IMPOSSIBLE ENGINEERING

PRINCETON STUDIES IN CULTURAL SOCIOLOGY

Paul J. DiMaggio, Michèle Lamont, Robert J. Wuthnow, Viviana A. Zelizer
SERIES EDITORS

A list of titles in this series appears at the back of the book.

IMPOSSIBLE ENGINEERING

Technology and Territoriality on the Canal du Midi

Chandra Mukerji

PRINCETON UNIVERSITY PRESS

Princeton and Oxford

Published by Princeton University Press,
41 William Street, Princeton, New Jersey 08540
In the United Kingdom: Princeton University Press,
6 Oxford Street, Woodstock, Oxfordshire OX20 1TW

Library of Congress Cataloging-in-Publication Data

Mukerji, Chandra.
    Impossible engineering : technology and territoriality on the Canal du Midi / Chandra
Mukerji.
        p. cm.
    Includes bibliographical references and index.
    ISBN 978-0-691-14032-2 (hbk. : alk. paper)   1. Canal du Midi (France)—History.   I. Title.
    HE445.M52M85   2009
    386'.4809447—dc22

                                                                                    2009004677

British Library Cataloging-in-Publication Data is available

This book has been composed in Minion

Printed on acid-free paper. ∞

press.princeton.edu

Printed in the United States of America

10 9 8 7 6 5 4 3 2 1

In memory of my father and to Cathy

CONTENTS

ILLUSTRATIONS

Figures

Tables

ACKNOWLEDGMENTS

I am deeply grateful to all those who have helped with this book. Perhaps the most important is the archivist at the Archives du Canal in Toulouse, Charles Vannier, who made the research both possible and often a pleasure. His patience and generosity over the last decade have been admirable. I am also indebted to Antoine Hennion for helping me to get access to the Archives du Canal through the Centre Sociologie de l'Innovation, and for both taking me into his family, and treating me to good conversation and good wine whenever I have been in Paris. Pamela Smith has been my intellectual guiding light and great friend as I worked on this project. At the Folger Library, she and Pam Long helped me think about distributed cognition and history, and at both Kew and Pomona, she encouraged my use of landscape history in the history of science as an extension of material culture studies. Karin Knorr-Cetina has importantly shaped my sociological analysis in this book, illuminating the role of epistemic authority and the struggles over it that animated the political history of the Canal du Midi. Michèle Lamont has not only encouraged my interest in French history but also usefully probed my ideas about the relationship of science and technology studies to cultural sociology. Steve Shapin served as a fine ally and intellectual friend while he was in San Diego and I was working on this book. His ideas about the social bases of credibility and the authority of gentlemen drew my attention to the political importance of impersonal measures of knowledge—detachment of ideas from their authors. Dissociating gentlemen from this vital source of power was crucial to the growth of technocratic authority and the development of modern state administration in seventeenth-century France. Steve Epstein has been particularly helpful in my efforts to think about peasant knowledge and technical expertise. We taught a seminar together on formal and informal knowledge that was not only tremendous fun but also clarifying about the politics of knowledge and social rank. Andy Lakoff was equally important to my thinking in this book, but in his case, about impersonal rule and technocratic authority. We taught a seminar on technology studies together in which he pushed me to think more clearly about modernism, rationality, and transformations of the natural world. Patrick Carroll has conspired with me for years to make sense of state formation and materiality, and provided endlessly interesting conversations on the subject. Jim Griesemer helped me consider the difference between formal collective memory (canonical stories of the Canal du Midi), and the practices of linking past and present in processes of innovation. Ed Hutchins, Carrol Padden, and Mike Cole have stimulated and shaped my thoughts on social cognition. Claude Rosental helped me think (again) about knowledge and practices of social authority, tracing the demonstrations and verifications of engineering efficacy on the Canal du Midi and their links to shifts in epistemic culture. I am grateful to Cathy Bacquet too. With her dedication to women's history, she pushed me to keep looking for data on the peasant women engineers, even driving me around the Pyrenees for weeks to look for water systems, old maps of the region, and old locals who would talk to me.

I would also like to acknowledge the support and intellectual help of my colleagues in science and technology studies at the University of California at Davis: Joan Cadden, Joe Dumit, Susan Kaiser, Carolyn de la Peña, Marisol de la Cadeña, and Jim Griesemer who were great friends while I was there. I also learned a lot from members of the cultural studies program and technology reading group, Patrick Carroll, Fran Dyson, and Caren Kaplan. In the period when technology was becoming more central to science and technology studies, we made it central to our conversations.

I have also been influenced by many students: Suzanne Thomas, Rick Jonasse, Charis Thompson, Bart Simon, Tarleton Gillespie, Sue Fernsebner, Fred Turner, Chris Henke, Matt Ratto, Lonny Brooks, Kate Levitt, Marisa Brandt, and Terra Eggink. And I am grateful to Veronika Nagy, who served as my research assistant in France for more than a year, and Etienne Pelaprat, who corrected my French on the manuscript.

Many individuals have given me good counsel, including: Kristin Luker, Fred Block, Tia DaNora, Jeff Alexander, Julia Adams, Deborah Davis, Gabrielle Hecht, Robin Wagner-Pacifici, Sharon Hays, Doug Hartman, Elizabeth Long, Ed Hutchins, Jeff Minson, Lesley Stern, Susan Smith, Sheldon Nodelman, Newton and Helen Harrison, David Bloor, Lissa Roberts, Simon Schaffer, Londa Schiebinger, Harold Cook, Simon Werrett, Carol Harrison, Ann Johnson, Michael Gordin, Kathy Pandora, Grace Shen, Cathryn Carson, Pam Long, Paula Findlen, Peg Jacob, Lynn Hunt, Mary Terrall, Sharon Traweek, David Turnbull, Helen Verin, Hélène Mialet, Luc Boltanski, and Laurent Thévènot; my colleagues, present and former, Carol Padden, Michael Schudson, Vince Rafael, Mike Cole, Leigh Star, Geof Bowker, Gary Fields, Lisa Cartwright, David Serlin, Morana Alac, Andy Lakoff, Naomi Oreskes, Nancy Cartwright, Charlie Thorpe, and Mary Walshok; and my French colleagues Bruno Latour, Luce Giard, Marie-Noëlle Bourguet, Ravi Rajan, Daniel Dayan, Dominique Pasquier, and particularly Louis Quéré, who was kind enough to invite me to lecture at the Institut Marcel Mauss.

I also want to thank my family and friends who helped with this—my father, Dhan Mukerji, and his wife, Marian, Kenneth Berger, Stephanie Berger, Bekah Fisk, Doug Van Horne, and Daisy Nieto. I also want to thank my "crew" who navigated down the Canal du Midi with me so I could take pictures: Cathy Bacquet, Rich Ortiz, and Marty and Stevie Malley. I want to thank my husband, Zach Fisk, too, not only for his interest in this project but also for buying me a good camera and (even) the Clément collection of Jean-Baptiste Colbert's letters. Others may love mystery stories, but a good letter from the seventeenth century is what I crave. It's amazing to be with a man who gets that.

ABBREVIATIONS

ABR   Archives Départementales des Bouches-du-Rhône

ACM   Archives du Canal du Midi

AHG   Archives Départementales de la Haute-Garonne

AHMV   Archives Historique de la Marine, Vincennes

BNF   Bibliothèque Nationale de la France

CCC   Cinq Cents de Colbert

EPC   Archives de l'École des Ponts et Chaussées

INTRODUCTION

The Canal du Midi today is a modest but pretty tree-lined waterway that threads through Languedoc in southwestern France. Most people who glimpse it from the nearby highway have little idea what was entailed to build it. Many assume that British canals predated it, or that it was contemporary to the Erie Canal. But it preceded them by nearly two centuries and was hailed at its opening in the 1680s as a wonder of the world. The work harbored a mystery. It was not technically possible according to the formal engineering knowledge of the period. Some called it a work of the Devil; the entrepreneur who built it, Pierre-Paul Riquet, attributed its success to God; and propagandists touted it as a measure of the glory of Louis XIV. It was otherworldly in its modernism, and huge by period standards. It did not just dominate the landscape but also haunted it with the foreignness of its powers.

Figure I.1
Canal du Midi in fall. Photo by the author.

IMPOSSIBLE ENGINEERING

CHAPTER 1

Impossible Engineering

When the Canal du Midi was first navigated to the sea, a flotilla of dignitaries crossed the waterless valleys of eastern Languedoc. The sight was uncanny. The strange apparition of boats in the continental interior seemed to turn the order of nature upside down, making palpable the powers of governmental administration. The royal canal to connect the two seas was mute, obdurate, and superhuman. For witnesses of this first crossing, it was something to see.

Originally named the Canal Royale des Deux Mers, the waterway linked the Mediterranean to the Atlantic through the Garonne River, crossing southwestern France just north of the Pyrenees. The canal extended across

Figure 1.1
Map of Languedoc, Nicolas de Fer, 1700. Courtesy of the University of California.

Languedoc for nearly 240 kilometers (150 miles), and at the divide between the Atlantic and Mediterranean watersheds, reached 189 meters (620 feet) above sea level. The passage required a hundred locks to manage the substantial shifts in elevation.[1]

The canal, as a silent demonstration of disciplinary power over the earth, pointed obliquely toward techniques of governance that lay beyond the visible and familiar practices of domination—war, taxation, and court life.[2] In its mysterious materiality, it suggested that political authority could also lie in something both spiritual and worldly—dominion over the earth and its creatures. The Canal du Midi officially represented the power of Louis XIV, but more memorably testified to the power of stewardship over the earth, using the human intelligence given by God in the act of the Creation to allow people to manage the earth and its creatures. The apparent impossibility of running water through dry hills and valleys across a continent made the canal seem evidence of God's work, and not just the will of the king. Here lay the cultural power of political territoriality.[3]

What made the waterway so mysterious and haunting was its unaccountable intelligence. No one knew in the 1660s how to do the elevation studies or estimate the water supply needed for such a canal. In the mid-seventeenth century, Jean Picard at the Académie Royale des Sciences was refining survey techniques, and Philippe de La Hire was working on elevations. Still, their methods were not adequate for determining precisely how to route a navigational canal through the complex topography of Languedoc. Picard's techniques were hard to use on uneven terrain because compensating for shifts in elevation could be tricky. Some academic geographers in the 1660s even believed Jodocus Hondius that continental divides were always high mountains, and doubted that there were passes low enough in Languedoc for a navigational canal. For all these reasons, respected geographers were ill prepared and not inclined to help build a canal in Languedoc.[4]

There were also elements of hydraulics that were not well understood by the learned men who dominated geography and engineering. For example, no one knew precisely how much water a navigational canal with so many locks would use. There were no reliable calculations for estimating water losses due to splashing, evaporation, or seepage into sandy soils. The Canal du Midi also needed reservoirs or holding tanks to assure an adequate supply during the long, dry summers of southwestern France. But how big would the reserves have to be? Where could they be located? How would they be filled? Currents could be a problem too, and depended on the inclines that directed water downhill toward the seas, filling locks as they emptied. Floods were a potential problem as well. The waterway was made mainly with pounded dirt and clay reinforced with timber, so storms could easily do damage to the walls of the structure both from the inside and out. In sum, there were myriad problems of engineering to address in constructing the canal that no one had quite realized or could properly appreciate when the project was initiated.

Hydraulics existed in the 1660s, but most of the literature addressed problems of mining and the design of pumps. Canals were used for irrigation and drainage projects as well as navigation, but the literature on them was scarce. Military engineers wrote on hydraulics after the Dutch flooded low-lying areas around their cities with seawater during sieges, drowning their attackers.[5] Still, vernacular knowledge of hydraulics was broader than what existed in written form in the 1660s. Cutting a navigational canal across southwestern France seemed feasible, but actually taking it over uneven terrain with multiple locks was not an enterprise that was obvious to achieve with the engineering expertise of the times.

Still, there were technical precedents to follow.[6] The Italians and Dutch both used locks; the Italians had built a number of canals on the plain of Lombardy, and also used locks to control rivers and flooding. The Dutch designed locks, too, to raise and lower ships as they moved between harbors and the sea where the countryside was below sea level. They also had locks on inland waterways where there were small shifts in elevation.[7] But the Canal du Midi was not and could not be built exactly on the same principles, because it was much larger and longer, needed many more locks, required

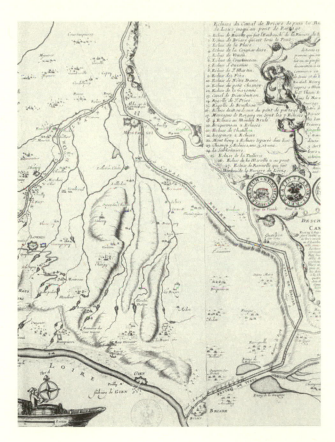

Figure 1.2
Canal de Briare,
Nicolas de Fer.
Courtesy of the
University of
California.

an extensive water supply, crossed drier and rockier land, and changed much greater elevations. So it remained a cultural enigma even though it appeared to be a technical possibility.

Entrepreneurs and political leaders had thought for a long time about cutting a canal across Languedoc, and their ingeniousness made this particular project seem more credible. France was relatively narrow in the region, so it was an obvious place to try to link the Atlantic to the Mediterranean. In addition, Languedoc was relatively wealthy and proudly independent from the Crown, so there were political motives for state intervention there. For many reasons and over many years, then, engineers and entrepreneurs made studies and plans for a canal in Languedoc during the sixteenth and seventeenth centuries, helping define the problems that had to be addressed and designing templates for their solution.[8]

There was also a convincing French technical precedent to follow: a working navigational canal near Paris completed in 1642. It was called the Canal de Briare, and with both a water supply and locks, was successfully carrying cargo from the Loire to the Seine.[9] The Canal du Midi was certainly a more ambitious and complex project, spanning a broader region and rising

Figure 1.3
Portrait of Pierre-
Paul Riquet.
Toulouse, Musée
Paul-Dupuy.

Pierre Paul Riquet, B.ᵒⁿ de Bonrepos,
Auteur du Canal du Midi,
Né à Béziers l'an 1604, mort le 1ᵉʳ Octobre 1684.

higher above sea level, but there was reason to think that the lessons learned from Briare would prove useful. The problem with the proposal for the Canal du Midi was less a matter of apparent feasibility than credibility.

It was easier to conceive of plans for cutting a canal across Languedoc than to engender public trust in Pierre-Paul Riquet. He was a tax farmer, not an engineer. He took contracts with the state to collect taxes on salt in Languedoc, guaranteeing a specified income to the treasury and profiting from additional revenues.[10] Although he would be hailed as a genius in future centuries, in his lifetime he was not treated so generously. At Versailles, Riquet was considered a provincial naïf without the authority to speak on matters of royal interest and without the mathematical skills to construct anything of consequence.[11] He was rich, but in the period, financiers of his sort remained something of a suspect lot. They could take on powerful roles in local politics, yet they were still socially disdained by members of the nobility. Noble opponents, who had no desire to have their lands or hegemony in the region disturbed by the project—particularly if this entailed transferring assets to a financier for a risky enterprise—were happy to cast doubt on the tax farmer's ability to do the work. No matter that Riquet had successfully collected revenue in a historically recalcitrant tax region, and had even made an additional fortune supplying the army with meat when France was fighting in Roussillon.[12] Neither enterprise could endow him with social dignity or make his word credible. He was no gentleman and was easy to paint as an opportunist seeking new ways to enrich himself. So constructing the Canal du Midi across the southwest of France required not only technical knowledge but also political and epistemological work to make the project legitimate.[13]

In spite of strong local opposition and the limits in knowledge of hydraulics, a navigational waterway of unprecedented proportions was indeed cut through Languedoc. Riquet did not have to become an engineer to make it work. The Canal du Midi was a product of a collective intelligence—the work of groups with both formal and vernacular expertise in land measurement, construction, and hydraulics. The minister of the treasury overseeing the work, Jean-Baptiste Colbert, sent experts from the north, and Riquet found skilled men and women in Languedoc who somehow worked together to realize this giant undertaking.[14]

The result was a modern "miracle" with surprising political ramifications. The canal demonstrated the efficacy of a new kind of power, engineering the land as a form of government. It was a model of impersonal rule, exercising a power over Languedoc distinct from that of local nobles and even the king. It was legitimated by stewardship principles, carrying the sanction of God's will. Those who excavated the channel and filled it with water, endlessly inventing their way around obstacles, produced more than new engineering knowledge or a piece of French infrastructure. They helped elaborate a modern logic of territorial administration based on "works" rather than "words."[15]

Their work was meant to be grand, but not as transforming as it became. Building a great infrastructure was supposed to make France more like the Roman Empire—a conservative, nostalgic vision of the future. Classical technical ingenuity used in pursuit of military power was the model of how to improve France.[16]

> The military Genius of the Romans made them trouble very little about Commerce, because all the Nations they had conquered traded to the profit of the Conquerors, who thus enjoyed every Thing useful and agreeable that the World produced; yet [the empire] produced a great number of magnificent Works [to uphold] its Advantage; how many Ports did they improve on all the Coasts of the Empire? On the other hand, the grand Bridges they made were not less designed for Commerce than for the Passage of the Armies and warlike Stores; nor did they only make those great Roads for the Communication of all Parts of their vast Empire, but also rendered most of the Rivers navigable, and joined them by Canals.[17]

France would seek power as the Romans had done, engineering an infrastructure for strategic advantage. This diachronic isomorphism was designed to align France with the classical tradition.[18] Change was a matter of restoration—of the glory and powers of Rome.

French literati of the period divided themselves into the ancients and the moderns, arguing whether the works of genius of the classical period could ever be matched or bettered by present or future generations.[19] The Canal du Midi did not fit the dreams of the literati.[20] Yet it was surely a modern effort—a claim about what was possible by combining new knowledge of nature with classical engineering. The ancients provided models of greatness, building ports, roads, bridges, and some canals. Now the French would design the locks needed to carry waterways across entire continents. The canal in Languedoc was a product of modern intelligence in the Roman tradition, and a monument to the capacity of the ancien régime to use and transcend the intellectual powers of the ancients.[21]

This vision of the "New Rome" had potent cultural purchase, but also disquieting political implications. The canal through Languedoc did not emerge from or serve the patron-client ties that dominated local political life; it succeeded in spite of the efforts of local elites to use social connections to defeat it. More than tactical skill, it depended on technical finesse that was not distributed according to rank.

A project like the Canal du Midi could allow a bourgeois entrepreneur like Riquet to reach above his station simply by mobilizing intellectual resources to further political territoriality. He could employ those with local knowledge of the land along with educated engineers to help make this part of the kingdom a visible part of the French state. In so doing, he turned himself into a new social type—a nascent technocrat—a man without station but with the capacity to exercise powers of government through the deployment of skills.[22]

The result was revolutionary. The waterway demonstrated the political efficacy of impersonal control, the stuff of modern state administration and technocratic government. Riquet recognized the importance of what he was doing, and to his credit and downfall, too, started to write to Colbert with naive enthusiasm about his contributions to the regime. The entrepreneur also started attributing his newfound abilities to God's will, not monarchical authority—a form of hubris at odds with the divine right of kings that made him seem a dangerous man.[23]

The improved countryside in Languedoc was hegemonically treated as a manifestation of Louis XIV's greatness, but the waterway remained impossible to reduce to the king's will. Excavating and routing the channel was dirty and delicate work that represented the monarch's authority, but also lay beyond it. It was an act of anonymous power that confronted normal politics in an oblique way, setting up the contrast between depersonalized administrative activity and patrimonial powers.[24]

Territoriality, as an administrative strategy and political culture, started being nurtured in the beginning of Louis XIV's reign by Colbert, the minister of the treasury and navy. It was based on *mesnagement* politics, an administrative and political philosophy developed early in the century under Henri IV. According to mesnagement thinking, the state was a great estate; its lands should be developed for the well-being of its people through the rational assessment of its virtues and vices, and the proper allocation of activities to appropriate sites around the countryside. Infrastructure was central to this approach. An estate needed roads and waterways to link diversified properties and the estate to marketplaces. It needed to have mills and other commercial ventures, too, and the water to feed them. To Colbert, this kind of land stewardship made sense as a program of state administration, and Riquet's canal seemed appropriate to it.[25]

But this logic of improvement flew in the face of entrenched political interests. Projects for improving the kingdom required lands, and their indemnification worked against the smooth flow of favors and information among nobles that cemented patron-client relations. Stewardship was a cultural tool of political change whose potency was greater than hoped and whose powers were more difficult to control than expected.[26]

The Canal du Midi became a peaceful eye in a contentious political storm, where patronage politics confronted territorial stewardship.[27] Land was requisitioned against the desires of local nobles, and the navigational waterway was excavated from the countryside in spite of hostility that only increased in passion from the 1660s to the 1680s. Nonetheless, the channel was cut; the canal's water supply worked; and boats started passing through the dry and rocky landscape of western Languedoc where no one had been able to navigate before.

Figure 1.4
Detail of a portrait
of Jean-Baptiste
Colbert. Réunion
des Musées
Nationaux.

Why did Colbert, normally a cautious man, ever agree to such an out-
landish plan that was sure to stir up local ire? The simple answer was the
potential value of heroic engineering to the king's plans for empire. A navigational canal connecting the Atlantic Ocean to the Mediterranean Sea
would allow French ships to avoid Gibraltar. The original plan was to make
the canal wide enough for naval vessels, and Colbert was minister of the navy
as well as the treasury.[28] The military dream had to be abandoned as the
canal had to be narrowed in the mountains, but the waterway still joined the
Atlantic to the Mediterranean and furthered the propaganda program of
building a New Rome. The canal positioned France as a land of moderns—a
triumphant possibility in the ancien régime.

The success of the Canal du Midi has traditionally been attributed to
Riquet's naive genius—a thesis questioned in this book. The canonical view,
first crafted in the eighteenth century and blown to larger proportions in the
nineteenth century, fit nicely with the entrepreneur's own assessment of
the situation.[29] Riquet claimed sole authorship of the canal, including some
technical details whose uses were commonplace.[30] He spoke in words that
served the myth of his isolated and beleaguered genius—a narrative he
probably believed unquestionably—but his view was not correct by historical standards. His claim to sole authorship of the canal belied the diversity
of techniques needed for the waterway, and the surprising story of collaborative engineering that made the Canal du Midi a work of technical art and
political transformation.

Still, Riquet's status as a local hero was not misplaced. He believed in what local people could do, even as he took credit for it. He was an advocate of Languedoc, and liked to imagine how to improve the province with canals, roads, water supplies, and a permanent Mediterranean port, recognizing in regional peasants, artisans, artists, and engineers the capacity for this work. Colbert insisted that he consult with experts to assure his honesty and the canal's efficacy, and Riquet learned from them, but he sought his own counsel from indigenous experts in hydraulics and construction who lived in Languedoc. In this historical moment, no educated engineer could say for sure how to carve a navigational canal through Languedoc, but a group could do the work. For including regional peasants and artisans in the collaborative work, Riquet became the personal instantiation of their collective achievement—the intelligence of the region.[31]

How are we to understand a work of "genius" in this context? Do we dismiss genius as simply a romantic notion that was particularly appealing to local advocates in nineteenth-century Languedoc? Riquet's status was certainly elevated in that time (along with fitting statues of him). We could simply argue they exaggerated the importance of the canal and its "author" in those heady days of French nationalism, and deny that the canal was a great accomplishment at all. But there *was* a kind of genius behind the Canal du Midi, a transcending of individual limits through collaborative work. It was a work of superhuman intelligence because it was the product of both formal expertise and a region's collective abilities, not just the mind of a gifted individual. And this gave the Canal du Midi its uncanny presence and power, too.[32] It instantiated an impersonal form of knowing that had surprising potency.

This explanation of the success of the Canal du Midi bears and veils its own mysteries. Drawing skilled people to work for the local tax farmer to build a canal across two hundred and forty kilometers of the province was not a sure thing, and getting them on-site did not in itself assure a good outcome. Riquet was not a credible engineer, so the project had no clear leadership and could have been a disaster. With the historical record littered with failed collective enterprises, why did this one work?

The question is all the more pressing because the collaboration in Languedoc required groups with differences in language, gender, education, and rank, which created social chasms among them. The people of Languedoc spoke Occitan, or the langue d'Oc, while experts from Paris spoke French; educated gentlemen who were sent by Colbert had courtly manners and other marks of social rank distinguishing them from bourgeois financiers and regional tax farmers like Riquet; and military engineers from noble families lived in social worlds far from those of Pyrenean peasants and artisans from the valleys of Languedoc. Gender played a role, too. Many laborers recruited to work on the canal were women. Why anyone in the period paid attention to their ideas is a mystery in itself. The brilliance of the Canal du Midi, then, lay not just in the success of the hydraulic engineering but also the social engineering. It was perhaps Riquet's greatest contribution to the canal and the political heritage it left behind.[33]

The power of collective intelligence as a political asset and the effectiveness of collaboration as a problem-solving strategy can be understood better from Ed Hutchins's theory of distributed cognition.[34] Hutchins contends that much of what we consider cognition or thought takes place not in individual minds but rather in conversations among individuals trying to accomplish a joint task. He uses the example of a large naval vessel coming into port to illustrate the multiple calculations and observations that can be linked to accomplish difficult activities. Some sailors check depth readings, others look for landmarks, and still others regulate the speed and direction of the ship. In a complex division of cognitive labor, they make separate calculations, but they get the ship to harbor by coordinating their activities around a naval chart. They relate their readings to these standard measures of the harbor, and make decisions based on the results. Their work constitutes a form of social intelligence organized for practical ends.[35]

Hutchins argues that much of thought *can* be deeply and complexly social, because it is physically distributed both in artifacts like charts and plans (with their memory and measurement systems) as well as among people with different roles in the problem-solving process. The cognitive result of these interactions, Hutchins claims, may not usually be called cognition because of its social nature, but it should be. Where calculations do not take place inside an individual's head but emerge from social interaction, they are not any less a form of cognitive activity. To navigate or engineer a canal clearly requires thought, although it is not mental activity in the usual sense (secluded in the mind of an individual thinker). It is an ongoing, social learning process—a pattern of socially distributed cognition.[36]

The Canal du Midi was clearly a product of such distributed reasoning, yielding both an impersonal intellectual power and a new culture of engineering. To cut the canal across southwestern France, diverse participants organized themselves around the plans and specifications for the canal that functioned like Hutchins's naval chart. The parameters for the work did not determine how to build the waterway but instead provided a common focus for the collaboration. Rather than inhibiting the success of the venture, the social differences among those working on the Canal du Midi actually increased the range of problem-solving strategies available for the work. Gentlemen, artisans, surveyors, military engineers, and women peasants all had different ways of reasoning, so when one engineering strategy proved unsuccessful, they were able to find others to try until they figured out what would work. They provided a rich cultural repertoire of modes of reasoning.

On the Canal du Midi, formal knowledge of hydraulics was no use without local knowledge of the region along with its terrain, soils, weather, and the sea. And formal measures for designing artifacts were useless without artisans who could realize the plans with local materials. Engineers, who could do mathematical calculations, had a limited range of relevant techniques and did not always know how to take advantage of the topographical detail they could measure. Military men who could design fortifications did not necessarily know how to build a water capture on a river. Reasoning from both principle and precedent, then, the men and women who worked on the Canal du Midi created an intellectual ability much greater than any one of them could achieve alone. They not only brought different knowledge and analytic ability to the problems but also learned from one another. The engineering became a product (in the mathematical sense) of different epistemic cultures—a learning system in which the participants acquired new capacities for reasoning from their interactions, making their accomplishments seem miraculous even to those who were involved.[37]

That a variety of social groups was needed to build the Canal du Midi can be demonstrated by using counterfactuals: recognizing what would *not* have been possible if any one of these groups had *not* participated.[38] Using this method, it is possible to see, first, that no single group of experts could have solved the engineering problems of the enterprise, and second, that a range of groups with different backgrounds were essential to completing the Canal du Midi.

The role of culture in social forms of cognition has been described in the psychology literature by Michael Cole and his colleagues.[39] They have argued that in school, individual learning is a group accomplishment that is deeply embedded in the culture. We may ask schoolchildren to take their tests as individuals, but mastery of subject matter is embedded in school systems, curricula, textbooks, teaching styles, and student participation—cultural forms. The thought of each student is a product of and serves to reproduce these sociocultural arrangements. They make schools carriers of intellectual cultures and bases of the intellectual "scaffolding" needed for effective learning.[40]

If learning depends on a base of common culture, this raises questions about the Canal du Midi. Given the social and cultural divides among the participants in the enterprise, how did they find common ground for learning from each other?[41] The answer is surprising. They shared remnants of the classical past; they all knew some elements of Roman engineering. Academics and military engineers studied literature written in ancient times, and scholarship derived from it. Locals, on the other hand, unknowingly reproduced elements of classical culture over generations, thereby making them taken-for-granted elements of daily life.

The Roman presence had been pervasive in the region. The city of Narbonne near the Mediterranean coast of Languedoc had been an important administrative center of ancient Gaul; Romans had also colonized the Pyrenees, building bath towns near hot springs with extensive waterworks and leaving behind a complex tradition of hydraulic engineering. The people of Languedoc did not see themselves as guardians of ancient knowledge but they also had not abandoned ancient techniques of building or hydraulics just because Rome had fallen; they continued to use the skills pertinent to their lives, simply forgetting their provenance and treating them as common sense.[42]

Some Roman techniques were clearly maintained by the Visigoths, who drained swamps and irrigated dry lands in the former Narbonnaise. Peasants in the Pyrenees who built domestic water supplies and public laundries in former Roman bath towns also transformed Roman hydraulics into commonsense practices. And artisans in the building trades reproduced classical methods as well. Those who dismantled the ruins from ancient Gaul learned lessons in construction methods. When they tried to reuse stones from old buildings in new ones, it was easier to reproduce the classical forms for which they had been cut than to recut them, reproducing some ancient building techniques over time. So a variety of ancient skills became "commonsense" practices and were available for building the Canal du Midi— carried by peasants, artisans, and gentlemen engineers alike. Their common heritage allowed participants to engage in a generative creativity at the interstices between traditions of formal knowledge and everyday practice, taking advantage of the distributed quality of their knowledge as well as its common assumptions. Even where the participants found themselves at odds with one another, their differences permitted an intellectual reflexivity that itself was productive. This was only possible because the debates were grounded in a culture of classical thought that might have lost its provenance but not its ancient logic.

Distributed reasoning would not have become so important to the Canal du Midi if there had not been a vacuum of authority as well as a deficit of formal knowledge for the work. Riquet wanted to be in charge, but his word as a financier had no credibility with either engineering experts or men of rank. He was continually supervised, spied on, argued with, and vilified by enemies, so his leadership remained mitigated to the end. Still, no one else could displace him.[43] In the absence of both sure technical authority and firm leadership, those trying to build the canal turned to each other to realize the daunting task of cutting a navigational canal across southwestern France.

These efforts produced a new technical capacity on the ground, and in participants, a thrilling ability to change the world. Apparently intractable problems over time became possible to address; technical innovations suggested other improvements or experiments to try. Participants started thinking in ways that were novel to them, and found new powers in their own common sense. Even laborers continued to work on the project toward the end when Riquet was broke and they were slow to be paid. The excitement of doing the impossible was intoxicating—explicitly so to Riquet and the military engineer, Clerville, who were able to write about it.[44] To conjure up the powers of Roman engineering by mistake and put this knowledge to work successfully on a modern project must have been a powerful experience.

The Canal du Midi was attractive as a political experiment because it fit with the propaganda campaign to define France as the New Rome, the proper center of a new Catholic empire to rival the Hapsburg one.[45] The waterway was hailed as Roman in spirit, and designed to reference and resemble the classical past. Still, no one dreamed how Roman the canal would become. No one expected or knew what classical techniques artisans and laborers would bring on-site, giving the Canal du Midi its daunting anonymous intelligence. And no one anticipated how French the structure would be. The techniques used to fashion it were honed against the countryside itself—French valleys, forests, hills, rock formations, rivers, and riverbanks. The knowledge of artisans and peasants was situated in France itself, and that made all the difference.[46] State territoriality was not imposed on the land and its people with this canal but rather built by locals from the ground up with knowledge that only they could deploy.

Place matters to technical innovation, and so it did in Languedoc in the late seventeenth century.[47] The social and material inheritance from Rome, the movement of diverse groups into Languedoc, and the mixed character of the land itself gave a vitality to the region that was consequential. Social life is always dependent on the material possibilities of places and gives rise to forms of intellect that make sense there. The intelligence in Languedoc about land and its uses was both deep and broad, providing rich cultural resources for engineering a canal there.

In seventeenth-century Languedoc, where the towns dotted diverse landscapes from high mountain valleys to the Mediterranean coast, people found a variety of ways of life, and nurtured multiple forms of intelligence. Some groups in the fertile valleys north of the Pyrenees grew grapes, olives, wheat, and vegetables, irrigating the dry, rocky soil from the Montagne Noire to the sea. Others herded sheep, cattle, and goats in high mountain meadows, making cheese for sustenance in the winter and using the abundant water supplies of the mountains to run mills for tanning the hides from the animals they slaughtered. In mountain forests, people collected firewood, made charcoal, or produced glass. On the coasts, they fished the sea, the wide rivers, and shallow salt ponds toward the Mediterranean, and even found ways to do agriculture in almost pure sand. Along the pilgrimage routes to Compestella, villagers witnessed miracles, made monuments to them, and served the believers who stumbled through their towns. Others were migrant laborers or engaged in trade across the high passes into Spain. This complex region gave rise to multiple logics of social life.[48]

The peasants of Languedoc certainly seemed at the time an unlikely source of state power. They were (as a whole) famous for their unruliness—a stubbornness that sometimes served tradition, and at other times refused attempts at control. In this land so rich in its topography and so variable in its productivity, bandits and rebels against the king's men found a viable niche. Those who killed tax collectors could disappear into mountains that they knew better than any authorities and where sympathetic villagers sustained them. They straddled a border region where authorities from both Spain and France had limited abilities to control the flow of people and things. Their knowledge of the landscape was a political asset as well as an economic resource, and one they both cultivated and used with agility.[49]

It was precisely this kind of local knowledge—the intelligence of bandits, fishermen, washerwomen, masons, charcoal makers, and women indigenous engineers—that made it possible to build a canal in Languedoc. These were the people who knew what you could do in this particular countryside, its topography, seasonal rainfall, rock formations, rivers, forests, and coastal sands. A navigational canal might have been long and wide, but it was not so different from the diversionary canals of the Pyrenees or the irrigation systems of valley estates. In some undefined way, it was an appropriate task for the people of this region, and Riquet knew it. He understood the stubborn and tenacious character demanded by the landscape, and recognized that the people who could make rocky hillsides or sand dunes yield a living, could realize his ambition to build a canal through Languedoc. What he and the other participants did not understand was that they were inventing a new kind of silent, insidious power.

CHAPTER 2

Territorial Politics

Before the Canal du Midi could become a technical accomplishment, it had to be an object of political desire. Prior to the physicality of measuring or digging, its form was ephemeral; its existence was an act and object of political imagination and strategic calculation; and its possible success or failure was weighed more in geopolitical terms than engineering ones. The canal had to be initially imagined as an act of state since it could not be built privately; it depended on the monarch's ability to indemnify land and employ treasury funds to improve his domain. It had to become an object of the king's will, competing as progeny of power with the minister of the army's

Figure 2.1
Colbert presenting the Canal du Midi to the king. Toulouse, Musée Paul-Dupuy.

plans for war—ambitions that were more tempting to Louis XIV for the glory they could bring the reign and the bloodlust they could inspire in the nobility against foreign enemies.[1] Although it was a useful piece of infrastructure, the Canal du Midi did not gain state sponsorship mainly for its economic usefulness; it was a gamble with the potential to change the material conditions of power, and this made it weighty enough to attract monarchical consideration. Water served travel in the period, of course, but was also difficult to control. Bending such stuff to the will of the Sun King and changing the geography of the continent to do it was a project worthy enough to compete with war in ambition and symbolic possibility.[2]

Making the Canal du Midi larger than life—like Louis XIV hoped to be—took substantial ideological as well as material work.[3] The enterprise had to edge its way toward becoming a wonder of the world and threaten France's rival, the Hapsburg Empire. Engineering itself had to be seen as glorious—a legacy of Rome and a sign of French capacity for empire. Then the canal could be a demonstration of power as well as perhaps an alternative route from the Mediterranean to the Atlantic for French ships.[4]

As minister of the treasury and navy as well as the administrator for the king's households, Colbert saw possibilities in the project for serving his ambitions. He was particularly intrigued by the strategic role the canal could serve for the navy by carrying warships across the continent. He wanted to make France a more effective sea power to feed the king's penchant for combat and gain status vis-à-vis the minister of the army, Michel Le Tellier (and even more so, his son, François-Michel Le Tellier, Marquis de Louvois).[5] Colbert also argued that the infrastructure would serve commerce, although that was only of local, not royal importance in his mind. At the same time, he could strengthen his ties with one of his own tax farmers—a man already under his control—while extending his ministerial reach into Languedoc. If it were technically viable, Riquet's project would yield glory, strategic advantage, and riches all at once.

Dreams on dreams on dreams: dreams of Rome built on a minister's dreams of naval and managerial power, a tax man's dreams of transforming his province and his family's fortune, and a king's dreams of legacy—all of them transmuting countryside into power. These ambitions were fueled by fantasies, heroic images of men willfully connecting the two seas, Atlantic and Mediterranean, etching a channel into the rural landscape to do it.[6]

Lines of translation from technical possibility to political significance and back again were drawn to connect the canal to these political desires. Riquet did most of the work, knowing enough to couch his enterprise in explicitly political terms. But the significance of his words also depended on their reception—how they fit the modes of political reasoning being nurtured at court. The canal's political effectiveness was also dependent on natural knowledge, immediately raising epistemological questions: When could anyone be sure that the measures were right, or the hydraulics sophisticated or precise enough for the task? Making the Canal du Midi politically viable meant making it convincing in all these terms.

Modes of Political Address

Riquet started the process on November 15, 1662, when he wrote awkwardly but baldly to Colbert to present his surprising proposition.

> Today I am writing . . . on a subject distant from [the salt tax]. It is about the design of a canal which could be made in this province of Languedoc for the communication of the two seas, the Atlantic and the Mediterranean. You may be astonished, sir, that I take the liberty to speak of something which I apparently know nothing about, and that a salt tax man [*homme de gabelle*] would dabble in elevation studies. But you would excuse my boldness once you know that I write you under the orders of Monseigneur, the Archbishop of Toulouse, [and] . . . not of my own accord. In any case, sir, if it pleases you to take the trouble to read the account [of the archbishop's visit to the site of the canal and my model of it], you will judge that it is true that the canal is feasible, that it is difficult [only] because of the costs, but recognizing what good it will bring, one should make little of the expense.[7]

Riquet's letter displayed a surprising political understanding of the canal's significance. Even in this short passage, his language addressed three different political strategies salient to the administration. By saying that the canal would connect the two seas, Atlantic and Mediterranean, he gave the enterprise grandeur, speaking to the propaganda of the period promoting France as the New Rome. Riquet was an insistent and engaging publicist, sure that a canal through Languedoc connecting the two seas would indeed be a wonder to behold. He spoke in his letter of high-minded things—the seas, the church, the truth, and virtue. He suggested that his enterprise would be a marvelous monument to the king, contributing to the legacy of his reign.[8]

Riquet also used the language of patrimonial or clientage politics to suggest his eligibility for this honor, invoking the name of his Catholic patron to both justify his breech of decorum in writing Colbert directly and suggest his reliability as a client.[9] Colbert already knew that the entrepreneur could serve the state well, since his record as a tax farmer was impeccable. But his sponsorship by Charles-François d'Anglure de Bourlemont, the in-coming archbishop of Toulouse, suggested that Riquet could connect the minister more closely with a powerful cleric in Languedoc who might serve him well. Colbert generally favored Catholic entrepreneurs with his contracts, and even more so in Languedoc, where Huguenots had become influential. Even Catholic nobles in the province were starting to put their children in Huguenot schools, displaying an independence that could spell trouble for the king. So

Colbert wanted loyal allies in the region, and the clergy in Languedoc was anxious to strengthen its ties to the court.[10] Riquet may not have had a credible voice about matters of engineering or any basis for addressing the minister about such a project, but he had a powerful Catholic patron who could vouch for him and be useful in other ways to Colbert. The canal held the promise of tipping the balance of power in the direction that the minister wanted.

Building a peaceful waterway would be an act of stewardship, too, demonstrating the virtue of monarchical power. When he spoke of the good the canal could bring to the region, Riquet certainly underscored the practical advantages it could provide for trade (as part of the logic of stewardship), but he implied even more. The canal was good in the sense that it embodied virtue and was an act of moral significance that could underscore the spiritual basis of royal rule.[11]

In these three ways, Riquet used vocabulary in his first letter to align his desire with that of the Sun King and his minister. As a bourgeois entrepreneur and financier, he was not and never could be a direct client of the king; he was too low born. But as a tax farmer, Riquet was in a position to take on civil engineering projects for Colbert.[12] Since the treasury had little money for domestic improvements, the minister routinely asked those with capital to take contracts for infrastructural work. Colbert used these contracts to not only serve the economy but also extend his patronage networks in strategic ways, including favoring Catholic entrepreneurs over Huguenots to make the kingdom more Catholic like the king.[13] Perhaps Riquet knowingly addressed the minister's religious bias when he wrote to Colbert, citing the patronage of d'Anglure de Bourlemont. Certainly, he presented his patron as a potential local broker for the administration.[14] In any case, with this rhetorical flourish, he brought more than a canal project to the attention of the minister.

In the rest of the letter, Riquet continued to interweave these three threads of political ambition and technical possibility. He outlined a genealogy of desire for the project, tracing hopes for a canal through Languedoc down a long line of distinguished French political leaders, emphasizing the Catholic clergy among them.[15] Henri IV was first in Riquet's genealogy, but since he had started his reign as a Protestant king, Riquet skillfully attached the idea of a canal through Languedoc more firmly to two of the most powerful cardinals in French history: Armand-Jean du Plessis, duc de Richelieu, and François de Joyeuse. This purified the project of reformed implications, designating the canal as a legitimate and dignified object of political desire for Louis XIV.[16]

Riquet also suggested that building the canal could undermine the Hapsburg Empire. He put it this way:

> The strait of Gibraltar will cease to be a necessary passage point, [and] the revenues of the king of Spain will be diminished and those of our king will be augmented as much from duties as from the export of merchandise from this kingdom, not to mention that the

tolls paid on the canal will add up to enormous sums.... [The] subjects of the king will profit in innumerable ways from a thousand new kinds of commerce and will gain great advantages from this navigation, such that I believe this plan will give you pleasure.[17]

Finally, Riquet presented the canal as a way to use knowledge of the earth revealed by God for exercising dominion over the earth. Now such knowledge was available, indicating spiritual sanction for the enterprise with Louis XIV on the throne.

Until now, people had not known which rivers to use for this project, nor knew where to set out the easiest route for the canal. For as much as one could imagine it, there remained insurmountable obstacles to building such a waterway and finding the machines for raising water, and these difficulties discouraged people from trying to execute the project. But now, sir, one has developed proper routes for the canal, and found rivers that could be turned from their traditional beds and diverted into this new canal. Using their natural flow and their proper incline, all difficulties cease except that of finding the funds for paying for the cost of the work.[18]

Colbert needed loyal clients as well as brokers in the provinces, and Riquet was volunteering for the job. The minister understood that although traditional noble officeholders were technically responsible for maintaining or improving the kingdom in the name of the monarch, such men usually served their own, not the monarch's interests with their political powers—particularly in Languedoc. Colbert had learned to distrust such "officers of the Crown," when he found them lying to him about their private exploitation of public resources. So now he was developing his own set of clients to fill the administration. Riquet seemed a fine candidate for this new coterie of political agents. He could improve France and undermine the king's adversaries at the same time, and as the tax farmer, he apparently had the stomach for enforcing the will of the king even when it was resented.[19]

Riquet offered Louis XIV and Colbert not only a grand project of infrastructural engineering but also spiritual justification for it. The enterprise could serve the navy, stimulate the economy, threaten Spanish control of the seas, give Colbert a new agent in a dissident region, and connect the Catholics in Languedoc more closely to the administration. But most of all, it was an audacious undertaking of Roman proportions and deep moral foundations. It was a project that even an ambitious young king drawn more to war than commerce could love, and it was an enterprise that Colbert could use to demonstrate the efficacy of territorial politics for bringing local nobles under more certain control of the king.

Niccoló Machiavelli explained well the problems with the nobility in France that turned the attention of Louis XIV and Colbert early to patrimonial politics.[20] He argued that French monarchs had never gained full control of great noble families. Local aristocrats often had loyal followers, who would obey them before the king. Since nobles of high standing depended little on their monarch, they could ignore his wishes or stymie his policies. Their loyalty had to be explicitly solicited with ongoing acts of respect, favoritism, and generosity: the tools of patronage politics.[21]

Patronage relations served weak monarchs as a means for accommodating autonomous nobles and powerful members of the clergy, but it was an unstable system. On the one hand, favors could build ties to leading aristocratic families, making patronage beneficial to the monarchy. Yet patronage could also add to the vulnerability of kings by dispersing powers in the form of favors or appointments to offices to the monarch's potential rivals or enemies. Sometimes the king bought his enemies; sometimes his enemies usurped his powers.[22]

In Machiavelli's terms:

> The king of France is surrounded by a large number of lords of ancient lineage who are recognized and loved by their subjects. They have their degrees of pre-eminence, which their king cannot deprive them of without danger to himself. . . . You can [conquer such a kingdom] with ease by winning over one of the barons of the kingdom, since malcontents and others who desire a change can always be found. For the reasons stated, such persons can open the way for you and facilitate your victory. But in preserving it afterward infinite difficulties will arise in regard to both those who have helped you and those whom you have oppressed. . . . Unable either to please them or to annihilate them, you will lose the state at the first likely opportunity.[23]

The last two kings before Louis XIV, Louis XIII and Henri IV had been effective soldiers, but failed at internal politics, and had faced premature death. Louis XIV came to power after a set of regencies had further eroded royal authority. As a child, Louis XIV witnessed the Fronde, an effort by powerful nobles to take control of the kingdom.[24] As Machiavelli would have advised on gaining the throne, Louis XIV actively cultivated noble connections, moving to Versailles, and providing those who followed him with presents and opportunities that they became loathe to give up.[25]

> The man who becomes prince through the help of the nobles will . . . be surrounded by men who will presume to be his equals. As a consequence, he will not be able to command them or control them as he would like. . . . [Such nobles] ought to be viewed in two principal

ways: their procedure will be either to join their fortune to yours or to hold back from doing so. If they join to yours and do not act rapaciously, you should honor and cherish them. [If] they hold back . . . through fear and natural lack of spirit, . . . these nobles [can still] bring you honor in prosperity and need not be feared in adversity. But if they avoid joining you, through cunning and ambition, that is a sign that they have more concern for themselves, . . . and [you] must fear them as though they were professed enemies.[26]

Patronage was a means of negotiating alliances among quasi-equals, reducing the incentives for becoming the king's enemies. Loyalty was secured through exchanges of favors and information, and the king's power depended on doing it well.[27] When Louis XIV gained the throne, he needed and cultivated these ties, presenting himself as the monarch who was bringing the nobility back to the center of politics after a period of church domination. By defining himself as a staunch Catholic, he cultivated the clergy, but by defining himself first as a nobleman, he capitalized on noble hostility to the church as well.[28]

The monarch in this situation had little other than patronage ties as a basis for rule. The "state" had surprisingly little "apparatus" of its own.[29] There was no central legal system, but rather many regional courts run by local elites; there was also no unified tax structure, and nobles often skimmed money from the revenues that passed through their hands.[30] They also routinely cut the forests without paying the required duties on timber, and charged tolls on those who traveled through royal lands. Colbert instituted reforms in all these areas, but he was working against the officeholders who wanted to buy themselves offices for their titles and privileges, not to serve the state. The result was a monarchy dependent on patron-client ties. Institutional authority was of limited importance. Getting something accomplished required personal influence and depended on finding allies of goodwill.[31] Military victories and a glittering court life helped draw nobles to him, but Louis XIV needed administrative powers, too, and it was Colbert's job to find a way to provide them.[32]

As minister of the treasury, Colbert managed the state patronage system—the distribution of offices, pensions, and favors. With some monies, he attracted artists and literati to Paris and Versailles to make the court a cultural center that nobles would seek for pleasure. He created a scientific infrastructure for the state as well by developing academies and pensions to attach learned men to the king. In addition, Colbert distributed economic favors to loyal nobles. They were formally restricted from engaging in trade by the constraints of their rank, so he made exceptions by offering them opportunities in royal projects, allowing landed nobles to enter the capitalist economy without losing their privileges.[33] This was a powerful job, but also in Machiavelli's terms, a losing game. The gifts empowered those who received them, creating degrees of independence in the nobility that could threaten the king. Something else was needed to disempower nobles without making them dangerous enemies.

Stewardship Politics and Administrative Reasoning

Colbert turned to stewardship politics as an alternate cultural basis for increasing the powers of the administration.[34] Noble estates were in principle the king's lands, given over to great families to manage in his name. This left most French land functionally in the hands of the nobility, although potentially owed to the king. The independence of French nobles that Machiavelli described made them eschew any effort by the monarch to regulate or intervene in their land-use practices. Colbert used the principles of stewardship to change this.

Extant literature on proper cultivation methods for the moral restoration of Creation provided a basis for questioning noble land-use practices on their estates and their rights to any land they had abused. Evidence of bad stewardship offered an excuse for taking parcels by eminent domain, or enforcing regulations from the king to restore the kingdom's beauty and fecundity. The result was a shift in power over the land, and a political strategy that put territory at the center of politics and intelligence about the natural world in the heart of political administration.[35]

Mesnagement ideas about stewardship as a moral duty started to develop in France around the turn of the seventeenth century during the wars of religion, when violence racked the interior of the country, making it seem anything but a haven of peace and rationality. Religious violence was particularly intense in the southwest, including Languedoc. As the fighting flared up, towns were sacked by Catholics, then Protestants, and then Catholics again (or the other way around). In the fighting, not only did thousands of people lose their lives but city walls, buildings, and bridges were also destroyed, making it necessary to reconstruct towns and regional infrastructure.[36]

The violence stimulated dreams of peace—of a utopian social orderliness and a physically restored environment. Hopes for rebuilding the countryside echoed biblical dreams of restoring Eden—returning the landscape to the perfection at the Creation. Attention to Creation also seemed a way to sidestep the religious intolerance that was shaking the kingdom. If there was no way to adjudicate differences of faith, then perhaps it was time to turn attention from words to works—from theology to the stewardship implied in the Creation. Good stewardship could bring both prosperity and calm to places as well as people suffering from warfare and the resulting poverty.[37] A combination of urban economic rationality, French humanism, and a deep spiritual yearning for restoration of the French countryside contributed to the formation of a mesnagement politics in France.[38]

Stewardship as a political philosophy in France had its roots in a humanist land-management literature based on the work of the Roman author Lucius Junius Moderatus Columella, who described techniques for managing estates.[39] In the mid-sixteenth century, the scholar and publisher Charles Estienne wrote one of these classically based books, *L'Agriculture et La Maison Rustique*.[40] It was finished and published after his death by his son-in-law, Jean Liebault in 1564. Estienne and Liebault argued that the lives of ordinary people could be improved by the rational use of the land. Well-studied and carefully tended estates would provide bountiful harvests, but only through serious thought and hard work. Those willing to approach agriculture with a studious attitude and purposeful discipline could produce the abundance that had sustained the civilizations of the ancients, and was implicit in the Creation itself.

Bernard Palissy, the Huguenot naturalist, ceramicist, and garden writer of the late sixteenth century, added a deeper spiritual dimension to the mesnagement tradition.[41] He said that his ideas about rational land use were stimulated by reading the 104th psalm, and its tender description of human dominion over the creatures.[42]

> He causeth the grass to grow for the cattle,
> And herb for the service of man:
> That he may bring forth food out of the earth:
> And wine that maketh glad the heart of man,
> And oil to make his face to shine,
> And bread which stengtheneth man's heart. . . .
> O Lord, how manifold are thy works!
> In wisdom hast thou made them all.[43]

Like a good humanist, Palissy advocated the direct study of nature, but his intellectual concerns were driven by a deep religious passion. The point of working the land was spiritual more than economic. "I also wish to build this admirable garden in order to give men occasion to become lovers of the culture of the earth, and to relinquish all other occupations or vicious pleasures and evil traffic, that they may instead delight in the tilling of the soil."[44] Soul and soil were fundamentally linked.

> I came to consider the marvellous deeds which the Sovereign has commanded Nature to perform; and among other things I contemplated the branches of the vines, of peas, gourds, which seemed as though they had some sense of their weak nature; for being unable to sustain themselves, they stretched certain little arms like threads into the air, and finding some small branch or twig, came to unite and attach themselves, never again to part thence, that they might sustain the parts of their weak nature. . . . [W]hen I had seen and contemplated such a thing I could find nothing better than to employ oneself in the art of agriculture, and to glorify God, and to recognize Him in His marvels.[45]

Palissy was also driven by the belief that God had given him the gift to recognize truths about the Creation, and out of respect, he needed to convey his knowledge to others. Intelligence was a human endowment that had to be exercised for the purpose of stewardship.[46]

During the reign of Henri IV, Olivier de Serres not only elaborated land-management techniques but also transformed mesnagement ideas into a more overt political philosophy. Land control for Serres was a source of both wealth and power.[47] Like Palissy, Serres wanted to use philosophy (intellectual rigor and spiritual guidance) to improve his farming methods, and sought to learn from plant cultivation and trade both how to improve gardening and understand God's Creation. He argued that rational estate development implied making the best use of the existing natural tendencies of the property by deciding through reason and surveys where to plant trees, cultivate gardens, lay roads, and set up mills. Serres described, in other words, ways of constructing a more managed infrastructure, a "second nature" to improve the abundance and value of estate lands.[48]

Although concerned with the market value of agricultural products, Serres defined estate improvement as a means to a happier *group* life, not personal profit. Mesnagement politics, in turn, was the route to a more contented kingdom of France. The well-run estate was for Serres the model of a good (virtuous as well as wealthy) state.[49] Stewardship was not only a profitable and virtuous quality in men but also the basis for good government, and the source and mark of effective leadership:

> As much as the father of the family is adorned in these qualities [of stewardship], and has made himself knowledgeable in all the aspects of rational land management, leading his workers firmly, who will follow him all the more willingly if they know by experience that his orders are reasonable and profitable. . . . Not only in estate management is such great solicitude and vigilance required, but also in all action in the world; neither are kings exempt from keeping knowledgeable about their own affairs, so they can so much more readily make things happen the more curiously they study and understand them; this maxim seems usefully verified in the establishment of this realm by the virtuous conduct of our king, Henri IV.[50]

Serres's political philosophy importantly shaped French politics overtly in the reign of Henri IV and covertly after it. His views on property, good government, and natural leadership helped codify and stabilize a Christian humanist approach to public administration in France. If the state was a great estate, the first step to improving it was to study the land and learn how to manage it well. It was for Serres as well as Palissy a pleasure in addition to a duty.[51]

Stewardship based on religious principles provided a rationale for administrative expansion that was hard for nobles to challenge. It was easier to ignore a king who had little immediate power over them than to completely ignore the arguments for stewardship if this was indeed God's will. So the adoption of this administrative practice gave Colbert a tool for extracting new powers for the government in the name of God as well as the king.

Patronage and Stewardship in Territorial Politics

The development of the mesnagement tradition in France would have been of less historical significance if it had not been so crucial to Colbert's administrative strategies. Colbert's territorial rationality gave stewardship a more secular purpose and aspect, making it the basis for government. For Colbert, mesnagement political philosophy provided a way around the interpersonal rivalries that marked patrimonial politics. It also gave the king powers that were hard for others to rival. Louis XIV could indemnify French lands, and he could also use his patronage power to gather to France those with the natural knowledge of geography, botany, and engineering to make these policies models of good stewardship. The land was obviously the kingdom's greatest political asset; now studying it and using it wisely could make the countryside better serve the monarch and his subjects through the exercise of impersonal as well as personal powers.

Stewardship principles had a variety of political applications. They supplied a moral measure of leadership that could be used to elevate the standing of the king. They provided bases for taking and also enclosing land. They could be used to discipline men of standing who had abused natural resources; they could also be used to extract money from nobles. Those who misused forests, for example, could be fined and required to pay duties on the profits they made. Autonomous nobles did not like these procedures, but none could question the king's right to exercise stewardship over his lands. Colbert had employed mesnagement practices on a smaller scale in administering the king's properties, making his forests and estates bountiful and beautiful. Then like Henri IV's surintendant des Finances, Maximilien de Béthune, duc de Sully,[52] he treated the state as a great estate, too, trying to bring the same kind of order and abundance to the kingdom as a whole. For the minister, there was nothing contradictory about serving his patron and tending his lands; it was his duty to be both client and steward, and his version of territorial politics was the way he tried to do that.[53]

In spite of his efforts to undercut noble independence with stewardship politics, Colbert was a creature of patron-client politics himself who had been groomed to serve Louis XIV by Jules, Cardinal Mazarin, the cleric who had ruled France during Louis's minority. The cardinal trained Colbert to serve him loyally and efficiently, and then recommended the aspiring bureaucrat to the young king. Colbert had helped the cardinal become the most powerful and richest man in France—richer than the king—and could bring to the monarch some administrative skills that he dearly needed.[54] So as much as he promoted stewardship administrative policies, he wielded most of his power through patron-client relations.

Because of his early proximity to power, Colbert was quickly disabused of the illusion that patronage ties were firm and nobles loyally served the monarch. He had begun a project of forest reform when Louis ascended the throne to learn about and manage the trees in the kingdom needed for ship-building. Colbert asked for information from the nobles and clerics who were forestry officials, but learned quickly that they had freely used the forests for their own purposes and were not inclined to divulge their indiscretions. They lied about the trees under their supervision, and when he recognized it and then realized that these officers of the state intentionally felled trees belonging to the king, Colbert was appalled. So he treated them according to Machiavelli's strictures "as though they were professed enemies" and used stewardship politics to manage them.[55]

Officeholders did not in themselves constitute an effective administration.[56] After learning this disconcerting lesson, Colbert encouraged Louis XIV to appoint his own political clients—from foresters and engineers to his regional emissaries, the intendants. These appointees were chosen for their loyalty, and measured at least in part by their material effectiveness or good stewardship.[57]

Colbert could not and did not seek a large-scale reform of state offices when he took over the administration. He was in no position to disturb large numbers of noble officeholders. Many were powerful men who needed to be allied with the king and enrolled into the administration, not alienated by an upstart minister. So Colbert added new offices to his administration and instituted new taxes, setting up a governmental structure parallel to the one that already existed. His was no dream of rational offices with distinct duties or a clear chain of command—the Weberian ideal of bureaucratic order based on nineteenth-century models.[58] Instead, he worked with clients of the king, and used principles of stewardship to measure them and identify his enemies.[59]

Colbert maintained some discipline over officeholders by accumulating public records—archives of regulations, judicial decisions, land claims, and claims of traditional rights. The minister used these public records for surveillance more than to make government rational. Many documented poor stewardship, and usurpation of royal lands or tax monies by local lords. They were means for making rogue nobles more accountable to him; they contained information that could be played like cards in patrimonial politics. In patron-client relations, the circulation of favors was one currency; the circulation of information was the other.[60]

Territorial politics as it developed in France under Colbert's guidance, then, depended on the interplay of patronage and stewardship politics: exchanges of favors and information, on the one hand, and land control and improvement, on the other. Political administration included personal rule of the king and impersonal rule over the kingdom—a powerful combination. This was the strategy that made the French state so powerful in the period, and the basis of what has been called "absolutism."

The Canal du Midi as Riquet proposed it seemed perfect for this political moment: a model of stewardship offered by a man already in Colbert's patronage network. The problem was only assuring that the waterway was a credible project to entertain by the king. It helped to have a clergyman of the stature of d'Anglure de Bourlemont as Riquet's advocate for the canal, but a work of stewardship also needed material demonstration. As Riquet put it, "In desiring to make visible that I am right in my thinking, if it is absolutely necessary that I provide proofs of [what I say,] and make them at my expense and peril, I do so willingly."[61] By convincing d'Anglure de Bourlemont to visit the sites for the proposed canal and view the techniques he had developed for the waterway, the highly esteemed clergyman could be his witness at court.

Riquet first took the archbishop up into the mountains to the area where he wanted to build the canal. The clergyman could see from the Lauragais ridge the valley where rivers flowed near each other, but in opposite directions. That is where Riquet wanted to build a canal to connect rivers that flowed down the Mediterranean and Atlantic watersheds, and join the two seas. Then the entrepreneur accompanied the cleric to his estate at Bonrepos, where he had built miniature locks and a water supply to illustrate the techniques that could be used to make such a canal work. The mock-ups spoke eloquently in the language of mesnagement politics. If the estate was a microcosm for the macrocosm of the state, then what could be built on a gentleman's lands could be realized for France.[62] The demonstration was persuasive to the archbishop and convinced him to advance Riquet's cause with Colbert.[63]

D'Anglure de Bourlemont did this by writing a "relation de la visite," telling the story of his excursion to the potential site for the canal and of the models Riquet had built at Bonrepos. They provided the king and Colbert with a means of "virtual witnessing" the plan and its demonstration.[64] Men at Versailles could look through the pure eyes of a high-status clergyman, a man of rank whose voice could be trusted, to see Riquet's dream not as a flight of fancy but rather as a vision of the truth.[65]

Patronage politics and stewardship logics met in Bonrepos. Patronage ties allowed Riquet to write to the minister, asking permission to engage in a massive project of territorial "improvement," cutting a canal through Languedoc to join the two seas.

Cartographic Tools of Territorial Power

Territorial stewardship depended on precise knowledge of the land, its properties, and its flora and fauna, relying above all on geography and engineering. By mesnagement principles, the first act of rational land governance was to measure the land and determine its qualities, so Colbert systematically developed and deployed a network of clients—geographers, engineers, and cartographers. He began early in the 1660s when he initiated his survey of French forests. The foresters charged with assessing the health of the trees, were accompanied by *arpenteurs géometres*, or land surveyors, who could mark the forests on the ground, and document on paper as well their extent and relationship to rivers and streams that could carry timbered trees to mills. Colbert extended his geographic networks in another direction, too, by employing military engineers for infrastructural projects., He also took an interest in academic geography, establishing both the Académie Royale des Sciences in 1666 and the Observatoire the following year. The minister was interested in connections between geography and botany, taking over administration of the Jardin du Roy and expanding the garden into a major intellectual center. To serve the colonies, he also connected the Jardin du Roy to a series of coastal botanical gardens for the acclimatization of new species and the cultivation of pharmaceuticals for the navy.[66] All these institutions promoted geographic study and provided France with expertise needed for "improving" the land.

Geography in the seventeenth century was beginning to change as the Atlantic economy eclipsed the Mediterranean basin as the center of politico-economic life for Europe. The intellectual uses of maps shifted with this reorganization of trade. Before this time, mapmaking had been centered in Italy and fueled mainly by needs for navigation, but as trade declined in Italy, skilled surveyors and mapmakers there sought out new work. They found it in the engineering projects that were multiplying in northern Italy for controlling rivers, draining marshlands, and extending waterways with canals.[67] Surveys of estates and the distribution of natural resources increased in the period as well. As geographers explored the different technical and social possibilities of their techniques, they elaborated useful practices of the sort that preoccupied Colbert.[68]

Experts in geography were not at all of a piece in the period—socially as well as technically. There were astronomer-geographers, military geographers/engineers, and civil surveyors.[69] Colbert worked with all three types, putting into his patronage system experts with the range of geographic knowledge pertinent to territorial administration.

Under Colbert's ministry, the French state acted as an intellectual "incubator" for geographic work.[70] The diversity of geographers patronized this way could and has been treated as an indicator of the immaturity of geography as a science in the period, but in fact the diversity of these "situated knowledges"[71] provided a basis for distributed cognition about land that proved useful to the administrative development of the modern state.

A system of distributed cognition could only arise if the different participants contributed problem-solving strategies not reducible to the expertise of others, and no one person in the group could dominate their activities and synthesize information from the others, functioning as the

Riquet expose son projet aux Commissaires du Roi et des Etats.

Figure 2.2
Riquet demonstrates his project to experts. Musée Paul-Dupuy.

single "thinker" for the whole. The lack of technical expertise in the administration itself and the variety of cartographic traditions available to serve the minister in late seventeenth-century together helped to assure this; the groups maintained different logics of representation and methods of measurement; and social differences helped to keep their work distinct.

The gentlemen from the Académie Royale des Sciences were mainly interested in mathematical techniques, particularly means of determining longitude and comparing elevations.[72] *Ingénieurs militaire* (military engineers) primarily did surveys to aid in constructing fortresses, building port facilities, planning sieges, and measuring frontiers. And arpenteurs or géometres marked property lines and made maps of land parcels, both royal and private (sometimes for civil engineering projects, and at other times for estate management, tax or legal assessments).[73]

Each group of surveyor/cartographers not only maintained a different tradition of practice but also generally came from a distinct social milieu, which kept the groups socially as well as intellectually isolated from each other. The cosmopolitan mathematicians of the Académie, the traditional gentlemen of the military, and the local civil/forestry surveyors of humble origins could communicate yet did not mingle. Fundamental distinctions both in social rank and ways of life helped to buffer their cognitive styles, and to prevent their becoming a single professional group. Army surveyors did not share information with civil engineers; foresters did not help gentlemen from the academy make their measures.

Colbert was the one person who could call on all of them, but he could not—any more than Riquet—synthesize their data or create a unifying technical vision of land improvement. He simply used the patronage powers of the state to cultivate and give himself singular access to the combined cartographic intelligence of all these groups. The result was that the administration could pull together (when needed) a social intelligence about French territory more sophisticated than the expertise of any one group, establishing an intellectual foundation for impersonal, territorial governance more "intelligent" than the strategic capacities of the king's "enemies." No wonder under Colbert, geography became an increasingly powerful tool for governing France.[74]

Military Engineers. The strategic importance of mapmaking was most obvious in the military where territoriality and legitimate violence met. Both the arts of war and fortress design were furthered by detailed knowledge of the land. So not surprisingly, the military in France had under its purview untold numbers of manuscript maps, and a long tradition of surveying and fortress modeling.

By the seventeenth century, the French became noted across Europe for their extensive use of military maps and geographic models. The visiting Englishman Ellis Veryard described it this way:

> I am inform'd, by such persons as pretend to know the Affairs of France, that this King spends more Money on Intelligence than all the *European* Princes besides, which proves extreamly Beneficial in time of War, and in support of a Government in time of Peace. . . . They never besiege a Town but their Engineers have been privately at work in the place for a considerable time before-hand, viewing the Fortifications, taking all the necessary Heighths and Distances, finding out the Magazines, the quantities of the Stores, the strength of the Garrison, sounding the Depth of the Ditches, and the like; so that they rarely fail of accomplishing their Designs.[75]

Military engineers were probably the most prolific cartographers in France during the seventeenth century and developed distinctive ways of drawing maps.[76] Military engineers were the experts on topography and elevation, both because they needed to know how to move troops through mountainous areas, and because they built and attacked fortresses using topographical differences for advantage.[77]

Successful fortifications exploited the natural attributes of particular places—extant defensive features, especially waterways and rock outcroppings—and were build with local wood, stone, and mounded dirt. Their fortresses were artificial landscapes, using walls, levees, ditches, and

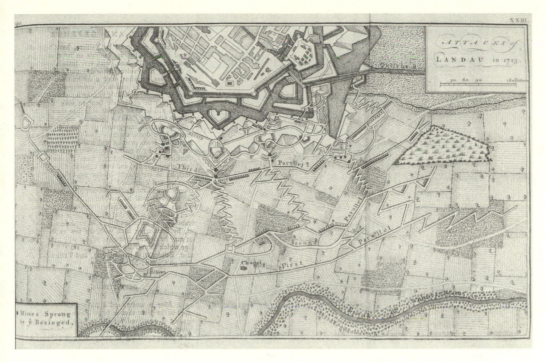

Figure 2.3
Map of the siege of Landau, 1713. Courtesy of the University of California.

water channels to create miniature gorges, valleys, ridges, and mountains. In sieges, attackers created countertopographies of tunnels, *sapes*, and landslides, again using engineering to try to utilize the powers of the earth for strategic effect.[78] Understandably, the details of the local landscape, its soil, rock formations, water systems, and the like, were at the heart of war and military cartography.

Soldiers also depended on nearby food and water supplies, so military engineers noted these kinds of features on their maps. They documented woods, orchards, fields, streams, and relevant topographical elements in relationship to fortresses, ports, or other military installations using common representational conventions so many people could read them.[79]

Although not minister of the army, Colbert was in charge of the king's properties, and this included the fortresses inside the traditional boundaries of the kingdom. That is why the Louis-Nicolas, Chevalier de Clerville, appointed as *commissaire général des fortifications*, worked for Colbert, building ports, dry docks, fortresses, and other infrastructural elements deemed important to improving the king's lands.[80] Clerville and other military engineers brought into the administration the tradition of land measurement and management from the army, adding to the mix of cognitive resources available for territorial politics.

Academic Geography. Colbert used his patronage powers to draw natural philosophers with geographic interests to the administration even before the Académie Royale des Sciences was established as an institution. He both admired and called on Abbé Picard, known for his precise methods for measuring latitude, and tapped his expertise for projects like the water supply for Versailles that depended on precise understandings of elevations over long distances. Since French geography lacked expertise on longitude, Colbert used the Académie and the Observatoire to bring both Jacques Cassini and Christian Huygens to help make the kingdom an intellectual center with practical tools for long-distance travel. Huygens thought he had found a solution to the problem of determining longitude at sea with his clocks. Cassini was studying the moons of Jupiter, which Galileo had argued could be used to measure longitude.[81] Helpfully, Cassini was also an expert on fortifications and collaborated with military engineers on their surveys.[82]

While academic geographers generally focused on measurement issues, they were less interested in the problem of elevation than longitude.[83] They treated land as a planar surface and often looked for the flattest routes through France to make their surveys, or they measured across mountain peaks that they assumed had similar elevations, ignoring topography except as an annoying source of error. With these methods, "the gentlemen" of the Académie made a series of new triangulation surveys of France, cultivating a form of geographic knowledge distinct from that of military men. Academic geographers were also more theoretical as well as more mathematical than their civil and military counterparts. The purpose of geography was to identify patterns in the landscape that were constant across the globe. The famous atlas publisher, Hondius, for example, argued that divides between watersheds had to be mountain chains, eliminating in principle the existence of low passes over the continental divide like those in Languedoc. Contradictions between theory and practice as well as between cartographic traditions added to the complexity of the geographic knowledge in France.[84]

Civil Surveyors. The lowest-ranking surveyors and engineers of the period were forestry officials, tax men, and civil engineers who made plot plans for estate planning, cadastral maps for legal cases, surveys of natural resources, or plans for or of civil engineering projects. These arpenteurs or géometres drew plans of land parcels, using chains of standard length for their measures, and locating places on their maps by name rather than by latitude and longitude.

The administration hired arpenteurs/géometres mainly for tax purposes. But simple "survey" and cadastral work was also used frequently in court disputes, and for land-use planning both for regional infrastructure and estates.[85] Arpenteurs with their measuring chains walked through the countryside and counted the standard lengths of chain (arpents) between property markers. Géometres generally represented on paper the arrange-

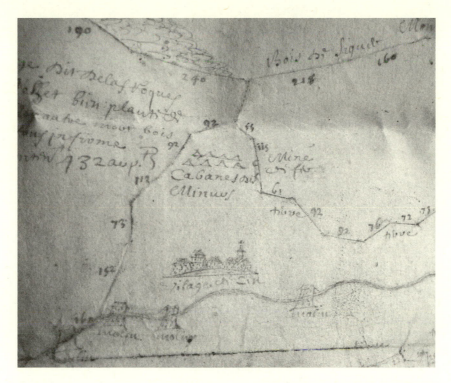

Figure 2.4
Forest survey map by Louis de Froidour. Courtesy of the AHG.

ment of property markers set into the land. Colbert deployed the greatest number of arpenteurs or géometres for his forest reform. They located trees remained in the forests that could be usefully set aside for the navy, and determined how to protect useful specimens for future use.[86]

These arpenteurs and géometres did not place their findings in grids of latitude and longitude like academicians, such as Cassini and Huygens. They did not locate land features within regional maps of France or study topography like the military surveyors, or try to make their maps into freestanding sources of information. The drawings referred both to property markers set on the ground and mental maps of the landscape around them.[87] The utter simplicity of their maps belied the geographic sophistication of many arpenteurs géometres because what they knew was not represented on paper, only the system of parcel measurement that they practiced.

What they knew about the land was local and deep. Because they traversed the local countryside on foot, studying it in detail, arpenteurs géometres developed mental maps with deep understandings of local features— where rivers flowed, meadows flowered, trees grew tall, and gorges cut through the mountains. They felt the topography under their feet, and understood ways to use natural forces and resources, since this was the purpose

Figure 2.5
Estienne arpenteurs.
Courtesy of the
University of
California.

Figure 2.6
Estienne arpent
chains. Courtesy
of the University
of California.

of most of their studies. They acquired knowledge about plants, ecosystems, watersheds, and river systems that was distinct from what military men and academics understood, and equally relevant to administrative engineering projects such as the Canal du Midi.

Precisely because of their simplicity, maps drawn by civil surveyors also could be particularly useful objects for distributed reasoning and collaborative work. On the Canal du Midi, simple drawings of the waterway were frequently drawn, subjected to questioning and redrawn without expense. They were used extensively not only for planning the canal at the beginning but also rerouting it, tweaking the water supply system, locating its dam, and otherwise managing the location of and flow of water through the system. These images documented ideas and agreements, providing temporary representations of engineering possibilities in a public form. They also could "explain" current thinking about the canal to men like Colbert who could not view it directly by focusing on a single, vital element. Some simply marked the locations of locks along the waterway; and others showed the shape of the seawall for the Mediterranean port at Sète extending from the shore into the sea. More finished maps did not have the same flexibility or tentative quality, and took too long to the draw to be useful objects of technical and political debate.[88]

Because of the differences in these traditions of surveying and in spite of the limited information in each genre of geographic practice, the whole array of survey and engineering conventions provided a surprisingly rich, if not always sophisticated, set of cognitive tools for assessing the feasibility of building the Canal du Midi. Colbert had at his disposal the experts who could be called on to collaborate in order to solve difficult problems of territorial governance. Not all of those working for this administration needed to be fine mathematicians or careful draftsmen; neither did all of them need intimate knowledge of woods and streams. They were not attempting to produce a single way of studying and representing the countryside. The strength of French cartographic expertise lay in the differences they maintained— the specialization in survey work that yielded not intellectual disarray but instead distributed knowledge of the earth and its creatures.

So Colbert was equipped to act after reading Riquet's letter proposing the Canal Royale des Deux Mers and studying the testimony of d'Anglure de Bourlemont. The minister turned to his patronage networks for advisers. He asked his fortress engineer, the Chevalier de Clerville, to head a commission of gentlemen assisted by a group of experts, who brought different traditions of surveying and engineering. The commission was charged with studying Riquet's proposal and assessing the feasibility of cutting a canal through Languedoc.[89] The Canal du Midi had become a matter of territorial governance.

CHAPTER 3

Epistemic Credibility

The Canal du Midi was a project that only a king could authorize and only collaboration could accomplish. It was so difficult to build that it tested the intellectual capacities of those who made it; it was risky enough to make its political status always problematic; and it was large enough in scale that it disrupted the traditional habitus of many communities, creating enemies as it stretched to the east.[1]

The scheme held alternately unquestionable and deeply doubtful authority. To the extent that the king supported the idea, his voice made the project real, but the ideas behind the proposed canal were not so robust, and opponents easily and insistently criticized them. It was not hard for them to do. The work was, after all, beyond the formal knowledge of hydraulics in the period, outside the accepted facts of geography, and proposed by a man without credentials. The design was based in part on peasant knowledge of the landscape and improvised into being by a politically forged collaborative network—not developed by a known leader of unquestioned intelligence. Colbert and Riquet may have assembled a group with the necessary expertise to address the technical issues required for the canal, but the project entailed epistemic work as well as engineering skill to give it credibility. Making the canal depended on ongoing struggles about true and false beliefs, until finally the canal itself became a measure of engineering knowledge.

Riquet had no independent standing or credentials to authorize his ideas. He wrote to Colbert, "I am convinced [of the plan,] but I see it in a different way than the others because I have studied it for a long time. I don't find in it the obstacles that others imagine."[2] In spite of his best efforts, serious study was not enough to give his opinions weight. No one in mid-seventeenth-century France was ready to "trust in numbers"; they only trusted gentlemen.[3] Riquet relied more on demonstrations than measures, but even when he used numbers, the entrepreneur was routinely faced with accusations that his measurements were flawed or, worse, faked. His fundamental problem was social credibility. As Steve Shapin has argued, truth in the seventeenth century was a matter of trust and rank.[4] Riquet was not a real gentleman (except in his own mind); he did not have a voice that carried gravitas even when he was heard. Reasoning was something he had to leave to his "betters," and this grated on him. Like a man of humanist France, Riquet thought reason and judgment should prevail over social authority.

"[If] you could give me the honor of listening yourself from my mouth my beliefs on this subject [of the canal,] you would surely find them reasonable . . . and not chicanery."[5] But reasoning on its own did not hold—at least in the patrimonial system of the day.

The canal project needed credibility because Riquet was asking for a large sum from the treasury and needed the king to indemnify huge swaths of land for his canal. So Colbert sent waves of authorities, advisers, experts, and spies to Languedoc who could verify the work or inform on the entrepreneur, creating ongoing struggles with Riquet over authority for the waterway. Still, rather than placing the enterprise on firmer intellectual ground, the proliferation of experts had the opposite effect. No one person was effectively in charge, undermining the epistemic authority of both the enterprise and entrepreneur. Given the conflicting accounts from those interested in its success or failure, it was even hard to know if a canal across Languedoc would hold water.

Adding to the uncertainties, the canal route itself was not fixed at first. Although Riquet wrote to Colbert proposing his canal to link the two seas as if its design was determined, there was in fact a range of possibilities for the waterway. Everyone who thought about it agreed that such a canal would have to connect the Garonne River on the Atlantic watershed to the Aude River on the Mediterranean side. But there were many ways to do it, and even Riquet had more than one idea of how to proceed.

Some things were clear. The waterway would have to cross the continental divide and needed alimentation at the high point from an even higher source. The Garonne River was reliably deep from Toulouse to the Atlantic, although it had strong currents coming out of the Pyrenees and rapids as far as Toulouse. But the Aude River flowing to the Mediterranean was dangerous, frequently shallow, full of rocks, and prone to floods. The rocks could be cleared, but floods would inevitably bring more down from the mountains. So it would be a weak link in the technical system.

How to do the engineering was in contrast quite muddy. It was not at all clear how to assure that the canal would hold water, how to build locks, how to design a water supply large enough for a navigational canal, or even where the canal should cross the continental divide. In light of the uncertainties, the problem of getting a plan that could go forward was in many ways more social than empirical; it entailed designing a project that made sense to state administrators, entrepreneurs, engineers, clerics, soldiers, and nobles.[6] It had to become credible to those who had the authority to know.

From the outset, the Canal du Midi required epistemological as well as technical work—ways to argue matters of fact as well as ways to reshape the land. It necessarily began with speculation and dreaming, but the canal had to become something more reliable and less imaginary—an object of concern that people could address as real.[7] If it had been unquestionably doable, given the formal knowledge of the times, it would not have been so mobile and mutable an object. But in the absence of definitive knowledge, Colbert and Riquet were left with a problem of government—of both things and knowledge.[8]

The problem of credibility was exacerbated by the low social standing of many important contributors to the enterprise. As a provincial financier, Riquet already had doubtful epistemic authority, but his credibility was further undermined when he turned to illiterate peasants and artisans to help design his project. Threading a waterway through the countryside, and giving it locks and dams, benefited from local knowledge of topography, traditions of building, and patterns of hydraulic engineering, yet this kind of expertise lay outside the realm of formal knowledge, and was carried by people of low status and no authority.

Just consulting peasants and artisans about land-management problems was not in itself discreditable. Soldiers of noble rank often worked with artisans and entrepreneurs to build fortresses. Gentlemen routinely relied on "their" peasants to make sense of natural features of their estates and advise them on issues of land improvement. Some locals were recognized as experts on irrigation systems or stonemasonry. Those who labored on the earth and built their communities from it honed their intelligence against the countryside where they lived, and knew the materials it yielded and the problems it presented. The specificity of their knowledge made them particularly valuable to those who had only abstract understandings of nature. Agricultural laborers knew qualities of topography, soil, water systems, weather patterns, and plant distribution—natural knowledge useful for constructing new buildings, developing water systems, and locating trees in the forests for different uses. This experiential knowledge could not be found in books, and it provided an essential underpinning for formal efforts at natural knowledge. That is why the gentlemen of the Académie Royale des Sciences routinely asked locals to help them do their surveys. People familiar with regional landscapes knew where there were high peaks to make distance measures or where there were broad valleys that surveyors could use instead. This indigenous knowledge had no authority, but many uses. So it was not surprising that Riquet would turn to local experts to help design a canal across Languedoc. The problem was that *his* low status gave their ideas little credibility, so his enterprise had an uncertainty that opponents to the canal tried insistently to exploit.

Riquet was more a man of experience than formal knowledge, schooled in finances more than philosophy.[9] Born in Béziers, he came from a family that had been a prominent part of the Languedocien bourgeoisie. As a wealthy financier, he remained part of the same social milieu. His father had been in the États du Languedoc, the provincial tax authority and governing body—in fact, when a similar canal project had been proposed and rejected because of questions about its water supply.[10] Riquet, as a tax farmer for the gabelle, took advantage of the contracts offered by the state. He had made a fortune supplying the army in Roussillon when the French army was fighting with Spain there. Riquet proved himself to be an effective manager, employing large numbers of people to get things done when they were needed.

Figure 3.1
Salt pans on the Atlantic coast. Photo by the author.

He also seemed adept at using the powers of government for the benefit of both himself and the state. This was experiential knowledge salient to building the Canal du Midi.

Riquet also schooled himself in the social and cultural life of Languedoc, developing a deep respect for local intelligence and skills. He traveled extensively for his work, and took note of the region's geography, demographic patterns, and habits of life. He socialized in an elite circle of financiers, but had much broader social reach because of his tax farms and investments. Riquet worked with engineers, artisans, and laborers to run mills, open up mines, and use the natural resources on his estates. In addition, as a farmer for the gabelle, he also learned the technical practices of salt production.

Salt tax farmers regulated salt making and distribution in their region in order to oversee sales as well as tax collection. This introduced men like Riquet to a tradition of indigenous engineering. Making salt was not just a matter of scraping up residues that nature left. Elaborate arrays of salt pans and canals were used to help form the crystals. Salt beds were made up of shallow pools made of pounded clay where seawater was held to concentrate. The salt pans were connected by complicated channels and sluices that kept water moving through the flats, sending more highly concentrated solutions into shallower pans where it was easier to dry and collect the crystals. Salt pans were works of hydraulic engineering with water supplies, reservoirs, canals, and pounded clay conduits—all necessary components of the Canal du Midi.

Riquet also knew intimately the mountains of Languedoc, particularly the area of Montagne Noire. He had lived for the first two years of his marriage in Revel, a town in a high bowl under the peaks of that mountain. He and his new bride had leased a house there before the young entrepreneur made his fortune and moved to Toulouse. Riquet remained attached to the mountains. He acquired properties around the Montagne Noire, both a mill near Revel and an estate to the north at Bonrepos. And he became *seigneur du Bois de la Pierre* at Revel, a forestry post for which he paid the town to collect taxes on timbering.[11]

To reach his properties and exercise the duties of his office, Riquet traveled frequently from Toulouse to Revel. The road up the ridge provided lessons in regional geography. It ascended the Lauragais hills, meeting the continental divide in the region southwest of the Montagne Noire. Windmills dotted the heights of the Lauragais, catching the winds that blew across Languedoc from the Atlantic to the Mediterranean, or in some moments, in the opposite direction. They helped mark the continental divide on the local landscape, indicating how much it rose and fell in elevation from place to place.

On the Toulouse-Revel road near St. Félix, it was possible to see down steep cliffs into the broad valley that lay beneath, looking toward the Montagne Noire.

The *seuil de Besombes* lay below near Graissens where the continental divide crossed the valley. The line separating watersheds crossed almost imperceptibly into the Montagne Noire. To the north, the Laudot River flowed toward the Atlantic. To the south, the Fresquel River headed from

Figure 3.2
Windmill on the Lauragais ridge. Photo by the author.

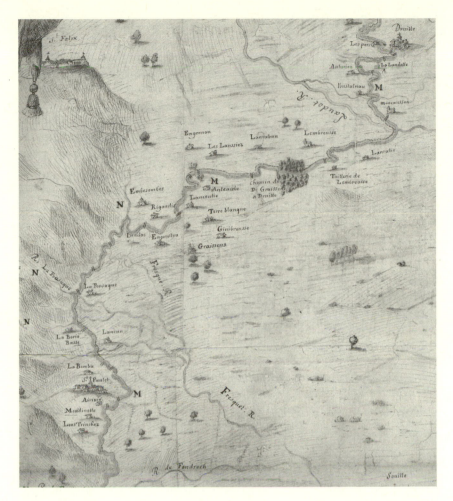

Figure 3.3
View from St. Félix to the Laudot and Fresquel rivers. Unsigned map detail.
Courtesy of the AHMV.

the base of the Lauragais ridge toward the Mediterranean. These were the rivers that Riquet first hoped to connect with a canal.

Riquet may not have understood all the formal topographical features of the area, but from this spot, he could see the rivers from the Atlantic and Mediterranean watersheds almost meeting below the ridge. The prospect provided "situated knowledge" of the region and a "God's eye view" of the local topography and hydrology.[12]

Louis de Froidour, a forestry official who also worked for Colbert doing forest surveys, was asked to assess the Canal des Deux Mers a few years into its construction. This is what he said of Riquet:

The sieur de Riquet . . . having heard that someone once developed a plan to connect the two seas, resolved to try to see if it was possible, and by what means it might be achieved. With the knowledge he had acquired of the region due to the diverse activities he had had in that province, and from the frequent voyages he had made all around the area, it was easy for him to recognize . . . where one could collect the necessary water for this canal . . . [and the divide] where the waters going to the ocean met those draining to the Mediterranean Sea.[13]

Riquet's understanding of the canal was clearly more experiential than technical. He could recognize the feasibility of taking a waterway across the continental divide in Languedoc because he saw a *place* to do it. The question he now faced was *how* to do it.

Collective Memory and Templates

A range of projects for a canal across Languedoc had been designed before Riquet considered his own enterprise. The earliest ones were proposed by engineers who came to France from Italy. When François I gained control of Milan in 1515, he visited Lombardy, and was struck by the number and usefulness of the canals there. He thought comparable structures could be built in France to further navigation. He returned to France in 1516 with Leonardo da Vinci in tow, hoping to profit from this gifted engineer's presence at court. He wanted a canal to connect the Garonne to the Aude River, and hoped it could be made in the Italian style.[14]

There was a long east-west valley north of the Pyrenees and south of the *massif central* that could—in principle—be spanned, using the Garonne to the west and the Aude to the east. It was not as flat as the Lombardian plain, but could conceivably carry a canal. Engineers already knew how to make locks to lift water over minor changes in elevation, and there were no large mountains to scale in this region. To those who saw the broad valley running from sea to sea, the route seemed obvious and easy.

The problem, according to Leonardo, was finding a water supply for the high point of the canal where it had to cross the divide, so that all the water did not just drain into the two seas.[15] He was correct. Locating a higher source of alimentation to feed the canal and bringing water to where it was needed turned out to be deeply vexing. But most early entrepreneurs focused more on the route for the main channel.

In the mid-sixteenth century, Nicholas Bachelier suggested connecting the Garonne at Toulouse to the Aude near Carcassone; twenty years later, Adam de Craponne developed a similar but more modest scheme; in 1598, a Dutchman and dikes master named Humphrey Bradley proposed a shorter canal to link the Lers River by Toulouse to the Fresquel, using these smaller rivers to reach the Garonne and Aude. Some ideas were left unfunded; some languished during war. The last two were trumped by a proposal from Hugues Cosnier for a canal linking the Loire to the Seine to serve Paris—the Canal de Briare. Sully chose Cosnier's canal to support in 1604, and left dreams of a canal in Languedoc to another generation.[16]

There was an additional flurry of interest in canals in the 1630s. In 1633, Etienne Tichot proposed a canal running from Toulouse all the way to Narbonne (the old Roman city near the Mediterranean coast), avoiding the Aude River. A plan for a shorter canal between Toulouse and Trèbes was also proposed, but without effect. Still in this period, the Canal de Briare near Paris that had been left unfinished after the death of Henri IV was bought by two entrepreneurs, Guilaume Boutheroue and Jacques Guyon, and was completed in 1642.[17]

From his letters to Colbert, it is clear that Riquet studied these plans and understood the difficulties of designing a canal with an adequate water supply. That may be one reason why he was drawn to a quite different approach developed by a Huguenot scientist named Pierre Borel. Borel had taught engineering at the university at Castres in the Montagne Noire before it was made a Catholic college. When he was ejected from the faculty for his reformed faith, he turned his attention to improving local rivers to facilitate navigation in the region. Familiar with the low area along the continental divide near Revel, he also thought of joining the Atlantic and Mediterranean ocean and sea with a canal there.[18] He was not able to realize this dream, and went instead to Paris to become a member of the Académie Royale des Sciences and a respected Cartesian thinker.[19] Given Borel's obvious abilities and the tax farmer's attachment to the region where Borel wanted to route his canal, this abandoned project must have seemed like a wonderful enterprise to Riquet.

Castres was a market town north of the Montagne Noire situated on the Agoût River, a tributary to the Tarn and Garonne rivers. Borel thought he could improve the Agoût and Tarn by dredging and removing rocks, making the rivers navigable to the Garonne. Once the Agoût was navigable, Borel wanted to improve the Sor, too, and join the Sor and Fresquel rivers with a canal, spanning the continental divide near Revel. The result would be a water system that could carry boats across Languedoc (fig. 3.4).

This approach was counterintuitive because his canal would not follow the main east-west valley north of the Pyrenees, instead veering north into the mountains and joining the Garonne beyond Toulouse. Still, the plan served Castres and was technically conceivable; in fact, Borel successfully completed a set of locks for the Tarn that allowed boats to ascend the river into the low valleys toward the Montagne Noire. But when Borel got to the Agoût, local landowners stopped the work. He wanted to straighten the river to facilitate navigation, but this required taking land from shoreline estates, and the landowners refused to cooperate, ending the work in 1662—the same year that Riquet wrote Colbert to propose essentially Borel's plan.[20]

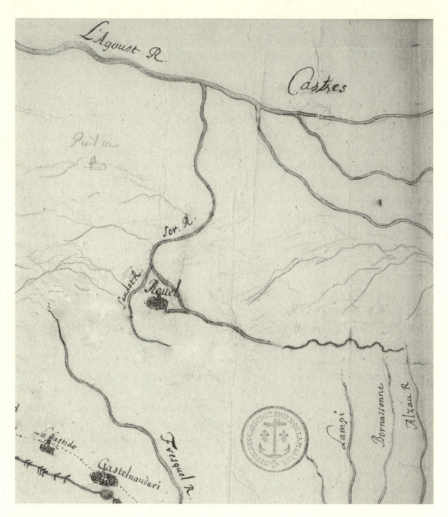

Figure 3.4
Detail of the area between the Sor and Fresquel rivers (retouched).
Courtesy of the AHMV.

Riquet thought he could complete the system that his predecessor had started, probably hoping that as a Catholic entrepreneur, he could gain the support of the king to override the local landowners. D'Anglure de Bourlemont was the bishop of Castres when Riquet first contacted him about the project and surely would have known (and perhaps worried) about Borel's work. The clergyman may well have feared a Protestant's taking control of the infrastructure in Languedoc, giving Riquet a reason to seek his endorsement. Riquet was not interested in Castres per se, but wanted to link the Agoût and Sor to the Fresquel below the Lauragais ridge to serve Revel. He also dreamed of connecting the canal system to a smaller river, the Girou, which ran by his estate at Bonrepos.[21]

Oddly, this plan was in some ways more practical than the canals proposed from the Garonne River at Toulouse to the Aude River near Carcassone in spite of the turn into the mountains. The continental divide in the valley east of Toulouse was lower, but there was no high source of water nearby, and no obvious way to bring alimentation from the Pyrenees. In contrast, the area between the Sor and Fresquel lay right at the foot of the Montagne Noire, where there was plenty of water at a higher elevation. Best of all, Borel's canal could be short and concomitantly easy to build. Riquet only needed the power of the state to indemnify land to sidestep the local opposition that had stopped Borel.

Collaborative Work and Collective Intelligence

To design a water supply on the Mt. Noire, Riquet quite logically turned to a man from Revel, Pierre Campmas, a young *fontainier*. Campmas and his father had constructed a millstream for Riquet, diverting water to power a mill that Riquet had bought near Pont-Crouzet on the Revel plain. *Fontainiers* like the Campmases designed water systems for towns, groups, and

Figure 3.5
Riquet's mill by Pont-Crouzet. Photo by the author.

individuals in this mountainous region full of mills and mines. The young Campmas was an expert on local water sources and their delivery—just what was needed for Riquet's canal.[22]

The millstreams near Pont-Crouzet provided a model of how to move water on the plain of Revel. Four millstreams were diverted from the Sor River, taking water from the Atlantic watershed across the low divide on the plain to power mills on the opposite watershed. Riquet could see that the Sor could in principle be carried over this same plain to join the Fresquel River on the Mediterranean side, using water from the Sor to fill a canal between them. Better, he knew that Campmas could design the channels to do it. The trick was getting *enough* water, and maintaining an adequate level throughout the year. Even if the Sor itself had enough for supplying Riquet's canal, local residents relied on it heavily for their own needs. So it was important to find additional sources to add to its flow. Happily, these were abundant in the Montagne Noire. A diversionary channel or *rigole* could carry water to the Sor, and that was all that was needed to complete the system.[23]

Figure 3.6
Rigoles de la montagne and de la plaine. In *Carte du Canal Royal de Communication des Mers en Languedoc* (1670). Courtesy of the AHMV.

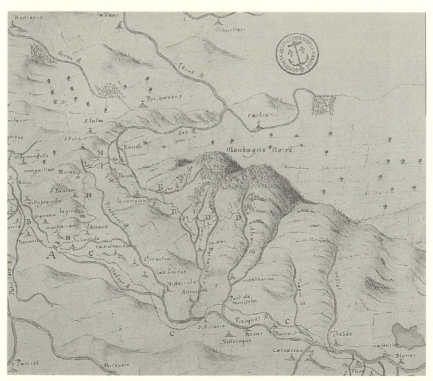

Figure 3.7
Capture of the Alzau,
detail from the "Plan
Géometrique de la
Rigolle faite de
l'invention d ... 1665,"
François Andréossy.
Courtesy of the
AHMV.

Riquet and Campmas turned to the streams on the southern face of the Montagne Noire to add new water sources to the system. These were on the Mediterranean watershed, and the Sor was on the Atlantic one, so they had to capture water high enough to carry it down and around the mountain to reach the Sor. Luckily, the Alzau River—the farthest one from the Sor—started high up the Mt. Noire in the forest of Ramondens.[24]

Riquet and Campmas imagined building small reservoirs near the captures from these rivers that could collect water in winter to dispense in summer when the river flow diminished. The rivers themselves would be connected by a rigole following the contours of the mountain to the Sor. The Sor riverbed itself would be "improved"—deepened and divested of rocks—to hold the expanded flow. The water needed for the canal could then be diverted from the improved Sor near the mills on the plain of Revel.[25]

Riquet and Campmas designed a viable water supply, but they could not produce a convincing plan for the water system without the help of a trained engineer who could assess water flows and map the route on paper. Happily, Riquet already had one in his employ: François Andréossy. This young man had sought out Riquet as his patron, tying his fortune to the wealthy financier. Apparently, he had been recommended to Riquet by d'Anglure de Bourlemont, raising the question once again of the cleric's role in this history. The tax farmer was an appropriate patron for Andréossy in any case, because Riquet had the land and money to be able to use a young engineer. In addition, both men came from Italian families that had migrated along the Mediterranean coast into Languedoc.[26]

Andréossy's family had settled in Narbonne, but François was born and educated in Paris, where he (apparently) studied engineering. In 1660, after finishing his schooling, Andréossy took a trip to Italy to both trace his roots and see firsthand the experiments in hydraulics that had been tried in Lombardy and Padua.[27] He clearly had a passion for hydraulics, and put it to good use with Riquet.

According to Andréossy's family, the young engineer came directly from Italy with thoughts of a canal in Languedoc on his mind, and was the real "author" of the Canal des Deux Mers.[28] But Campmas designed the water system; the route from Toulouse to Carcassone had been discussed for at least a century; and the canal that Riquet wanted to build to Revel was the plan by Borel. Unless Andréossy had brought news of Borel's project from d'Anglure de Bourlemont to Riquet, this claim does not make any sense.[29] Still, the young engineer was important to the project. He calculated water flows from mountain sources, measured the lengths of each section of rigole, drew maps that included topographical features relevant to the canal, and worked on technical specifications. His abilities to map and measure along with Campmas's knowledge of water flows on the Montagne Noire were essential to designing as well as giving credibility to Riquet's proposal.

In the end, the canal was a product of collective intelligence that benefited from a diversity of people and forms of problem solving. It was a child of earlier attempts to build a canal through Languedoc, Borel's technical imagination and political failure, Andréossy's experience in Italy, and Campmas's intimate knowledge of the Montagne Noire. Each of these participants helped to design the waterway, working at the interstices of formal knowledge and vernacular expertise, high concepts and the local terrain. Riquet could imagine an improved Languedoc, Campmas knew what rivers to tap, and Andréossy studied their flows. Their collaboration produced a social expertise that was adequate to the task. What made the canal plan feasible was distributed cognition and collaborative work.

The social origins of the canal's design were obscured in Riquet's letters. He never acknowledged his intellectual debts to others for technical ideas; he claimed to have thought of the water system in a dream even though he clearly derived "his" plan from Borel, Campmas, and Andréossy:

> As to the success [of the water system], it is inevitable, but in a manner completely natural and which no one has ever thought. I count myself among this number, and I assure you that the road that I follow now has always been unknown to me, as much as I have tried diligently to discover it. The thought came to me at Saint Germaine. I dreamt of the means, and no matter how distant I was then, the measures proved true on site: the elevation studies proved true what my imagination had told me. . . . My measures are so well taken that I could not fail.[30]

Riquet denied Campmas and Borel when he said that the plan for his canal was "unknown" to him before he started imagining it. And he ignored the importance of Andréossy when he said his dreams were confirmed by measures without giving credit to the man who made them. Andréossy's descendants saw this as an injustice, but what Riquet did was to present himself as the credible voice for the project—like a true nobleman. The ideas of peasants and servants indeed had no independent authority, so he took the role of author. Still, his claims about inspiration and sole authorship of the project did not in fact *make* him the lone genius behind the canal, as later historians assumed. Neither did they make the ideas more credible at the time.

Collaborative design may have made the Canal du Midi technically possible, but epistemic work was required to make it politically viable. Because no one was clear what would assure a good outcome, and few were convinced that Riquet could judge the facts, developing a plan for the Canal du Midi posed questions of epistemology: How could one be sure that the facts of the plan were true?[31] It was not just that Riquet got ideas from peasants; his knowledge of the divide between watersheds, for example, was based on folklore, a product of neither a canonical flash of genius nor carefully considered measures.[32] Both the *seuil de Besombes* by Graissens and the *seuil de Naurouze* were well-known in the region as *points de partage* or divides. When it rained at Graissens, people said, water flowed in both directions. At Naurouze, the same phenomenon was described in a local legend. The spot was marked with a pile of boulders that, according to myth, had been brought up from the underworld by giants who were doing an errand for the Devil. As evil henchmen often were, these giants were clumsy. Somehow, they lost hold of the boulders, and in a fright, dropped them in a pile. Where these evil-saturated stones landed, water fled in both directions.[33] This was the seuil de Naurouze.

Riquet did not need to find the continental divide with measurements; he had stories of water flows and windmills on the Lauragais ridge to guide him. But what did it mean to "know" the continental divide this way? Folk narratives were charming, but not credible. Local observations had to be verified with measures. So Colbert appointed a commission under the leadership of Clerville, the commissaire générale des fortifications, to study the project. The commission consisted of men of standing partly appointed by the king and partly by the États du Languedoc. The king chose Claude Bazin de Bezons, the local intendant, and the trésorier de Montpellier, M. Bourdon, who oversaw the funds collected as taxes in Languedoc. The États du Languedoc appointed a long list of political and church leaders, representing most of the major towns in the region:

> The bishops of Montauban, Mende & Saint Papoul; the Barons of Ganges, Lanta & Castres; the vicars of the abbey at Chambonas, vicar of Béziers; sir d'Angrain, vicar of Le Puy; sir de Caunes, representative of Consoulens; the Capitouls [city fathers] of Toulouse, the Consuls of Carcassonne, Narbonne, and Le Puy; le sieur Toche-Pierre, Syndic from the area of Vivairais; the Consuls of Castres; the Syndic of the Diocèse of Toulouse, le Syndic of the Diocèse of Saint Papoul; les Synics generaux of the Province of Languedoc, & the sirs de Roquier & de Guilheminet, Secretary & Notary of the États du Languedoc.[34]

The highborn gentlemen of the commission had the credibility to speak about the truth of things, but were hardly in a position to provide engineering advice of their own, so they were assisted by "experts" who could study the plans, survey the route, identify engineering problems, and develop workable specifications for the enterprise. The experts included Hector Boutheroue de Bourgneuf, heir to the Canal de Briare; Etienne Jacquinot, the salt tax farmer for Province and Dauphiné; Marc de Noé, a lieutenant for the king's army stationed at Aigues-Mortes on the Mediterranean coast; and Joseph Avessens, sieur de Tarabel, described as a man of experience from the Midi-Pyrenees, south of Toulouse.[35]

These experts were in turn supported by a large contingent of engineers, surveyors, and artisans who had more practical experience in assessing land and building on it. The surveyors included *ingénieur-géometres*, who could do triangulation studies of large areas of land; *nivelleurs* who took elevation measurements; and arpenteurs who measured lands by placing markers in the ground and using chains of even lengths to count the distances between them. The ingénieur-géometres included Pons, sieur de la Feuille, a trusted client of Colbert who would return to the canal later in its history; Andréossy, Riquet's engineer; and Jean Cavalier, a *géographe du roi* and *contrôleur de fortifications du Languedoc*, who was the most celebrated cartographer of Languedoc, having drawn the canonical map of the region.[36] The other two surveyors, unknown elsewhere in cartography, were nivelleurs and arpenteurs.[37] Other experts assigned to help the commissioners were masons and carpenters-artisans from the building trades. Master builders were routinely asked to judge the viability of civil engineering projects, and had assessed in the past plans for a canal across Languedoc.[38]

The commissioners were charged as follows: "Being transported to the relevant sites, [they] should give their advice about the possibility for constructing such a canal, and the form in which it should be constructed; [and provide] plans . . . and an accounting of the work . . . necessary to do for the construction of the aforementioned canal."[39] This they began in fall 1664.

Designing a Credible Canal

In spite of the language used in the documents, the canal that the commissioners "studied" was not a predetermined object. It was a concept: a canal to cross Languedoc to link the two seas. Its form shifted radically as they examined it—particularly as the head of the commission, Clerville, and the manager of the Canal de Briare, Boutheroue, looked closely at the plan and alternate possibilities.

Clerville was a fortress engineer, who had worked previously with Colbert rebuilding fortresses as well as constructing ports and other infrastructure for the navy. He worried mainly about architecture, sites, specifications, and budgets, and like other engineers with formal knowledge in the period, knew much less about hydraulics. Military engineers used water in fortress defense systems, constructing channels and sluice gates to fill and drain moats; they also thought about water supplies to serve fortified sites, designing storage and delivery systems; but they did not try to carry water over vast distances or complex terrain.

As a structural engineer, Clerville applied himself to the design of ports and considered the hydrodynamics that affected coastal installations. In this, he acquired more knowledge of hydraulics than most engineers of the period, but he still treated port construction as a building problem more than a hydraulic one. In spite of his work on coastlines and harbors, including a new Mediterranean port at Sète, Clerville was no hydraulics expert, so he referred Riquet to Boutheroue to vet the entrepreneur's plan.

Boutheroue was appointed to the commission to study the Canal du Midi because of his experience as manager of the Canal de Briare. The opposite of Clerville, he knew a great deal about practical hydraulics, but little about canal construction, since his father had finished the Canal de Briare twenty years earlier. Still, he knew all the nuances of running a canal that yielded lessons in building one. He knew where the banks ruptured or where drains were needed. Boutheroue was well aware, too, of traffic problems that could be avoided by proper design. The connection between the Canal de Briare and the Loire River, for example, often had to be cut because of high water, strong currents, or dangerous debris; this left barges backed up on the canal waiting for days or even weeks for the river to stop running high. This made him skeptical of using the Aude River in connection with the Canal du Midi. Also traffic was slowed by locks, so Boutheroue recommended using as few as possible, and clustering them in staircases.

Boutheroue was most concerned about the number and size of the locks Riquet wanted for the Canal du Midi. He worried about instabilities caused by their depth and length—concerns that proved later to be well-founded. He urged Riquet to use staircase locks rather than the proposed large ones and argued against routing the canal through the mountains because reaching the higher elevation at the continental divide near Revel would require more locks. He wanted the canal to cross the divide at its lowest point—Naurouze—shifting the route of the waterway from the mountains to the central valley of Languedoc.

Boutheroue could change the route of the Canal du Midi both because he had gained Riquet and Clerville's confidence, and he had the obligation to use his experience to make the canal work. Clerville had specifically *asked* Boutheroue to improve the plan before the commission formally verified it to give Colbert what he wanted: a dramatic infrastructural project that would serve the regime. To do this, Boutheroue walked the entire route with Riquet.[40] For twenty months starting in April 1663, the two men took careful measurements of the proposed canal. This gave them both time to talk and figures to talk about.[41] Boutheroue convinced Riquet that the canal should be routed through the central valley of Languedoc, using Riquet's proposed canal to bring alimentation from the Sor. The result would be a more central and broadly useful waterway, starting at the

Garonne in Toulouse and heading over the divide at Naurouze toward the Mediterranean.[42]

In his letters to Colbert, Riquet seemed excited by Boutheroue's suggestions. He was happy to change the plan; he just wanted one that would work. Boutheroue's ideas for the canal and the credibility of his measures gave Riquet new confidence in the project. He wrote his patron, d'Anglure de Bourlemont, on May 29, 1663:

> For a month, I have been working on the verification of the canal project, but with such care and exactitude that I can speak with assurance at this moment and tell you in all truth that the thing is possible. I will bring you perfected plans and estimates with a calculation of what this work should cost. I have gone everywhere with [the tools for measuring] elevation, compass direction and distance. So that I know perfectly the route, the length of the channel and of the locks, the disposition of the land, whether it is rocky or clay, the drops in elevation and the number of mills that can be found on the routes. In a word, sir, I am ignorant of nothing pertaining to this affair, and the plan that I will bring will be right, being made on site and with great skill [*conaissance*]. I will soon be in Paris about this subject.[43]

The commission met briefly on February 5, 1664, just to agree that placards should be posted on church doors and other public places to announce the project, and ask those who wanted to make competitive proposals for the canal to appear before them in Toulouse on September 30 (postponed to November 7).[44] Then Clerville asked Boutheroue and Riquet to set out stakes along the canal route from Toulouse to the seuil de Naurouze, and asked two local surveyors, Pierre and Roux, to set out similar markers for the water supply along the ground on the Montagne Noire.[45]

The Commission: Social Authority and Technical Assessment

On November 7, 1664, the commissioners finally met in Toulouse and entertained proposals from four different entrepreneurs. Not surprisingly, they found Riquet's plan the most promising and decided to survey the route from the port du Bazacle near Toulouse to Naurouze. They began on November 8 and finished on November 13, declaring that the canal to this point was possible to build—if enough water could be brought to this site.[46]

To assess the feasibility of a water supply on the Montagne Noire, the experts started surveying up the valley under the Lauragais ridge and into the mountains, checking the inclines all the way to the Revel bowl. They reached the Sor River at the town of Durfort, and measured the flow of the river. Determining that the Sor was too small as a water supply by itself, they asked to see the other sources on the Montagne Noire.[47]

Since there was no lodging in the forests, the highborn commissioners did not climb onto the peaks of the Montagne Noire; they returned to Carcassonne to wait for the results of the experts' studies. Their underlings dutifully persevered, trudging uphill to measure the inclines and distances between the Sor and Alzau rivers in the forest of Ramondens.

An unsigned manuscript plan of the canal from 1665 displayed one possible arrangement for the captures in the Montagne Noire, using two rigoles rather than one. It separated the problem of taking water from the Mediterranean watershed around the mountain to the Sor River on the Atlantic side from the problem of capturing water from the high rivers. It even showed a large reservoir (a circular form) on the Bernassonne River and lines across other rivers—perhaps suggesting where earthen dams could be located to create the small reservoirs Riquet hoped to build in the mountains. The map was unsigned but obviously drawn by one of the experts from the commission, since no one else knew enough to do it. It was not by Andréossy, since it had neither his handwriting nor the rigole design from Riquet and Campmas.[48] More likely, it was drawn for Clerville and perhaps Boutheroue, too, trying to reduce the scale and cost of the water system in the Montagne Noire. In any case, it was rejected for the single rigole proposed by Riquet, but still reveals something about the decision making that went into the water supply.

The commissioners met in Carcassonne on November 26, and declared the water system in the Montagne Noire feasible, if difficult to build, and the alimentation available from the high rivers double what was needed. Water for the canal could indeed be directed into and later diverted out of the Sor, the commissioners agreed. The only question left was how to carry alimentation down to Naurouze. This was a problem they postponed considering further until they could verify the viability of the canal proper.

The commissioners continued their studies of the canal route the next day. On November 27, they examined the rivers running east from Naurouze on the Mediterranean watershed that might be used as extensions of the waterway. They described the Fresquel as *"peu de chose,"* or too small to be worth trying to improve. They also declared that the Aude was too wild to use, since anything they did to deepen it could be ruined by a few storms, and the riverbed could easily refill with rocks and debris dangerous to shipping. Riquet explained to Colbert:

> [The Aude] is inconstant, full of rocks, badly set, and of a sort that one could not render navigable except for small flat bottom boats, so that we are all agreed that it would be better to dig a new canal along its shores, at enough distance and elevation to not fear floods from the river. I have gone to see sites where one should route the new canal and what means would allow us to take enough water from that river and not overwhelm [the canal].[49]

From December 1 to December 3, the commissioners studied the possibilities for an outfall of the canal on the Mediterranean—an area of intense

Figure 3.8
Route of the Canal du
Midi to la Robine from
the manuscript map of
1665. Courtesy of the
BNF.

interest to Clerville. He was particularly attached to the idea of improving the old Roman canal, la Robine, to connect the Canal du Midi to the ancient Roman city of Narbonne, where he had lived and still held property. From there, he wanted the canal to turn north to Sète, where the new port was being developed.

But the commissioners found la Robine too silted up, turning instead to the *grau* and small harbor at La Nouvelle as a possible port. But this was too shallow for seagoing vessels to use. So the commissioners considered routing the canal farther north into the étang de Thau to arrive at Sète, the terminus that Clerville preferred. On December 9, the commissioners verified the feasibility of the route from Béziers through the étang de Vendres to Sète.

Following an idea from Bezons, the intendant for Languedoc, they also considered extending the waterway north of Sète through a set of shallow salt ponds along the coast to the Rhône. This would allow trade between the Canal du Midi and Lyons in central France. Colbert was very much in favor of this program, and Clerville was intent on making Sète an important trading hub. This decided, the commissioners and their experts worked their way back through to the Aude valley, taking measurements to see where the canal could be placed in this region to avoid using the Aude as part of the navigation system.[50]

Determining the proper route for the main canal and its alimentation channels still left open the question of reservoirs to ensure the canal could continue running in the dry summers of southwest France. Clerville thought a single, large dam was the most efficient way to proceed. He dreamed of building a reservoir in the mountains by closing the mouth of a valley where the Laudot River flowed.

The valley by Saint Ferréol seemed an ideal spot to Clerville. Few people lived there, the river could fill the reservoir, and there was a solid rock base at the mouth of the valley that would serve as a good foundation for a dam.[51] The dam itself could be constructed like a massive battlement with a frame of stone walls and pounded dirt around and between them. The structure

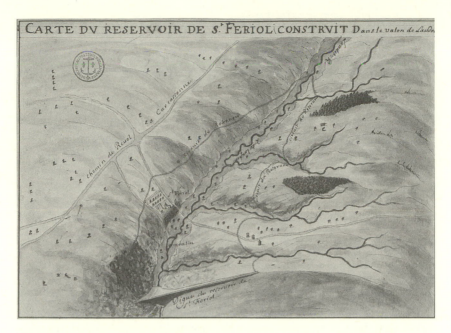

Figure 3.9
Carte du Reservoir de St. Feriol [Ferréol] construit dan le valon de Laudot.
Courtesy of the AHMV.

would have tunnels piercing it near the valley floor to let out water when it was needed for alimentation or when the reservoir had to be emptied for repairs. Clerville thought the water supply channels on the Montagne Noire could be integrated with the reservoir by digging a tunnel through the Cammazes ridge to the *rigole de la montagne.* This would bring all the water from the captures on the Montagne Noire directly into the reservoir, not only leaving Sor River water out of the system but also allowing Saint Ferréol to serve as a settling pond for the water supply.[52]

Riquet was strongly opposed to this approach for two reasons. First, he did not want the water system to bypass the Sor and his beloved Revel. Second, the engineering needed for the dam seemed too ambitious. He wrote Colbert: "These sorts of things are not at all what I know."[53] The water system he and Campmas had designed was simpler to build and probably less expensive. He continued to agitate for multiple small reservoirs and seemed to convince the commissioners. When Boutheroue wrote the estimate for the work for the commission, the "Devis et Estimation des Ouvrages des Experts," the budget only included money for small dams.[54] On the other hand, the document called for ten thousand livres for *twenty* small dams, not just the eight or so Riquet had wanted. So perhaps Clerville purposefully inflated the budget to cover the costs of the Saint Ferréol dam without mentioning this audacious proposal in the report.

Surprisingly, Riquet finally included the dam in his bid for the contract, but not the tunnel through the Cammazes. According to his proposal, the rigole de la montagne would still flow into the Sor by Revel, but the *rigole de la plaine* diverted from the Sor near Revel would then connect to the Laudot River below the dam.[55] What made Riquet change his mind and Clerville give up the Cammazes project? Perhaps it was a compromise so the canal could move forward. Building the dam pleased Clerville and dropping the tunnel perhaps made the project seem more viable to Riquet. Clerville agreed to write the specifications for the bid that Riquet tendered once the commission's proposal was put up for bid, and Clerville appointed a number of his clients to important positions on the canal. Once again, the Canal du Midi was a product of collaborative work.

Constituting the Object

The commission's report finally constituted the Canal du Midi as a stable object of political imagination and technical desire. It was made to seem real with numbers and drawings—not only specifications and cost estimates in the *devis*, but also a map of the canal crossing Languedoc, joining the two seas. Andréossy's "Plan Géometrique de la Rigolle faite . . . pour la conduire des Rivieres and Ruisseaux de la Montagne Noire au Poinct de Partage, Affin de monstrer la possibilité de la Communication des Mers par Garonne et Aude, En Languedoc" gave the canal a form. Circulating this image on paper allowed people at court and beyond to "see" the yet-untried enterprise as though it already existed.

By calling it a "plan géometrique," Andréossy asserted that the map was based on surveys—scientific measurements—witnessed by the commission. It represented the routes of the rigoles through the Montagne Noire and down from the Sor River to the divide at Naurouze. It was a virtual demonstration of the possibility of joining the seas by connecting the Garonne and Aude rivers in Languedoc.

In spite of the commissioners' general support for the canal and the convincing detail of Andréossy's map, some commissioners doubted that water from the Mt. Noire could be directed to Naurouze. As Riquet put it, they said: "I would not receive any of the contracted money if I did not make a convincing demonstration, or more explicitly, [show] by a little rigole, that it is possible to carry [water from] all the rivers [on the Mt. Noire] to the pierres de Naurouzes."[56] To satisfy the skeptics, Riquet proposed to cut a *rigole d'essai*, a ditch along the proposed route to demonstrate its efficacy.[57] Finally, after some political wrangling about funding, the entrepreneur received the *lettres patentes* for the exercise on May 27, 1665.[58]

The rigole d'essai was a tough test. If the water for the rigole had come solely from the Sor, it would have been easier. The mountain's waters could have crossed the continental divide on the Revel bowl near Riquet's mill and

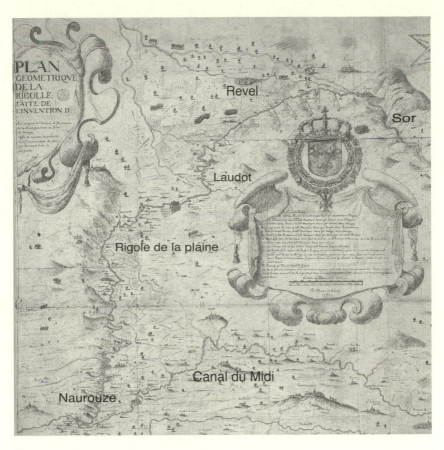

Figure 3.10
Left side detail of the "Plan Géometrique de la Rigolle faite de l'invention d . . . 1665,"
François Andréossy (retouched). Courtesy of the AHMV.

gone south along the Mediterranean watershed to Naurouze without any
major impediment. Designing the rigole to receive waters from Saint Ferréol
complicated the situation; the channel would have to intersect the Laudot
riverbed below the dam, where releases from the reservoir would flow. The
Laudot River was on the Atlantic watershed south of the Sor, so the rigole
now had to stay on the Atlantic side of valley below the Revel bowl. Farther
south of the Laudot River, the next crossing of the divide was by Graissens.
Unfortunately, the *seuil de Graissens* rose above the valley floor, so to reach
the Mediterranean watershed, the rigole had to be elevated on a levee.[59] This
made the "simple ditch" anything but simple to build.

Torrents of rain slowed construction as local rivulets carried away embankments as they were being built. Sandy soil in some regions did not even
hold the water in the rigole. Nasty rumors started to circulate about Riquet's
incompetence. But Riquet wrote reassuringly to Colbert in September 1665:

Figure 3.11
Right side detail from the "Plan Géometrique de la Rigolle faite de l'invention d . . .
1665," François Andréossy (retouched). Courtesy of the AHMV.

> The places by which I pass are favorable to my design, since over
> about 39000 toises of length [about 75,660 meters], one encounters
> only three or four small sandy sites. A quarter [of the test rigole]
> rests on stone, and three quarters [on] the best clay in the world, as
> when waters pass through a marble conduit, these would be no less
> sure [water tight]. . . . I resolved before that in order to conserve el-
> evation to pass the water through conduits on wooden scaffolding
> for about a thousand toises [1,940 meters] . . . but considering what
> that would mean in wood, it might be better anyway to make it in
> stone in spite of the expense and time . . . [but even] without this
> artifice, I am making the conduit with ease and do not need the
> wood.[60]

Finally, by late October 1665, the waters arrived at Naurouze in quantities
that even surprised Riquet, making the alimentation system seem feasible
and no longer just a product of the imagination.[61] In the entrepreneur's
mind, nature itself had become the final judge of his work:

> Bad-intentioned people and the most incredulous will have to avow
> to this proof that what I have made is a good thing. Few people have
> faith in my success, and now that one can no longer look at it in
> doubt, most say that what I've done seems like a miracle and that it

could not be without the help of God or the Devil. I am convinced of the first, sir, and for the rest, they would do me justice when they say of me that I have a modest nature, no artifice, and that I am no magician.[62]

With its moral status in question, even this demonstration needed authoritative witnessing. On November 8, the intendants Bezons and Jacques Tubeuf were sent to the area to examine the rigole, verifying its efficacy to the commission and Colbert.[63] The Canal du Midi started to shift epistemological status, moving from the dream of a tax farmer to a state-sanctioned experiment in territorial engineering. Riquet's bid for the project was finally accepted; the contract was awarded in 1666, and work began on the Canal du Midi in 1667.[64]

CHAPTER 4

New Rome Confronts Old Gaul

The Canal du Midi was not only a piece of difficult engineering that linked the Atlantic and Mediterranean; it was also part of a propaganda effort to promote France as the New Rome—the rightful center of a new European empire. As a simulacrum of Roman Gaul, the project was meant to be a fine counterpart to the grand divertissements in the gardens at Versailles with their excesses of classical symbolism. Through engineering projects such as the canal, the kingdom of France was meant to take on the perfect orderliness of the royal gardens. The landscape was expected to dignify the reign with its glorious subordination, but achieving this was no small task.[1]

Versailles, the model of the New Rome, was decked out for Louis XIV as the palace of the sun, using imagery from Ovid; the *petit parc* was a model of territorial control filled with triumphal arches, classical statuary, and friezes of Roman helmets, arrows, and spears, claiming continuities between the ancient empire and the current regime.[2] The divertissements played out in the gardens included mythological spectacles, enrobing nobles in togas to play gods and heroes in stories scripted by Molière and dances choreographed to the music of Jean-Baptiste Lully. This dream world was conjured up on stage sets with ancient precedents, and in a garden with a vast and complex water system partly modeled on the aqueducts of ancient Rome, and built by the military like the Romans to keep up the army's physical discipline during times of peace.[3] France was made Roman at Versailles, but to become a great empire, it had to exercise this kind of control across the kingdom and beyond.

Colbert was determined to remake France in this image, working in the interstices between art and identity, stewardship and patronage, regional power and state authority, infrastructure and propaganda. He honed the political agenda with his *petit académie* of writers and scholars, and enrolled engineers and artisans to build new fortresses, roads, bridges, and manufactures to make the kingdom more powerful. He dug new ports, too, for trade, colonial expansion, and shipbuilding—all to transform France into a seaborne as well as land-based empire using Roman methods. Colbert could not make France a great *military* power like Rome; that depended on the minister of war, first Le Tellier and later Colbert's rival, Louvois. But the minister of the treasury could make France more Roman on the ground, where social obedience was expected to follow material control.

Into this world of political imagination and administration stepped Riquet, requesting the authority to build a canal where Julius Caesar had crossed Languedoc between Toulouse and Narbonne.[4] It was perfect—a large-scale display of engineering finesse with a classical flavor. A canal through Languedoc was considered (incorrectly but routinely) a goal of the Romans, but beyond their abilities.[5] The claim made intuitive sense, since Roman engineers did link major rivers with canals, and Languedoc was partly straddled by the Aude and Garonne rivers. Only a little political spin and mythological dreaming was needed to turn Riquet's version of the Canal du Midi into a project to prove the efficacy of the "New Romans." Louis XIV's empire would pick up where the old one had left off, using engineering for domination; the moderns would surpass the ancients.[6]

Riquet's project was a crucial test, a critical political experiment. But the dash toward modernity in Languedoc was fraught with missteps—errors that were expensive both politically and financially. The risks of this ambitious program were enormous—all the more so because of the dramatic scale of the Canal du Midi and its distance from Versailles. The threat of failure left the ever-anxious and overbearing minister with the daunting task of juggling resources, expertise, and patronage ties to try to force a positive outcome. No other outcome was permissible to the king.

The obvious risks were technical or fiscal, but reengineering the landscape in Languedoc soon became a struggle of the New Rome with Old Gaul. Languedoc had a long-standing pattern of elite as well as popular resistance to the Crown now activated against the canal. The nobles of the province had no strong reason to further the king's territorial ambitions, and many reasons to keep his hands as far as possible from Languedoc. The powerful families of the region were exactly what Machiavelli saw in France—political actors with entrenched powers and strong local followings. The king's canal tested the loyalties of the provincial governors, the États du Languedoc, threatening local practices of quotidian power, and hijacking the Roman heritage that was a strong element of local identity.[7] The work endangered the countryside, too, turning the lush valleys of the former Roman Narbonnaise into a set of ugly and socially disruptive work sites. From a local elite perspective, the Canal des Deux Mers turned interest in the classical past in a destructive and dangerous direction.

Men of standing in Languedoc were not about to risk monarchical intrusion into their traditional realm. Using their patrimonial networks to carry their views to the court, they attacked the legitimacy of the enterprise with negative gossip about Riquet and the canal. If they could tarnish the image of the enterprise, making the project a political embarrassment to the king, they might be able to stop it. To them, technical risks were political opportunities—just the opposite of what they were to Colbert.

Noble opponents to the Canal du Midi also tried to use their powers over state finances to stop the canal. The king depended on the États for tax collection, so the treasury money for the project had to pass through their hands. By law, nobles were obedient to the king, and Louis XIV was determined that they would subordinate themselves to him. But provincial elites did not have to oppose the wishes of the king directly to undermine them. They could simply fail to execute his orders effectively or put up conditions for obedience that limited their obligations to the Crown. This is what they did in Languedoc with the canal funds. The New Rome and Old Gaul played out their power struggle over the province around matters of technical competence and finances—impersonal matters of importance to the exercise of personal power.

The commission under Clerville was meant to mitigate these problems; that is why it contained local notables who knew nothing of canals. Yet the commission had been charged by the king to determine if the canal was *feasible*, so the commissioners were not free political actors. They could not deny what their studies indicated. So the commission's report was positive, but did nothing to quell the hostility of regional power brokers. The dangers of political subordination to the Crown remained potent in the political imagination of many a local gentleman.

Early opponents of Riquet's project probably hoped that the rigole d'essai would fail and invalidate the enterprise. But the entrepreneur had triumphed in spite of the difficulty of the task. So the detractors remained on guard, watching like hawks for evidence of incompetence, and the États made the enterprise more risky by blocking Riquet's access to treasury funds. The local game was set.

Not surprisingly, local chauvinism had its contradictions, and the canal had a few regional investors and proponents who wanted closer ties to the north, and better connections of Mediterranean Languedoc with the growing Atlantic world. The center of European trade had shifted perceptibly to the Atlantic, and Languedoc had an opportunity to benefit by linking the two seas with the canal. But there were fewer regional advocates than enemies of Riquet's plan among the ruling elite of the province, since they lived more from their estates than trade.[8]

The ordinary people of Languedoc, in contrast, had a different relationship to the project. Peasants and artisans with no land to lose had more to gain from the Canal du Midi. Locals were notoriously hostile to tax collectors and had no reason to love Riquet, but this did not prevent them from wanting to take the tax farmer's money and become his New Romans. Suppliers and carters happily gouged Riquet while serving the canal, charging for their services double or triple what was normal. Meanwhile, artisans and laborers were attracted by the allure of high wages—one and a half what itinerant peasants earned during the harvests. The same autonomy that made nobles so hostile to Louis XIV also gave them vast powers over the peasantry. Working on the canal provided an alternative to the routine poverty that haunted many peasants bound to the land and landed families. Apparently these factors outweighed the hostility toward Riquet, and workers joined the brigades of laborers as well as ranks of artisans and engineers who began to cut the Canal du Midi across the province.

Local laborers made Languedoc more Roman not only with their willingness to work on Riquet's waterway but also their tacit knowledge of engineering that proved directly useful for a navigational canal. The ancient empire had helped form their habitus, and some of its practices remained living parts of local life.[9] Roadways, bridges, theaters, aqueducts, baths, and arenas from the Roman period were ubiquitous in Languedoc, supplying an environment with and against which people shaped their communities. Rome was both a proud memory and an extant feature of the province.[10]

Ancient remains in this part of France had helped keep classical engineering alive by serving as tutorials in Roman technique. Many structures had been abandoned after the fall of Rome, but others continued to provide infrastructure (bridges and roads) or usable walls and buildings for towns of the region. Monasteries had picked up ancient methods of irrigation and craft techniques, too, and so did peasant communities. Even when ancient structures were torn down by local communities during the Middle Ages to build their churches and towns, classical techniques were ironically sustained. Pulling apart ancient buildings taught lessons about classical building methods. The inner secrets of construction were divulged in a pattern of plunder that was also a dissection process.

Figure 4.1
Ruins of a Roman town by Saint Bertrand. Photo by the author.

The desire to reduce quarrying and stonecutting by reusing Roman materials also helped reproduce classical construction methods, since the stones and bricks extracted from antique buildings were easier to reuse with comparable building methods. Remnants of ancient structures, then, became primers of classical practices and helped sustain them as living traditions in Languedoc.

As Elting Morison has argued,[11] Roman construction could be reproduced long after the end of the ancient empire in part because it was efficient and made sense to builders. Techniques for doing masonry described in the *Ten Books* of Vitruvius[12] remained culturally robust even after the collapse of the empire itself. Artisans who carried these traditions into the seventeenth century did not need to know their provenance to value them. The methods of construction "worked," and they served their local cultures.

For example, Roman buildings were generally constructed with walls made of two exterior shells of stone or brick masonry filled with a mixture of cement and rubble. When the cement dried, the result was a strong structure built without precision stonecutting or good contact between most surfaces.[13] The mortar was made with lime or calcium oxide mixed with

Figure 4.2
Roman-style wall.

Figure 4.3
Roman-style wall near the *pont-canal* de Répudre. Photo by the author.

sand and water—the calcium oxide coming from sandstone or marble heated in hot furnaces until it turned soft. This lime could be produced from chips from quarrying, making the method economical as well as efficient.

Roman architects and engineers could build walls this way with both bricks and stones, and could use materials that were irregular, filling the cracks with mortar. The practice produced wide and strong structures by almost literally throwing materials together. These were precisely the techniques used for the bridges, locks, drains, and dam for the Canal du Midi.

Walls were held together with special tricks as well. On arches, some of the stones forming the interior of the vault were also made longer, creating a pattern of radiating spokes inside the structure that helped connect the walls above to the arches themselves. Long keystones hinted at these interior structures. The corners on buildings were also made with rectangular stones alternating between the two walls. They created a sequence of interspersed "fingers" that helped secure the two faces of the building together.[14]

Building this way was flexible and fast.[15] Along the Canal du Midi, mills, spillways, bridges, intake channels, warehouses, aqueducts, locks, laundries, and small harbors were created with these techniques. The New Rome benefited from this reservoir of ancient skills and served the modern effort to "recover" the classical tradition.[16]

As elite opponents of the Canal du Midi fought the king's efforts to build a New Rome in Languedoc, then, peasants from their estates became New Romans, bringing experience with waterworks and construction methods with classical provenance. The result was a political struggle of enormous complexity played out on the landscape. The Canal du Midi started shifting the focus of political dispute away from interpersonal contests and toward material efficacy. The logistics of exercising do-

Figure 4.4
Detail of a bridge over the Canal du Midi with a Roman arch and wall. Photo by the author.

Figure 4.5
Mill for the Canal du Midi with Roman corners and arched doorway.
Photo by the author.

minion over the earth took on strategic importance, raising the stakes of the engineering and heightening the risks of the technical experiments for the Canal du Midi.

Managing the Risks of Building a New Rome

The risks of realizing such a modern enterprise were both multiple and enormous, and much of the political jockeying was over the dangers and responsibilities of the experiment. The task was daunting. The channel had to be 111 kilometers long just from Toulouse to Trèbes.[17] The trench itself had to be 1.8 meters deep and 9 meters wide on the bottom of the channel, sloping out to become 17 meters wide across the top.[18] In the area of first enterprise, there was a 54.57-meter rise in elevation from Toulouse to Naurouze, and a 113-meter drop from Naurouze to Trèbes. To make these shifts in elevation, the canal needed at least forty locks.[19]

Negotiation over the terms for the 1666 contract or "l'Arrest d'adjudication des ouvrages à faire pour le canal de communication des Mers" was a high stakes game, and played with care.[20] For the administration, the risks were massively reduced simply by contracting the project out to the entrepreneur.

The contract gave Riquet full responsibility for the technical success of the enterprise, promising him only limited funds from both the treasury and the États du Languedoc to fulfill his obligations.

Colbert also cut potential losses to the administration by dividing the Canal du Midi into two parts. The "first enterprise" covered the area between the Garonne River at Toulouse and the Aude River near Trèbes, including the basin at Naurouze and the water supply in Montagne Noire. This part of the project could prove the efficacy of the canal system itself, and was about the size of the canal originally proposed by Riquet. The second enterprise from Trèbes to the Mediterranean had been added by the commission to bypass the Aude, but was an extra expense that made no sense to take on if the first enterprise failed. By giving Riquet a contract only for the first section of the project in 1666, Colbert could see if the canal could indeed cross the continental divide using water from the Montagne Noire.

Colbert also tried to reduce the financial drain on the treasury by requiring Riquet to shoulder more of the costs and offering him as compensation the contract for a new tax farm in Roussillon. That province had recently

Figure 4.6
Map of the area of the first enterprise. Courtesy of the AHMV.

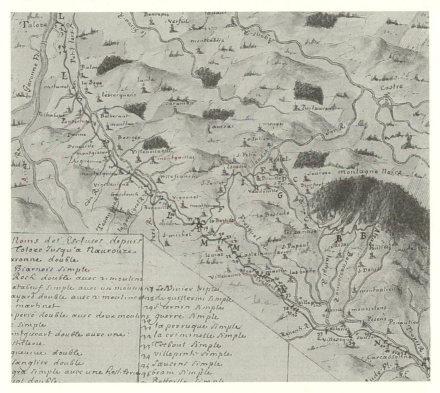

been acquired by treaty from Spain, and the minister wanted to impose French taxes there immediately. Riquet had been successful with the gabelle or salt tax in Languedoc in spite of a history of rebellion against that tax in his province. The inhabitants of Roussillon would surely not want to pay a salt tax either, but as part of France, they were subject to the same laws. Of course, trying to impose the new tax would be a gamble, too. Riquet would have to promise money to the king, yet there was no guarantee that he could collect it from Roussillon. Worse, since tax funds were deposited in Montpellier under the nose of the États, any failing would be evident to his enemies. This was a gamble that Riquet embraced but would regret.

Without doubt, Riquet carried the greatest burden of the technical, financial, personal, and political risk for the Canal du Midi. He gambled with the new tax farm, his family's fortune, his relationship with Colbert, and his reputation among provincial leaders. He hoped to place his sons in good offices through Colbert and raise his family's standing in Languedoc because of his connections to the court. He also assumed that tolls from the canal would bring his family substantial new income. In return, the family would commit to maintaining the canal. It was a huge risk for a massive potential gain.

Riquet tried to manage the risks of the canal project by asking that his family be raised to the nobility, and insisting that the land for the canal become his domain or fief—land held by the family in perpetuity. He claimed that his Italian ancestors had been nobles, so the king could restore the family's nobility, giving him legal authority as well as property rights over the indemnified lands. In this way he could protect what he built and leave his family assets of lasting value. Financially, he would take his chances.

The other party with the most at stake was the États du Languedoc. By opposing the project, the États entered into a struggle with the Crown not only over the canal but also over the region's political autonomy. They had refused to fund any bit of Riquet's plan, but the king insisted, claiming that the benefits of the waterway would go mainly to the people of Languedoc. The États had no way to refuse the monarch outright, since they nominally served him, so they had to agree. But in turn, the États demanded control of the budgetary process.

The canal was given a construction budget of 3,630,000 livres to be allocated over eight years in sums of 453,750 livres per year. In addition, the king would pay the indemnities owed to local landholders for taking parcels for the canal by eminent domain.[21] The funds for either purpose were contingent on verifications overseen by the États, and the monies were held in the bourse in Languedoc.[22] A permanent commission was set up to oversee the verifications, but the costs of these inspections were left to the entrepreneur.[23] Finally, if the project was not completed in eight years within the prescribed budget, the treasury and États de Languedoc had no further responsibility to the entrepreneur.

The deadline for completing the canal, the ambiguities about selecting experts for the verification process, and the lack of funds for doing these studies all provided the États with ways to impede work on the canal without directly contradicting the king's orders. They had to pay part of the costs for the canal, but only when they thought it proper. They retained the power of the purse and had a time limit to hold over Riquet's head.

Almost immediately after the 1666 "Arrest" was published, Riquet found that his worst problems lay not in the terms of the contract for the canal but rather in the new tax farm he had acquired. Rebels in Roussillon were killing tax collectors—encouraged and protected by Catalan peasants of the mountains who were disdainful of the French occupiers. Their violent reaction to the gabelle was not unexpected, yet it was more effective than anticipated. Riquet worried about his contract with the treasury, so he started making frequent trips to the provincial city of Perpignan to consult with local authorities about what to do, leaving work on the canal to his employees, who quietly took up their role as architects of the New Rome.

The New Roman Construction

Colbert hoped and Riquet expected that building the Canal du Midi as a monument to Rome would be a matter of following the recommendations of the commission and specifications of the "Devis et Estimation des Ouvrages des Experts" that provided technical guidelines for the work. But while the construction on the Canal du Midi could sometimes be quite straightforward, often it was not—and frequently in surprising ways. Those following the specifications soon found that much important technical detail had been left out of the devis, and some of the designs were inappropriate to the tasks at hand.

Clerville had obscured the extent to which the engineering was too novel to anticipate precisely. He said that work should be done according to conventional standards—terms familiar in contracts of the period. But for new engineering, there were no conventional standards to be followed. No one knew how to build deep locks. No one knew how to maintain a large supply of water from the mountains for alimentation. No one knew whether anyone could in fact build a dam strong enough to make a reservoir out of an entire valley and one that would be reliably watertight over time. For the novelties that the canal required, designs had to be improvised on the spot, forcing the participants to confront the depth of their ignorance, and scramble to extrapolate solutions to problems from their extant repertoires of knowledge and practice.

Even just walls could be problematic. The dam at Saint Ferréol was impressively difficult, but succeeded because it could be built mainly with known principles of military architecture. The same was not true of the much smaller walls of the locks. Riquet worried more about Saint Ferréol, but the lock walls proved much more unstable, wobbling enough as the water moved into and out of them to become loose over time.

As problems multiplied on the Canal du Midi, the lack of clear social or technical authority to guide the work became palpable.[24] The project was supervised, of course, but no one had the expertise to say what to do. The lines of authority were clearly drawn though functionally misleading, and served more for bookkeeping purposes than engineering supervision, since no one could say for sure what to do. Clerville had a network of clients who had collaborated with him previously on infrastructural projects for the Crown, and he recommended many of them to Riquet as supervisors or *contrôles* for the Canal du Midi. These men had experience in building structures according to period rules of military architecture, and Clerville asked Colbert and Riquet to trust them to make the project work, writing the entrepreneur: "I've avowed [to Colbert] that if there is any fault [in the work on the canal], it is mine for giving you the men who supervise it. . . . I gave him my word and promised his satisfaction."[25] Clerville presented the success of the canal as a matter of supervision and trust, an artifact of patronage politics, but the Canal du Midi was a modern experiment that surpassed the experience of his men, requiring that the canal be built through improvised, collaborative work.[26]

If technical limitations provided one barrier to erecting a New Rome in Languedoc, the regime's conception of Rome itself was another. The ancient empire was not just a historical entity but an ideal of perfection as well. So to serve the politics of empire, the works in Languedoc had to "stand for the power of a grand king who had them made . . . [and] they must, if it is possible, endure until eternity . . . [and be] done to such perfection that the men ordered by his majesty to inspect them will find nothing there to redo."[27] The New Rome had to be monumental and robust—not simply grand—to carry the regime's legacy through the ages.

To make the canal *seem* Roman in this context meant inventing a legacy through technical innovation. For example, the main channel was dug as straight as possible like a Roman road, but routing a canal this way (rather than seeking the most level ground or gentle slopes) made it necessary to use locks to take the water down steep hills. In this case and many others, making France seem Roman entailed the use of modern contrivances.

Some ancient techniques also made the structure seem *less* rather than more Roman. For instance, Vitruvius recommended using wood rather than stone and concrete for construction in water, so wood was widely used at first in the channel and locks for the Canal du Midi. The walls of the waterway were stabilized not with stone facings but rather wooden palisades and *chandeliers* or *corbeilles*, thick bundles of twigs tied into log-shaped forms. This method was Roman but did not assure that the waterway would become a monument to "endure until eternity," and became all the more suspect when some parts of the channel built with *corbeilles* failed in sandy soil.[28] Worse, the walls for the locks on the first enterprise were made with timber and soon threatened to collapse. The Roman advice to use alder for underwater construction had made sense to Clerville and pleased Riquet because of the lower cost, but ancient advocacy had not made the practice worthy of the New Rome. Colbert, thought wooden structures were flimsy, and took the lock problems as corroboration for his view. The classical provenance of the practice was not enough to recommend it for this regime, or to make it appropriate for the Canal du Midi.

Other unglamorous techniques with classical provenance were not flashy enough to sustain French dreams of empire. Using clay to line the water channel was one example. Romans had routinely dug their canals down deep enough to reach a layer of clay to assure that the channel would hold water. On the Canal du Midi, the depth of the trench was specified in the devis, so where the underground layer of dense earth was too deep to reach given the specifications for the canal, workers applied a thick layer of clay, and pounded it into the base and sides of the channel. This quasi-classical technique was effective, but clay was not a material for the New Rome. Lining salt pans with clay was a familiar method used on the salt flats that Riquet knew well, and that local workers could easily apply to canal construction, but it was also understood as a peasant technique, not a defining element of the New Rome.[29]

Conversely, some design elements on the Canal du Midi with unknown provenance were "marked" as classical by using names that linked them etymologically to the Roman tradition. For example, since Italian military architecture was associated with ancient technique, when artifacts on the Canal du Midi were named after elements of battlements, they took on the aura of Rome. Ditches on the inside and outside of the canal, for instance, were called *fossées*, and *contre-fossées*, and interspersed with levees to resemble the complex contours of battlement walls derived from principles of classical architecture. They stood for a military power associated with Rome.

The point of the levees, ditches, and platforms along the canal—like that of fortresses—was to create an artificial topography for defensive purposes. The flanking levees protected the canal from the floodwaters that periodically ravaged the region, and ditches along their exteriors carried excess rain away from the Canal du Midi. Along the canal itself, a wide ledge provided a horse path for pulling barges. The alternating trapezoidal

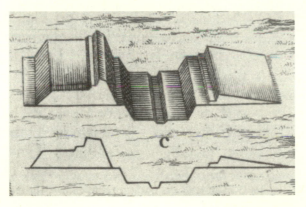

Figure 4.7
Fortress wall profile,
Mallet. Photo by the
author.

Figure 4.8
Levees on the uphill side of the Canal du Midi. Photo by the author.

forms of levies and ditches looked like a battlement wall, and justified the military terminology that associated it with the military power of Rome.[30]

If Roman grandeur was hard to represent in the trenches for the Canal du Midi, it was realized at the Saint Ferréol dam, where dirt and stone combined to make a giant structure strong enough to hold up against the force of a valley full of water. Dams and reservoirs were in fact more familiar elements in Moorish water systems than Roman ones, but the dam at Saint Ferréol was grand enough to be worthy of the ancients.[31]

Work began on the dam in January 1667. Delays in indemnifying estate lands in the valley near Toulouse turned the attention early toward Saint Ferréol, where land was quickly available. As a result, the dam was one of the first structures to be started. Beginning construction on the water supply made technical sense, too. As Louis de Froidour put it:

> Since the success of the canal depends on supplying adequate water to Naurouze, they also built a reservoir in the mountains to capture the water from all the little streams there. The location is a valley below the small town of Campmases, which is just below Revel, which they call Saint Ferréol because of a *Metrairie* nearby with this name. The Audant [Laudot River] passes through the valley, and this stream brings water from the storms and snows of the area. Mountains run along two sides of the valley. At the end of the

Figure 4.9
Wall and tunnel system for the dam at Saint Ferréol. Courtesy of the AHMV.

valley is a stone outcropping of such height it could be called a third mountain. The foundation of the dam is of solid masonry, all set onto this rock. On the bottom is a small vault for letting water out of the reservoir. . . . Above this they have erected a large wall.[32]

The dam was an extraordinary structure. Its longest wall spanned 975 meters and rose 49 meters high. The set of three walls plus the earthworks between them created a barrier across the valley 119 meters deep pierced by a tunnel (*voûte*) for releasing water from the reservoir.[33] There were two outlets for the tunnel: one on the valley floor to drain the reservoir completely for cleaning or repairs, and one higher up the wall for releasing clear water for alimentation.[34] The tunnel itself was 3 meters wide and 4 meters high, the painstaking work of masonry taking most of 1667 to complete.[35]

The three main walls of the dam all had different proportions and purposes. The wall on the reservoir or interior side was short and squat, providing a sturdy barrier where the force of the water was most dangerous to the structure. It was roughly 20 meters high and 12 meters deep, and was reinforced with a wall of dirt that sloped into the reservoir.[36] The highest wall at the center of the dam—set 64 meters back from the first—was tall and thin. It was only 6 meters wide, but nearly 49 meters high, defining the high point of the dam.[37] The third wall on the outside or downhill side of the dam—60 meters from the central wall—was 8 meters wide and roughly 30 meters high at the center. Its wide foundation was set into the rock of the valley floor to anchor the whole structure and contain the pressures from the water against the void of the vale beyond.[38] The wall system straddled the valley like a small ridgeline, mimicking the hills behind the reservoir that rose into the Montagne Noire.

The walls themselves had specially designed masonry. Artisans knew that water could find small fissures through a structure, particularly as the walls settled over time when mortar was prone to crack. Leaks could undermine the dam, so they built double walls and filled them with a thick layer of pounded fine clay to keep water from seeping through. Workers pounded another thick layer of clay into the huge cavities between the three walls before they were filled with stones and dirt. The result was a huge reinforced mound, with a set of walls multiply reinforced with clay to be watertight and stand up against a huge valley full of water.[39]

Riquet visited the dam site only periodically because he was called away so often to Perpignan about the Roussillon taxes. When he came, it was mostly to make financial arrangements for materials or authorize new work, such as the construction of buildings to hold the tools needed by the stonecutters, masons, carpenters, and cement makers. Clerville was also infrequently on site, serving the king elsewhere, leaving his devis and supervisors to give the work direction. In this unlikely situation, the dam at Saint Ferréol started to rise successfully from the valley of the Laudot River—a work of hands that carried the tacit knowledge of how to proceed.[40]

Figure 4.10
Reservoir at Saint Ferréol. Photo by the author.

The small workforce came from the nearby towns. In February, a master carpenter named Élie Nègre from Sorèze was hired to supervise work on the walls and discharge tunnels. He helped with the foundations, meanwhile ordering materials for the voûte. In July, masons started to erect the arches for the tunnel. The massive amounts of timber ordered at this point suggested that the discharge tunnel was made in the Roman style on timber scaffolding.[41] The dam was an example of what trained engineers and artisans could do together, using classical building techniques to address new problems of engineering. The resulting hybrid was grand enough for the New Rome.

The octagonal basin at Naurouze was a more modest structure charmingly designed to evoke Rome. Riquet wanted to build a town around the basin in the ideal city tradition—a type of urban design derived from Vitruvius and developed by Italian military engineers around the turn of the sixteenth century.[42]

Ideal cities were generally set out as pentagons, hexagons, or octagons, and contained streets designed either in grids or concentric iterations of the larger geometric form. They were oriented geographically as well as geometrically—ideally aligned to the wind directions to encourage the

Figure 4.11
Octagonal basin at Naurouze with a street plan, Louis de Froidour.
Courtesy of the New York Public Library.

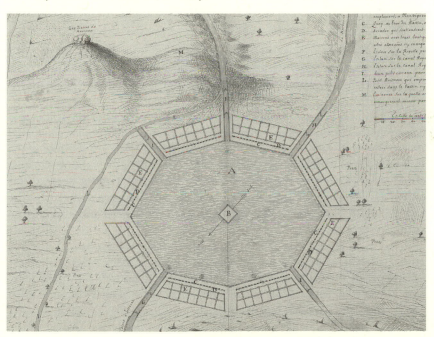

healthful circulation of air, but also to prevent gales from debilitating the inhabitants.[43]

Naurouze could be imagined as an ideal city because the holding tank was shaped as an octagon, and was in a site where there was an abundance of good stone to quarry. The *pierres de taille* removed from Naurouze provided such good building material for the canal that the excavation was more extensive than originally intended, and produced a huge holding tank—twice as large as the port on the Garonne.[44] The scale of this grand basin fed the entrepreneur's dreams of turning Naurouze itself into a great port at the intersection of the rigole de la montagne and the Canal du Midi. The abundance of stone on-site also made it seem financially feasible to construct a town there with the adornments that Riquet desired. According to Froidour:

> There is little left to do to make [the basin at Naurouze] usable, but there is lots to do to embellish it in the way one has resolved to do. Primarily, [the holding pond] must be finished with good masonry, and surrounded with a beautiful quay. And one must build a town that has geometrical streets and has a measured exterior and pavilions like those of the Place Royale in Paris. All the houses should be equal in size and built one like the next. There should be beautiful and imposing arcades along the quay, so one could walk there in the shade. There must be a parish church or convent with an arsenal or magazine, where one could always repair boats inside and would have the materials needed to build new ones and to equip them for service. They have made a plan in relief of [Naurouze like those made] of the fortresses of [the king's] new conquests, and of other places in your part of the country.[45]

The port at Naurouze never met these ambitions, but the plans for it testify to Riquet's determination to set along the Canal du Midi grand monuments in the classical tradition that were at once usable pieces of infrastructure and markers of the New Rome.

The Roman theme of the enterprise was not only registered in structures but also in the organization of the workforce. Laborers for the Canal du Midi were placed in brigades or military-style ranks echoing those of the Roman soldiers who built the infrastructure for the ancient empire. Colbert could not offer Riquet a labor force of soldiers, since he was not the minister of war, but Riquet could provide the minister a labor force that evoked the military. Each brigade had a standard size and a rigid command-and-control structure.

The work groups contained two hundred laborers. These ateliers were managed by a *chef d'atelier*, and were broken down into five brigades of forty laborers each supervised by a brigadier. The brigades could contain men, women, or both, but with a conversion formula by gender: three women counted as two men. This meant a brigade could contain forty men or sixty women, and would get the same pay. (The lower cost of women laborers was one of the reasons that women were later recruited in large numbers for the project.) Two ateliers—in all, four hundred workers—were overseen by a *contrôlle sédentaire*, whose job it was to record who was present or not for work.[46] Books from each contrôle were collected by the *contrôleurs généraux ambulants*, who would check that the work matched the expenses and get the money owed to workers from the *contrôleur général*. Skilled laborers like engineers, locksmiths, smithies, and stonemasons were assigned to brigades, yet were not counted in the group of forty. Such skilled workers were fewer in number and better paid, but under the same chain of command.[47]

The strict lines of authority, and the use of precise numbers of laborers in each group, created a beautiful illusion of order and control at the work sites even though this approach was not efficient for getting the work done. Some sites needed more workers than others, particularly the dam in the Montagne Noire. So Riquet began to pay extra laborers by the basket to carry dirt uphill to fill the huge pans between the walls, abandoning the command-and-control structure. Still at most work sites, the military organization of labor was sustained. The tight system of control countered any questions about supervision of the construction, and the regulated methods of payment rationalized labor costs, while also bringing classical patterns of discipline to the work site.[48]

The Canal du Midi was most vividly and explicitly evoked as Roman in rituals, particularly at the opening ceremony at the Garonne River by Toulouse in fall 1667. This event was organized by the intendant of Languedoc, M. Bezons, ritually transferring the land from the province to the state and from the town to the canal.[49] The workers were arrayed in their military-style ranks like the soldiers in Roman times, and they were led not by Riquet but rather by an array of notables, representing the men of power in the region. All opposition to the project was momentarily suspended during this ceremonial event, when France edged closer to Rome. Père Matthieu de Mourgues, *inspecteur du canal*, described it with these words:

> On the 17th of November, the notables of the town, the ancient and new Capitouls dressed in red and black, the clergy and Parliament in grand attire with all the attributes of their rank, paraded to the walls of the capital city of Languedoc. They met there the workers for the canal, without whom nothing would have been accomplished. The *chefs d'atelier* [stood] in front of their brigades of workers; there were close to six thousand *terrassiers* set out in battle order, drums beating. It was a powerful site: the [notables of the town were] lined up behind the cross [parading] along the still-dry basin of the canal. The procession of authorities . . . flowed like a current into the middle of the enthusiastic crowd massed on the open banks of the channel, the people of Toulouse and workers mixed together.

The canal itself was portrayed as an arena like those of the ancients. God even appeared to bless this New Rome.

> They shouted out cries of joy, "Vive le Roy," or in the words of the author of the Annales de Toulouse, "[They] formed a kind of amphi- theater and provided a sense of the spectacles of the ancient Ro- mans." [This] perfect assembly of the powerful and honorable mem- bers of society . . . came to the place for the foundation of the locks [to link the canal to the Garonne]. The archbishop of Toulouse took the first two stones in his hand. He blessed them, giving one to the president of Parliament [the legal system that constituted part of the importance of Toulouse] and the second to the Capitouls [the mu- nicipal fathers]. A little mortar was taken with a trowel of gold from a silver plate, and the stones were placed. To the joy of the people, commemorative medals were thrown into the crowd, and Riquet had wine and liquor distributed, as the artillery massed on the banks of the Garonne fired as did the musketeers. God was present. In this ceremony marking the opening of the canal, it was the 17th of No- vember, but it was like a spring day, which people took as a good sign for the project.[50]

The November 17 ceremony was also commemorated on the site with plaques written in both Latin and French, equating the accomplishments of the moderns with those of the ancients. As Mourgues suggested in his com- mentary, this ceremony put memory, politics, ambition, and power all into play, with France starring as the New Rome.[51]

The Ancients and the Moderns

All the evocations of Roman greatness, as noble as they tried to make the enterprise seem, were undermined from the beginning by the technical troubles and political struggles that would plague the Canal du Midi. Angry local opponents denied the evocations of the ancients with "modern" talk of natural knowledge and material competence.

The gentlemen of the États voiced their opposition to early attempts to indemnify land in 1667—the same year as the opening ceremonies at Tou- louse. The political motivations for their complaints were clear: no land, no canal. The valley between Toulouse and Trèbes was fertile and well popu- lated, and landowners there did not want to give it up. So the États stalled the indemnifications as long as possible, while also insisting that the terms the king would offer for land were at least generous. They argued that the monies originally tendered for the parcels were insufficient because the tech- niques for assessing the land were faulty. They had a point. Land with mills, orchards, roads, or other amenities on it had been assessed at the same value as undeveloped properties. Thus surveyors started again, this time inventorying "improvements" that affected land value.[52] Given the eight-year limit to the contract, these calls for new verifications were multipurpose political tactics.

The route of the canal to the Garonne River by Toulouse also became a flash point for the struggles between the old Gaul and New Rome in 1667, when technical language was clearly mobilized as a political weapon. Riquet proposed to run the canal into the city moat to reach the porte du Bazacle on the Garonne River where the entrepreneur was preparing to build a port for his canal. M. Arquier, the senior member of the governing Capitouls, presented this proposal on Riquet's behalf, and the opposition fought back led by an "expert in canals," Jean Nivelle.[53]

Nivelle was contemptuous of using the town moat for the canal, arguing that it was feasible but would threaten the city with floods. He ridiculed Riquet, suggesting that the entrepreneur lacked both the insight and ability to take on such a great project: "[Riquet] presents fictions, which have more in common with fables [*romans*] than with facts [*histoire*]. One doesn't know how to respond to proposals in cases like this where measurement and

Figure 4.12
Map of Toulouse with possible routes for the Canal du Midi, Bellefigure.
Courtesy of the AHMV.

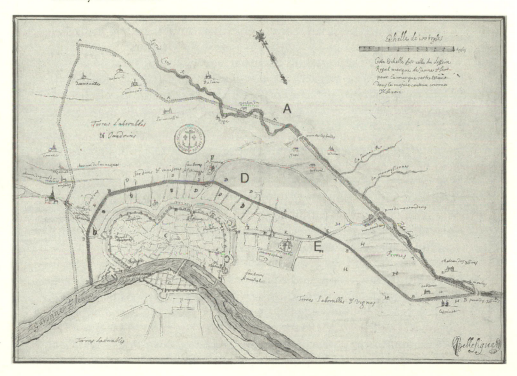

verifications should be worth more than discourse."[54] Although Arquier insisted that the plan had been "demonstrated" physically on the land itself, Nivelle called for a higher level of technical sophistication, attacking Riquet's lack of precise measurements to verify the plan.

Nivelle treated Riquet's proposal to build a new port by the Garonne even more dismissively:

> *This is nothing but an idea, and one of enormous expense.* It is better to ignore it than to talk of it.... [The entrepreneur] does not understand what other men comprehend that *our knowledge of engineering does not extend to this.* There is no knowledge of how to make the water go . . . this way. . . . There are no machines or tunnels adequate for this.[55]

Nivelle attacked the proposal on technical grounds, but for political purposes. The Capitouls had wanted Riquet to run the canal into the Lers River, a small river north of Toulouse, improving the riverbed for navigation. This would have routed the canal through the cheaper land far from the city, not the more desirable properties on the adjacent hillsides.[56] Nivelle not only ridiculed Riquet's counterproposal but also questioned his ability to understand the technical issues, characterizing the entrepreneur as more a propagandist than an engineer.

> It is fair to say that the author of this proposal is *full of zeal for the public good*, but he apparently *does not know anything*, nothing of water, nothing of canals, nor does he know anything about locks, or even that the canal would have to cross the river Lers to get to Toulouse. . . . *The arguments presented here without verification are bad ones, because they lack the most important basis, that is, telling the truth.* The entrepreneur for the canal has reason to advance his projects without resorting to fictions; if he continues this way, *he will never see the end of this work.*[57]

Riquet and Nivelle were interested parties to the project; still, they argued about the truth of things. When Nivelle contended that Riquet proposed a canal that was beyond the technical capacity of engineers in the period, he was using the epistemic vulnerability of the enterprise for political advantage. He could only protect his patrons in Toulouse from the Canal du Midi by posing valid questions. And Riquet could only promote his project by making it seem real in spite of his lack of authority over natural knowledge.

Not surprisingly, the Canal du Midi was routed around Toulouse, finally meeting the Garonne at the porte du Bazacle, as Riquet and Boutheroue had planned, but arriving there through the suburbs west of Toulouse. Local opponents had succeeded in changing the route, but not stopping the canal. But Nivelle warned the entrepreneur what to expect in the future: "If he continues this way, he will never see the end of this work."[58] Old Gaul could not face down the king but could make it difficult for Riquet to build the New Rome within the time specified by the contract by demeaning every proposal he made on technical grounds.

Failures of the Modern

Worse for Riquet, there was mounting evidence that he could indeed be incompetent. In the first year, his "engineering" near Toulouse seemed to produce no more than mud-filled ditches and tipping walls. The project spoiled the land more than it seemed to improve it and politically, it exacerbated the struggles between Louis XIV and local nobles, threatening to splinter and weaken the state rather than empower it.

Mistakes multiplied and became political lightning rods. Locks were the most obvious and dangerous example. The locks on the Canal du Midi were designed poorly with dimensions that were too large for their shape. The deepest ones near Toulouse began to fail as soon as they were built, bulging and tipping inward.

Riquet and Clerville had based the specifications on precedents, although it might not have seemed this way. The length of the locks was based on dimensions already used on local rivers like the Tarn, where shallow locks carried boats through rapids. The depth of Riquet's proposed locks was based on the height of the doors that the Dutch built to the sea. Without a good understanding of soil mechanics and the effects of water on soil, there was no way to calculate the forces on lock walls. So precedents rather than principles were used for the design. In themselves the dimensions were not preposterous, but the combination of proportions made them huge and unstable.[59]

Although the connector to the Garonne was the only lock required to be made of masonry, according to the devis,[60] even this structure could not stand up to the force of the earth on its walls and the changing levels of water in the interior. In a bit of bad judgment, Riquet originally built this lock even larger than specified, yielding proportions for the interior that were indeed impressive: roughly 5 meters long by 4.3 deep. The walls soon started bulging and tipping—and did so where all of Toulouse could see it. It was a technical and political disaster.

The forces on lock walls were quite different when they were full or empty, and working the locks kept changing the balance of forces on them. As long as the locks were full, the inward force of the earth was countered by the outward pressure of the water, but when the locks were empty, the earth had no counterforce other than the strength of the walls. The longer and deeper was the lock, the larger the wall surfaces and worse the problems.[61]

Riquet quickly tried Boutheroue's solution, and built staircases or flights of locks where the walls became unstable. He quickly had the lock at the mouth of the Garonne rebuilt as a double staircase of shallower locks, but even this was not enough to stabilize the walls. The wet and sandy soil

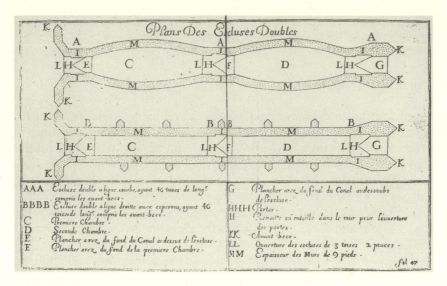

Figure 4.13
Plan for double locks with curved walls for the Canal du Midi. Louis de Froidour.
Courtesy of the New York Public Library.

provided poor footing for the construction of any lock, so the workers used a Roman technique for laying footings for bridges in a river: building cofferdams.[62] They pounded long wooden stakes or pilings deep into the ground close to each other, tying them together to create a barricade against the water. The interior space was then pumped dry, and the earth beneath excavated to reach rock or at least deep heavy clay. After the excavation, the whole interior was filled with stones and cement, creating a foundation that was both heavy and deep.[63]

The combination of the lock staircase and cofferdams finally yielded a functional connection from the Canal du Midi to the Garonne River. But inland, it was (oddly) less clear what to do. The wooden lock walls were rebuilt with stone, and the deep locks were redone in staircases, but still some started tipping. Only in 1670 did someone think of changing the shape of the lock walls themselves, curving them out to the sides.

Locks with an oval shape used the strength of the arch to resist the inward pressure from the earth. This design solved the fundamental problem, made the waterway feasible, and gave the locks on the Canal du Midi their distinctive shape.

The use of horizontal arches in retaining walls was a Roman technique developed at Fréjus—a distant town in the Narbonnaise—where a hillside over the new port required reinforcement. The butte Saint Antoine was held up with a retaining wall made of arches that resembled the curved lock walls used on the Canal du Midi.[64]

Figure 4.14
Curved lock wall at Fonseranes. Photo by the author.

The oval locks were not only more stable but also had more commodious interiors. They could be made long enough for the boats that navigated the rivers of Languedoc, and they could be set in staircases where needed. The only problem was that they shifted the inward forces from the earth through the arches to the ends of the locks. This focused the inward pressure on the door frames, so lock walls were made particularly thick there, and the span between the doors was maintained with a huge, single piece of cut stone set as a footing for the doors.

The result was a lock large enough for local shipping, strong enough to withstand outside pressures, and designed in a way that would be effective in a variety of locales and soils. The oval locks were both beautiful and innovative forms of engineering derived from Roman precedent and worthy of the New Rome, but not celebrated at their inception. The years of lock failures that had preceded their adoption had made this development anticlimactic.

If the lock failures near Toulouse plagued Riquet in part because they were so public, there were other technical disasters that were veiled and successfully hidden from critics of the project. Again in 1667, Riquet contracted with an entrepreneur to construct a machine for lifting dirt into the dam at

Figure 4.15
Roman retaining wall at the Butte Saint Antoine, Fréjus. Photo by the author.

Figure 4.16
Stone footing under the lock door at Montgiscard. Photo by the author.

Saint Ferréol—a failed mechanism that remained successfully hidden from view until Michel Adgé recently found evidence of it.[65]

From the sixteenth century, books of machines had been popular in Europe, presenting images of classically derived technologies for lifting massive stones from quarries or raising ore out of deep mines. In spring 1667, Riquet received a proposal from a M. de Combes to design a dirt-lifting machine for the dam; it was a tempting possibility, so he agreed to try it.[66]

No plans were left from the project, and only glimpses of it in the account books. In June 1667, a shipment of timber was sent up the Montagne Noire, and a carpenter began work on the basic structure for the machine. Shortly later, a cabinetmaker, a wheelwright, a locksmith, a farrier, and a smithy all came to work on it. They built (in all) sixteen wheels with brass grease reservoirs—one of them the "great wheel" powering the rest of the machine. This central mechanism was run by a "squirrel cage" or treadmill of the sort illustrated in Giorgius Agricola's *De re Metallica*.[67] The many small wheels built for the machine were probably part of a pulley system for hauling containers of dirt. Four wheels were apparently made for a cart that may have run along rails toward the top of the dam. There was a dumping lever for the machine, too, to tip the cart and empty its contents.

The machine was finished on August 23 of the year it was started, and apparently only tried once. The story of its development was buried by skillful bookkeeping. Combes was paid directly by Jean Gasse, sieur de Contigny, the controller for the project and Clerville's client, suggesting that Clerville was involved in the secret. In any case, the experimental machine, another leap by engineers into new conceptual territory, died silently in the mountains.[68]

Failures and Credibility

If arguments and rumors about technical competence had been his only problem, Riquet might have weathered the storms of 1667 more gracefully. But there was also trouble with the salt tax in Roussillon where rebels were killing Riquet's tax collectors. When he was unable to deposit his taxes on time with the États, Riquet was forced to give his enemies new ammunition to use against him. They immediately stopped payments for the canal until his taxes were paid and spoke disparagingly of his character to Colbert.

Throughout 1666–1667, the armed uprisings in Roussillon put Riquet into terrible debt. He had a contractual obligation to supply tax revenue to the treasury, but no way to collect money for the payments. Efforts to stop his tax losses kept him far from the work sites for the Canal du Midi, and his tax debts prevented him from obtaining the contracted funds for it from the États. He kept work on the canal going with his own fortune, as he wrote to Colbert in February 1667:

I have not yet touched funds other than my own, and that I dare expect no more this year than the sum of 50,000 livres of the 300 million to be given to the province by the king over eight years for the canal. . . . I am so passionate about seeing a happy and quick ending to my enterprise that I continue to work vigorously. So much so that I have spent great sums [of my own] this year. If it were possible for me to continue this way, I would achieve in four years what I have promised in eight. I wish I had the proportionate finances. . . . If it pleases you, sire, send me back . . . with a letter from you to the treasurers of the bourse obliging them to pay me without delay . . . [and demanding] that I receive payments from the funds designated for this work at the beginning of each year. Because without such a letter, the messieurs [of the États and bourse] will hold onto the payments until the end of the year . . . and I am too far into debt not to pay up. . . . I have already spent twice as much on the canal this year as was allocated. . . . I need your protection and help.[69]

Colbert was quick to reassure the entrepreneur, but still looked into the État's reasons for keeping the funds from Riquet.[70] On April 27, Colbert wrote Riquet:

Sir, having examined for a little while the monies which are due to the Royal Treasury by the tax farmers, and other charges about the recovery of taxes, I have found that you are in arrears by the sum of 349,784 livres 5 sols. And since the summer months that we are entering are near and moreover the amount due is considerable, I ask you to acquit promptly the [missing] part [of the taxes], being persuaded that having paid regularly until now and the king being satisfied [with your conduct so far], you would not want to put an end to his [goodwill toward you]. In awaiting your response, I remain, sir, your very affectionate servant.[71]

Riquet wrote back on April 12, 1667, asking for debt relief on the basis that Roussillon was a war zone: "Sir, you would give me a substantial relief from payment of this tax at the Royal Treasury, if you would please assign me some of the extraordinary funds for war for the payments that need to be made in Roussillon and in this province."[72]

The entrepreneur wrote again a week later, more penitent but still insistent, reassuring Colbert that his financial dealings with the treasury were in good order, and he was selling properties to pay off his debts. With the gabelle payments clearly pending, he again asked the minister to help him get the canal funds released from the bourse:

[In March,] you told me that to save me, you were going to expedite the freeing up of funds [specified in] my contract [for the canal]. I rest in my hope from that letter, and don't doubt that you will give me the grace that is just. I am already in great debt from my work because of all the provisions of material. I have, beyond that, great debts in the salt tax [with] all credits having been lost and the contracts impossible. . . . You know, Sir, that I have left 50,000 escus in the hands of the king that you wanted me to deposit until the sale of [some properties] to serve as insurance for my debts. . . . [What I ask, however,] remains absolutely a matter of your will, Sir.[73]

In another week, Riquet again wrote:

Whatever diligence I have known to make, it is still not possible to get a single penny of the money that the province owes me. . . . [M]y intention [has been] never to mix the salt tax accounts with those of the canal, and to find expedients and measures so I might pay the gabelle punctually, and nonetheless advance this current year my work on the canal so that I can finish in the three upcoming ones. . . . But this will not be without difficulties because [M.] Hurez [who holds the money] loves chicanery, and will find all the ways imaginable to extend the process, and avoid as effectively as possible paying what he owes me. And it's for this, sir, that I humbly ask for your protection . . . [and] submit to any outcome.[74]

The États now wanted verification of the work on the canal, and were arguing about who should do the necessary studies. Colbert wrote Riquet on March 10, 1667, that Clerville was the appropriate man for the job, and that the fortress engineer was coming to Languedoc.[75] Riquet at first balked, but then on April 12, wrote that he was expecting Clerville and was looking forward to the verification of his work. There were problems to fix, and Riquet needed time for the repairs, but he also needed money. Verification was a politically tricky process that he would have to accept. He claimed that he had work to show, in spite of the bad weather and holidays that had slowed him down. These impediments, he maintained, were simply mortification for his sins.[76]

After apologizing for delays, Riquet nonetheless bragged that he could achieve in four years what he had promised to deliver in eight, if the funds were made available.[77] The entrepreneur was in a desperate situation; he had to find new ways to get money to repair his mistakes; and he was afraid to confide in Colbert about his problems. He needed a new advocate at Versailles.

Patronage and Damage Control

Unable to exercise authority over the construction of the Canal du Midi from a distance, hearing incessant stories of failure, and now worrying that Riquet could be involved in graft that could threaten the king, Colbert turned to Clerville to keep a closer eye on the tax farmer. The smooth flow of information and favors had clearly broken down between the minister and the entrepreneur. Riquet was having problems that he was keeping secret from his patron. Clerville was a loyal client as well as technical adviser to Colbert. He knew the project intimately, had appointed his own clients as local supervisors, and could find out what was really happening in Languedoc.

It was a great opportunity for Clerville to become a beneficiary of patrimonial politics. He was delighted because he had some "joint enterprises" in mind that he wanted to pursue with Riquet. Clerville had been rebuffed by the entrepreneur at first when he had proposed this, but Riquet needed him now as a patron and confidant. There was no way Riquet could continue to evade his overtures or try to misinform him about the canal, and Riquet knew it. With the minister's encouragement, the entrepreneurial soldier was prepared to keep an eye on Riquet and was happy to do it in exchange for favors.[78]

So on June 6, 1667, he wrote to Riquet, emphasizing that the entrepreneur needed a good word from him to the minister. He asked only that Riquet consider his earlier proposals for joint enterprises. Now that Clerville had been asked to verify progress on the canal, he held the future of Riquet's project in his hands.

> I had hoped that as I traveled from one end of the earth to the other, you would do me the favor of entertaining a little commerce with me, through men in my service who are in Languedoc, and that not only would you let me know that the friendship that you promised me during my last visit at court had not diminished; but also that you would take the trouble, following a bit the desires of the king, [to include me in] what you are doing for the communication of the two seas. Nonetheless, since you have neglected me up to this point, never asking for news of me, at the least you might have found means to give me yours [news]. I beg you to take as a good thing that I speak to you of your silence that it has given me unhappiness, and that *it has disturbed M. de Colbert, making him disquieted that I have had no correspondence with you, and that I have nothing to tell him about your work.*[79]

Clerville made it clear that he knew about Riquet's problems and the slow progress on the canal. There was lots of work under way, Clerville observed, but much of it was repair work, and all of it was unfinished. He did not judge Riquet for his failures, he only asked him to consider how much he needed the military engineer to be his patron.

> I know well that you have started nothing there, that nothing has been completed as one would want; but with all that, if you had done me the honor to confer with me, by letter, and had asked me a

little advice, that perhaps I could have given, following what you and I had done in concert before, I could have assuaged all doubts that [Colbert] expressed in the presence of his Majesty. But if after all, things are in such a state as I hope to find them, when I am on site [verifying the work], I would forget easily the little care that you have taken to communicate with me by way of my business adviser [homme d'affaires], whose address I have made known to you, or by the four or five of my servants that have not left Langue-doc. Finally, as I leave this town tomorrow, to go visit your works, following the three express orders that I have received, I write this letter to let you know that, with God's grace, on the 10th of this month I will be at Montauban, where I will spend two days with M. Pellot, and according to plan, go to a place where you can meet me, that you can give me your news.[80]

Riquet received a lesson in patron-client relations that was chilling and effective. He was in trouble and could not refuse Clerville's offer of help. The rebuffed soldier was warning Riquet not to make him an enemy, and added that he would continue to be available through his homme d'affaires.[81] As a man of noble standing, Clerville could not engage in commerce directly, so he needed this kind of third party to exercise his entrepreneurial interests. Clerville ended the letter, "Notwithstanding the little care you have shown in communicating with me about your designs, I remain your obedient servant."[82]

True to his word, Clerville did assert in his verification that Riquet was making progress on the canal, and Riquet's homme d'affaires suggested in November 1667 that there was some possible work Clerville might do with Riquet's son. What came of this is not clear.[83] But the forced alliance between the two men had been established by this point, and would turn into a deep friendship over the course of the twenty years they collaborated.

Things started to go better for Riquet and the Canal du Midi in 1668. After haggling over prices, land was finally confiscated for the work. In spite of the rumors from opponents that Riquet was just pocketing the money for himself, more workers were indeed digging ditches through the Montagne Noire, packing fill dirt and stones into the dam at Saint Ferréol, and excavating the main channel of the canal.[84] Riquet wrote to Colbert from Montpellier on January 14, 1668, after meeting with members of the États de Languedoc, optimistic that the project was now well started.[85]

Perhaps with Clerville's encouragement, Riquet renegotiated his contract with the state, and on August 20, 1668, was allowed to extend the period of the contract for the first enterprise. The new "Arrêt de Conseil" gave him another four years to complete the work to Trèbes, not the two years left on the original contract. With Clerville on his side, Colbert and the king had been convinced to give Riquet an extension.[86]

The first enterprise began to impose itself on the countryside and in the political imagination—making the canal a political asset at Versailles, although difficult to complete and locally contentious. This impressive chasm began interrupting traditional patterns of life, making an unsightly mess as it extended across the province. Wide swaths of countryside were torn up and dirt was heaped to form the embankments. In the rain, work sites turned to mud and walls of the channel collapsed. The project was big, new, and in many ways out of control. Accustomed to dominating life in the province, the powerful men in Languedoc vilified the entrepreneur and minimized his progress so that his credibility would be reduced. This they did with relish and skill, but finally not great effect. Reviving and surpassing Roman Gaul was too important a political agenda to be easily abandoned by the king and Colbert.[87] In spite of Riquet's failures and lies, the king was determined to win the battle with the États de Languedoc and subordinate these members of his kingdom. By forcing the États to pay Riquet and insisting on the indemnifications for the work, he would make this recalcitrant province into a visible, if not enthusiastic, part of his New Rome.

CHAPTER 5

Shifting Sands

The New Rome was meant to dominate the landscape, turning a natural world of mountains, rocks, and trees into a place of human power. But in Languedoc, this was not so simple. Nature would often play the trickster; land could become a shape changer. Rivers would shift their beds; cliffs would fall into the sea; and sands would start to migrate along the shore, turning ports into ponds and then beaches. Like a trickster, nature in Languedoc did not confront power directly but rather periodically and predictably turned against it, showing the vulnerability of human efforts to control the natural world. Nowhere was this clearer than along the route of the Canal du Midi.[1]

Mediterranean Languedoc was famous for its storms. When people spoke of the Aude as a wild river, they thought of it in relation to the weather patterns of southwestern France. For much of the year the Aude seemed a lazy, meandering stream more than a threat to structures or human communities. It was the weather that changed it—the same storms that tore through the Montagne Noire. In moments when the sky let loose, visibility would go down to nearly zero, and the world would become everywhere wet, perhaps pooling more on the ground than in the air. Hills would slide; rivers would flood; and silt and debris would be carried across the countryside, relocating bits of the land, taking down trees, and moving topsoil to reveal stone beneath.

The political terrain in France changed its shape, too, soon after Riquet initiated work on the Canal du Midi. In May 1667, after marrying the Spanish princess, Marie-Thérèse, Louis XIV began the War of Devolution, invading the Spanish Netherlands on the pretext of claiming his wife's inheritance. Colbert hurried to supply new monies from the treasury to expand the army. French soldiers under Vicomte de Turenne moved quickly, grabbing money, lives, and land from the empire.[2]

A rapid series of sieges quickly gained France major victories, but a surprising reversal changed the war and limited French expansion. The panicked Dutch worried that Turenne would continue the drive north, and joined with England and Sweden in the Triple Alliance, overwhelming the French and forcing the king to sign a treaty in May 1668.[3]

Riquet's stumbling engineering efforts in Languedoc paled in the face of battlefield drama, and left the entrepreneur struggling just to find workers to sustain the project. In early 1668, Riquet's workforce had dwindled to

eight hundred laborers, but the numbers rebounded rapidly in November after troops were disbanded, and the harvest and *vendange* were both in.[4] There were seven to eight thousand workers on the Canal du Midi, and a thousand more came in the following year. The new labor force was not only large but also contained women in substantial numbers.[5]

Although progress resumed on the waterway, the luster of the enterprise had tarnished. Digging a large trench across Languedoc was slow, painstaking work that just did not have lasting political purchase. The king had turned his attention to the nobles who had served him well at war, planning divertissements and other pleasures at court to reduce their battle weariness and bind them closer to him.[6]

Riquet's inability to conjure up the New Rome swiftly was troubling to Colbert, and strained their already-complicated relationship. The minister was struggling to recover his standing vis-à-vis the minister of war, now Louvois, who was basking in the glow of battlefield successes. Colbert wanted victories of his own, and Riquet could not provide them. There was nothing the entrepreneur could do about it, so he began paying more attention to engineering than politics, stewardship more than patronage, lock doors more than the royal will. Territorial politics had two currencies: stewardship successes and patrimonial powers. As the political landscape shifted, the two men found themselves playing opposite ends of the system. Riquet sought material improvements while Colbert looked to regain stature at court. Not surprisingly, their political differences had unsettling effects on their fragile social ties.

Territorial politics was necessarily most difficult at the margins of the kingdom—at the political borders where land met sea, rivers, or mountains. These were precisely where Riquet tried to fulfill his contracts with the state. He sought to extract tax revenues from the most difficult of sites: the border and mountainous province of Roussillon, where populations newly subservient to France quite justifiably opposed state interference in their traditional ways of life. He also fought to extend his canal toward the Mediterranean coast, where rocky countryside and local elites both resisted his efforts. Storms followed storms in these volatile environments. Storms of protest against the gabelle and storms off the sea both stalled work on the Canal du Midi.[7]

Building a New Rome was proving difficult in the trenches of war and water. The pursuit of territorial politics through military action had reached an impasse; the effort to improve French infrastructure reached a limit of its own. Even as the administration pressed nature to be complicit with human political ambitions, the countryside still eluded full dominion. Cartographers might draw clean lines on paper to provide an illusion of stable geographic control, making the engineering of wars and waterways seem a reliable means of achieving political desire and territorial domination, but nature itself periodically mocked human efforts, avoiding capture.

All the foundations of the Canal du Midi—political, social, natural, and personal—were shifting in 1668–1670. The longer the channel became, the

more of it was eroded by the rain; the more land was seized to the distress of locals; the more Riquet needed money from the États; the more the États resisted giving him any; the more Riquet went to Perpignan to quell the uprest in Roussillon; the more opponents claimed that Riquet was not supervising the work; the more Colbert suspected he was being defrauded; and the more Riquet rhapsodized about the miracle of his growing engineering abilities. The canal became more substantial as the situation became more unstable.

A Second Enterprise

In the face of political, social, and technical troubles, Riquet dreamed about the future. Throughout 1667, as though oblivious to the war and his lack of progress on the first enterprise, Riquet repeatedly asked for the contract to finish the canal.[8] In January 1668, as the War of Devolution was reaching its crisis point, Riquet sent Colbert a full-blown proposal for building the second enterprise. The canal was beginning to take form, but still the timing seemed awful.

The language in the proposal seemed politically out of touch, too. Riquet took an oddly modern tone, speaking more like a bureaucratic functionary to his supervisor than a client to his patron at court. The choreographed manners and high-blown rhetoric of previous communications with Colbert were gone, replaced by a modern pragmatism, coolness, clarity, and specificity. He offered no favors, and asked for none, claiming that the second enterprise could pay for itself. The language in the proposal was surprisingly guileless. Riquet seemed to assume the second enterprise was his for the asking because he had the proven expertise to get it done. He provided facts and figures, a plan, and funding sources.

Riquet began by confessing that he had been avoiding the minister, taking time to think about the future of his enterprises. He had three goals: determining what additional work he could still do in Languedoc to glorify the king and serve his people, identifying new sources of funds to realize these projects, and finding ways to reduce the opposition to his projects. He did not ask if the king wanted the canal extended or express any gratitude for what the monarch had allowed already. Displaying a stunning narcissism, he wrote to Colbert about *his* ideas, *his* reflection, *his* project, *his* problems, *his* funding, and the program of engineering that *he* wanted to pursue for the province.

Riquet admitted that he had failed to serve Colbert as he had promised. Although he had agreed to finish the work in four years, it would now take eight. Still, Riquet maintained, only building the waterway between Toulouse and Trèbes would be a mistake. To make the project useful, the canal needed to cross Languedoc to the port at Sète—just what the commissioners had suggested. He did not claim to deserve special consideration or offer new service to the king in exchange for this new favor. Riquet simply argued that finishing the canal made sense. It was a matter of good stewardship.[9]

He also asked to take over construction of the port at Sète. As the terminus of the Canal du Midi, he thought it should fall under his purview, and Clerville concurred. Although the new harbor at Sète had been started and the design was well crafted, Riquet argued that the construction was shoddy and progressing too slowly. With the snowstorms arriving around Christmas, the entrepreneur claimed, about twelve to fourteen boats on the local coasts needed a safe port, and he could provide it for them.[10]

Riquet also wanted to include in the second enterprise construction of a permanent inlet [grau] by the Hérault River near the town of Agde and the canal system to the Rhône that Bezons had advocated to the commission.[11] An ancient waterway, the canal de Silveréal (or Sauvereal) was used in the salt trade to cross the Rhône delta and could be improved for general navigation. It could be connected to a small canal called Bourgidou near the town of Aigues-Mortes (Ayguemortes) and then to the La Radelle that discharged into the set of coastal *étangs* where Sète was located. This work would connect the Rhône to the Canal du Midi.

The Bourgidou was presently filled with sand, Riquet observed, even though the king had previously paid around fifty thousand ecus for its maintenance. The entrepreneur contended that it made more sense to use these funds to expand and rebuild the waterway to ensure its effectiveness in perpetuity rather than continuing to make useless annual repairs. This he said he could do.[12]

Riquet conceded that while there was lots of work to do to improve transport in the region, the costs mounted up to enormous sums. That is why he had figured out the costs of the second enterprise, and determined how to raise the money for it.[13]

Riquet first listed the costs. For excavating the rest of the Canal du Midi from Trebes to Sète, the work (at 42 livres per toise) would require 3,570,000 livres. To maintain the canal level, it would require thirty locks at 15,000

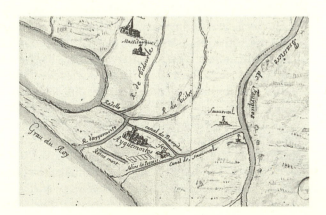

Figure 5.1
La Radelle, Canal du Bourgidou, and Canal de Silveréal by Aigues Mortes, detail of the map of the second enterprise. Courtesy of the AHMV.

each, amounting to 450,000 livres in all. For building the canal on top of conduits to cross the Aude River in two places, and span the Hérault and Orb rivers as well as transect some additional small streams, the cost would be 300,000 livres. And for additional unanticipated work, he added 100,000 livres. This would total up to 4,420,000 livres simply for completing the Canal du Midi.[14]

In addition, there would be costs for the port of Sète. Riquet would need 15,000 livres to finish the seawall and connect the étang to the sea with a canal. But he would also need to build two additional jetties (one from the beach and the other off the seawall) and construct a fort to protect the port. This whole enterprise would cost 1,500,000 livres.

The improvements on the canal de Silveréal would come to 50,000 livres, and annual repairs and maintenance could be assured with a reserve of 100,000 livres. The cost of acquiring land and building bridges from Toulouse to Marseillan would amount to 600,000 livres, bringing the grand total for the second enterprise to 6,620,000 livres of additional expenses.[15]

Riquet wanted to organize the enterprise in a new way. He would divide the canal into different sections and subcontract the work, setting out standards for the contracts. This would help solve the problem of obtaining funds from the États, since if the terms of the contracts were met, the États would be obliged to pay the subcontractors. The most powerful members of the assembly, Riquet assured Colbert, had already agreed to this.[16]

To fund the second enterprise, Riquet continued, the États de Languedoc already had on hand 1,200,000 from the original contract, having paid him only 800,000 livres for the first enterprise. Taxes on the fabric dyers of Toulouse who benefited from the canal in their region could yield 1,400,000 livres; and a tax on *cabaretiers*—the hotel keepers, cabaret and tavern owners, and wine merchants who profited from the influx of canal workers—could generate another 1,000,000. The Office of Receivers would be asked to contribute 1,000,000; one could sell the rights to the salt trade through the canals to the Rhône for 600,000 livres; and Riquet himself would contribute 200,000 for the rights to the canal land as part of his seigniorial estate. He claimed these sources yielded income of 6,400,000 livres, which was only 220,000 short of what was needed.[17]

Riquet offered to propose other sources of funding to make up the difference, but he also argued that if he could obtain from Colbert this additional support, he could bring the canal to completion in the eight-year period he had promised.[18]

There were reasons to believe his timetable. It seemed possible that the second enterprise would proceed more quickly than the first because laborers had already worked out ways to improve the channel walls and artisans were learning to design staircase locks. The reorganization of the project also held out the promise of ameliorating tensions with the États and speeding up construction by reducing the number of verifications they would require. The canal would be defined as a series of "works" (material units)

Table 5.1

1,200,000	États de Languedoc
1,400,000	Teinture des estoffes
1,000,000	Cabaretiers tax
1,000,000	Office of Receivers
600,000	Salt trade rights
200,000	Right to land in seigneurie
6,400,000	Income
6,620,000	Costs of project
−220,000	Difference

rather than a large, unmanageable enterprise.[19] It would become more a struggle with nature than a political battle between the administration and the États de Languedoc—a matter of stewardship rather than patrimonial jockeying.

Riquet did not ask to present the new enterprise at Versailles to promote himself along with his project. He proposed to send the papers with (Jacques?) Cambacérès,[20] a gentleman whose word might be trusted at court. At the same time, Riquet insisted that the project be approved before the end of the current assembly of the États, so work could begin during good weather. Then he ended abruptly, "I have nothing more to say to you now."[21] Only in closing did he switch to the form of address appropriate for someone writing to his patron: "I do beseech you to protect me when it is just, and be persuaded that in all this my personal interest has no standing. I have only one goal: increasing the glory of the king, [and] the advantage of his subjects."[22]

He no longer begged Colbert for help as he had in the past or tried to inflate his accomplishments to look better in the minister's eyes. Riquet did not complain about his enemies or blame them for his failures. He focused on nature, not politics. His proposal spoke of facts and measures. Riquet took for granted that the minister would see him as a good steward of the king's lands and indispensable for completing the canal. But he misunderstood how much he had fallen in Colbert's eyes by failing to provide him easy victories like those of the army. In the French patronage system, Riquet had become a liability.

Putting the Contract out for Bid

Rather than accepting Riquet's offer to finish the canal and improve its terminus, Colbert insisted that the plan be put out to bid. It was easy to justify this because it was normal procedure for civil engineering contracts, and what had been done for the first enterprise. But Riquet was shocked. He expected the minister to allot the second enterprise to him as an extension of the first; it was the kind of favor his patron could do.

In demanding that the contract go out for bids, Colbert was not simply trying to force Riquet to reduce his estimate for the second enterprise.[23] He wanted to make it clear that the entrepreneur was dispensable. Riquet had been striving to please the minister by becoming a good steward, but he had become a bad client—insubordinate, secretive, demanding, technically unreliable, and evasive and manipulative about money. The minister continued to hear rumors from Languedoc, intimating that the tax farmer was hiring nonexistent workers, charging for failed or incomplete work, and pocketing the money rather than using it on the canal. In fact, Riquet never defrauded the treasury, but Colbert was sure he was cooking the books. Riquet's word did not have the social standing to counter these charges, particularly after he was caught lying to Colbert about using tax revenues to pay bills for the canal (even if only temporarily). So in many ways, he was less attractive as a contractor for the second enterprise than he had been for the first.

The minister wanted to assure himself of the status of the first enterprise before even entertaining the new contract. So Colbert had the intendant Bezons, Clerville, and M. de Saint Papoul do a verification of progress on the canal. In February 1669, they prepared a document listing the work that still needed to be done, the expected cost, and who should pay for it—the treasury or the entrepreneur.[24] Clearly, there was much to be done, but the verifications suggested that the canal did seem feasible now. It made some sense to take it to the sea.

Riquet was surprised that the canal's successes did not give Colbert more confidence in him, but the minister was more worried about graft than he understood. The entrepreneur thought he had been clever by using his own fortune for the canal to evade the États' attempts to stop the project by withholding funds; he even felt martyred by his financial losses, and more virtuous because of them. But this strategy did not make him seem more virtuous to Colbert, only unfathomable in a disturbing way. Even in using money from the gabelle to pay his bills, Riquet was not trying to steal from the treasury but only wanted to juggle funds temporarily to keep the project going during 1667–1669. He made matters worse, too, by asking Colbert for permission to take on new projects that would increase his fortune, building roads, digging mines, erecting forges, and charging locals tolls for their use.[25] This work was easy to discredit as profiteering.

All the entrepreneur's strategies to keep the work going while the États withheld funds missed the fundamental practices of patrimonial power that Colbert valued and used himself. Riquet was not humbling himself and attempting to secure his place in patrimonial networks as his technical and political troubles mounted. Instead, he was ruining himself and his family both financially and socially. For Colbert, this was unimaginable; no one would do such a thing. So the minister was disturbed by Riquet and reluctant to keep working with the tax farmer—sure there was dirty dealing in the entrepreneur's weird ways of managing his finances.

From the point of view of Versailles, indemnifying more land for the canal was a politically dangerous act, and not one that Colbert was prepared to take without reconsidering how to manage the risks to the treasury and the king. Assigning the second enterprise to a new entrepreneur probably seemed like a reasonable thing to do. After all, Riquet was no engineer. And as a tax farmer, it was easy for him to mix his tax revenues with treasury funds, but it would not be for others. On this basis alone, it might have appealed to Colbert to keep him away from the second enterprise.

Riquet found the idea inconceivable because he and his collaborators had succeeded in making so many of the inventions that were now working. Yet to Colbert, these "inventions" were not Riquet's property and could be used by any contractor for the second enterprise. The utility of what had been learned did not imply any debt to the tax farmer at all. Others could continue where he left off.

So a devis for the second enterprise was posted by intendant Bezons, and a set of bidders responded with proposals. Jean Farrand of Montpellier offered to do the work for 6,232,000 livres over eight years. Louis Ponthier proposed a bit less: 6,182,000. Riquet, realizing that he might lose the contract, came in with the lowest bid—5,832,000 livres—but excluded from his bid two items in his original proposal: repairs to the port of Sète, and the opening of the étang de Vendres with a grau to the sea. By not adding new enterprises to the project, he could complete the canal proper with the lowest bid.

Begrudgingly, Colbert awarded him the new contract, and the "Arrêt of 1669" was posted in June of that year.[26] Riquet was given the honor of bringing the enterprise to a successful conclusion. Colbert wrote that the entrepreneur had acquitted himself well so far, and it made more sense to allow him to execute the second contract than to divide the enterprise in two.[27] The Canal du Midi would continue to the sea, and Riquet would take it there, although not under the same terms.

The Arrival of Pons de la Feuille

In June 1669, the same month in which the "Arrêt of 1669" for the second enterprise was posted, Colbert appointed Pons (Ponce-Alexis) de la Feuille as Clerville's second in command to supervise the Canal du Midi. De la Feuille was an engineer that the minister trusted who had served on the commission under Clerville. The new appointee was to be the minister's "eyes and ears" for the Canal du Midi. On June 6, 1669, Colbert explained to Riquet:

> I have sent to Languedoc the Sieur de la Feuille who will bring you this letter to [authorize him] to be incessantly at your work [sites],

to take care with you for their conduct, and to make sure that all the estimates and specifications of the chevalier de Clerville and the [innovations] that you have made with him would be well executed, so once more I can have complete faith in you about the success of this great enterprise. It will be always good and advantageous for you to have someone from the side of the king on the work sites to be an ocular witness of all you do and to take into account the warmth and zeal with which you execute this great enterprise. In this way, the king will be more precisely informed, and I assure you, this will augment the satisfaction his majesty already has of all you have done until now. Give [de la Feuille] your complete confidence and let him know all that you do as you would to myself, so that his efforts and presence might be useful.[28]

Colbert wanted to keep constant track of the canal project and put someone in Languedoc to monitor the entrepreneur's activities. Riquet had been a poor source of information about the Canal du Midi for some time. Clerville's visits had been intermittent since the War of Devolution, and would continue to be since he was needed to rebuild fortresses in the north. So de la Feuille was appointed to stay permanently in Languedoc to do verifications of the engineering and oversee the accounts. To Riquet, de la Feuille was a spy.[29]

De la Feuille's appointment finally made it clear to Riquet the weakness of his patronage ties to Colbert. Riquet was loath to accept new oversight and even more horrified to learn that de la Feuille had been sent at the request of the intendant Bezons, who he had counted as one of his supporters. But Riquet had alienated Bezons immediately on receiving the contract for the second enterprise. The entrepreneur had fired the men in charge at Sète, some of whom must have been Bezons's clients, and did so without consulting with the intendant. As a matter of courtesy alone, Riquet should have discussed his plans with his political superior in the region, displaying his loyalty and subservience to Bezons. Yet he did not, so the intendant complained to Colbert, and de la Feuille was sent to Languedoc to hold Riquet in line.[30] Once more the tax farmer proved himself too independent to be trustworthy.

Riquet was crushed, and after meeting with de la Feuille, he wrote to the minister:

I have received so many caresses from [Bezons], so many assurances of the good state of my work, that I thought him to be my friend, even as others assured me otherwise. I have always prayed him to tell me if he knew any fault in my work, that I was a man without contradiction and that I had no reluctance to do whatever he told me. . . . In truth, it is a great shock if he is my enemy. . . . What obliges the sieur de la Feuille to tell me one day that he was charged with being my constant companion [*bride*] to put an end to inventions that I have discovered and have cost me money. The memory of this conversation persuades me that he undermines me in every way around you.[31]

Riquet suffered personally from the loss of trust and support from Bezons and Colbert. The entrepreneur finally had to face the fact that the minister did not accept his word on faith and did not recognize his successes on the canal as contributions to the administration. To Colbert, he was just a man in arrears with a contract for the king whose character was entirely questionable.

Even Clerville's good reports had not saved his reputation. The verifications themselves were now under suspicion—perhaps because of Bezons. Colbert ordered de la Feuille to study both the canal and its entrepreneur, making his own verification of the project, assessing "the determination with which the work is being done, and whether one finds evidence of success in these visits in relation to the contract with the sieur Riquet, salt farmer of Languedoc."[32]

De la Feuille first was asked to inspect the water supply with Riquet, verifying the observations already made by Clerville of the state of the work on the rigoles and reservoirs. Subsequently, he was to check the work on the Montagne Noire again four times a year until the alimentation system was perfected. He was to report, too, on the navigational channel from Toulouse to Trèbes, paying particular attention to the solidity of the locks along with their placement, number, and construction. After checking the progress on the first enterprise he was then to take charge of the second, making sure that Clerville's maps and plans were properly authorized by the intendant Bezons.[33] Clearly, Clerville and Riquet had worked too independently of the intendant. Now nothing would be free from de la Feuille's and Bezons's scrutiny.

Finally and crucially to Colbert, de la Feuille was asked to check Riquet's account books to make sure that the payments did not exceed what was appropriate for the work done.[34] He was told to "write an exact memoir of the contracts that have been made with le Sr. Riquet, of his income and expenses up to the present, and . . . verify by visiting the work [sites] if the sums that he has received are proportional to the work he has done."[35] Colbert counseled de la Feuille to do this with tact, managing Riquet's feelings with care and understanding that the entrepreneur was more useful working on the canal than being cut off from it.[36]

Such tenderness and tact were not part of Cobert's own approach to the entrepreneur. Colbert warned Riquet that he wanted better accounts of the costs and expenditures, knowing full well that de la Feuille would ferret out the figures if Riquet did not. He wrote on June 27, 1669:

> In response to the letter that you have written me this month, I tell you that it is not enough that you tell me in general that the cost for the canal for the communication of the seas, ocean, and Mediter-

ranean amount to 17 or 18 thousand [livres] up to the day of the verification that is to be made. It is necessary that you send me an account of the receipts and expenses with which I could see in detail the use of funds.[37]

To Colbert's surprise, Bezons and de la Feuille found that Riquet had been honest in his financial dealings with the treasury. At the same time, they discovered that he had been less candid with Colbert about technical problems on the canal. In 1669 when Bezons and de la Feuille visited the locks by Toulouse, they found that the walls were still not plumb.[38] Colbert was exasperated, and exhorted Riquet to be more forthcoming about the problems and careful about the quality of construction. On the other hand, he was clearly relieved that the entrepreneur had not been stealing, so he defended Riquet's motives for continuing the work to the Rhône.[39]

I am pleased with the hope you have for rendering the navigation easy between the cap of Sète and the Rhône River. Apply yourself carefully to find the means to succeed in this enterprise which will be very advantageous for commerce. And after, I will give you my advice about what you think will serve for the execution of your plan. Above all, apply yourself to establish greater solidity in your works so that we can be sure that they will be long-standing, being sure you will redouble your zeal.[40]

Ironically, Riquet was partly redeemed in Colbert's eyes by de la Feuille—the man who was replacing him as Colbert's primary interlocutor on matters concerning the Canal du Midi. Still, the findings did not repair his relationship with the minister. The problems of credibility and trust that had plagued Riquet from the start remained too great for either Colbert or the entrepreneur to overcome; they were structural, not personal. Even if Riquet had never used any tax money to pay bills for the canal and even if the locks had never failed near Toulouse, he did not have the social standing to engender trust. He was still not a gentleman whose word was authoritative, and Clerville's support or de la Feuille's findings did not change this.[41] De la Feuille was a gentleman, so the latter became the voice of the Canal du Midi at Versailles.

Trust and Surveillance

It is difficult to know why Colbert chose de la Feuille in particular to be his eyes and ears in Languedoc; there were other gentlemen he could have appointed, including the great military engineer Vauban.[42] We know that de la Feuille had been part of Clerville's commission and had acquired background knowledge of the project at that time; he may also have been or become a client of Clerville, recommended by the soldier as his second. We know that Bezons had spoken well of him to Colbert, too. Certainly, de la Feuille was a man whose word he would trust. But there were other trustworthy gentlemen. Why de la Feuille?[43]

One argument is that Bezons wanted another engineer to do the necessary verifications with Clerville away, and that Clerville did not want to give the job to Vauban. Although Vauban had begun his career as Clerville's protégé, he was by 1668–1669 quite critical of his superior. Vauban had formed an alliance with Louvois in 1668 that would last for twenty years, and put him in conflict with Colbert and Clerville. Fortresses inside the traditional borders of France were under the purview of Colbert, but those in new territories came under control of the army. By 1668, Vauban was already on record opposing Clerville's plans for rebuilding Lille, and by 1669, the young engineer was being asked to finish projects that Clerville had started.

Given their rivalry, there is good reason to believe that Clerville did not want Vauban in Languedoc. And given Colbert's rivalry with Louvois, it is likely that the minister did not want him either. De la Feuille was a reasonable second, then, because of his neutrality, engineering competence, and ties to Colbert.[44]

There is scant and conflicting evidence of who de la Feuille was or even where he was born. The historians who should know disagree. Jean Mesqui says he was from Merville; Anne Blanchard says Marville.[45] Both towns were along the border with the Spanish Netherlands, a region where de la Feuille's surname was common. A Jacob de la Feuille was a well-known Dutch cartographer in the period.[46] And a M. de la Feuille was also the source of some paintings in the Louvre acquired for Louis XIV from the Netherlands. De la Feuille left few papers, including little of his correspondence with Colbert. His letters were not stored in the Canal du Midi archives—perhaps because the repository was set up by the Riquet family. He was not a member of the Académie Royale des Sciences, and did not leave any army documents, although he was sometimes called a military engineer. He continued working on Sète and the adjoining canals toward the Rhône after Riquet's death, and designed buildings for Beaucaire and Nîmes, before he died in 1684.[47]

According to Michel Adgé and Blanchard (who should know), de la Feuille held the title *ingénieur ordinaire*, and they list him among the military engineers who worked on the canal.[48] Riquet described him as a man with fine manners and courtly graces—ones that probably made him an amiable client for Colbert. As an *honnête homme*, he likely adhered to rules of civility and honesty that Colbert would have understood and appreciated. He knew Latin; kept good account books and could write specifications for structures; made precise engineering drawings; worked as an architect for nearby cities; and even designed in 1670 a decorative cornice with a Latin inscription to go on the Saint Ferréol dam.[49] Just as his social and political skills eased his interactions with the minister, so his literacy and numeracy gave Colbert confidence in his views on the canal and Riquet's account books.

Even though de la Feuille technically remained Clerville's second, and was cautioned to consult both Bezons and Clerville in all his actions, for Colbert he was the most reliable source of information and ideas about the canal. On July 12, 1669, Colbert explained: "You understand of what consequence it is to employ only the most useful persons and that no consideration be it mutual affection, family ties, or other personal considerations should prevail over what is the best service and moreover you must take control so that nothing is done more than what is usefully needed."[50] Clearly, de la Feuille would not supervise the canal as Clerville had done.

Once de la Feuille began examining the engineering and checking the accounts, Colbert expressed his confidence in the future of the canal. Its success was assured:

> I am glad to learn from your letter from the 19th of this month that you have started your visits in Languedoc with the mole [ar Sète], but I am not satisfied to see that there are only twenty-three men who are working there. I would believe that it would be better for the work if you took charge of the twenty or twenty-five thousand pounds that are destined [for that work] each month [so the funds] are entirely employed to the project and with all the economy possible.[51]

While most details of his life are a mystery, it is clear that de la Feuille knew something about engineering. The book he produced while traveling to the Netherlands for Colbert displayed technical sophistication both in his written descriptions of Dutch hydraulic techniques and the technical drawings he made of them.[52]

In his sketch of the "Escluze de Planshandael," he emphasized the thickness of the masonry near the doors used to maintain the gap. He also documented how the Dutch used diagonal supports on lock doors to keep them from sagging over time. Furthermore, he inventoried an array of sluice-opening mechanisms that the Dutch had devised for lock doors to help empty the chambers before opening them.

De la Feuille took particularly careful notes on a lock from near Ypres in Flanders—the Ecluze de Bousinghe—that had an unusual water supply. Two lateral small reservoirs outside the locks helped quickly fill the interior of the large structure. The walls of the lock also had powerful exterior stone buttresses that were almost as wide as the lock itself. Suggestively, this wall design was similar to what was used on the Canal du Midi for the lock staircase at the Garonne. Even though exterior buttresses mainly offset the effects of outward pressure from the water, not inward pressure from the soil, they still may have served as anchors for the walls.

Other technical details depicted by de la Feuille had practical uses on the Canal du Midi. For example, de la Feuille described techniques for

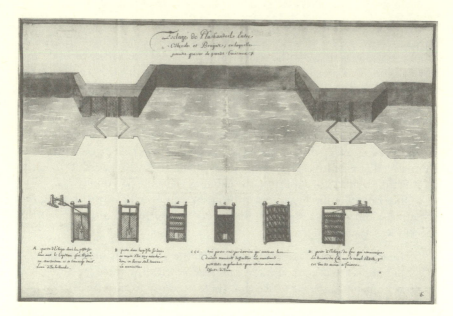

Figure 5.2
Ecluze de Plashandael, Pons de la Feuille. CCC, no. 448. Courtesy of the BNF.

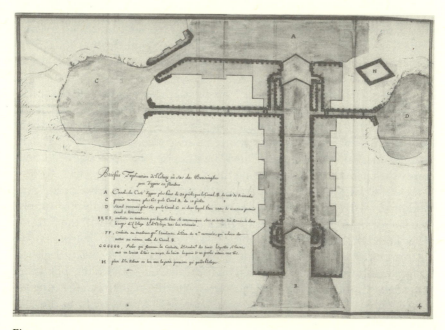

Figure 5.3
Escluze de Bousinghe, Pons de la Feuille. CCC, no. 448. Courtesy of the BNF.

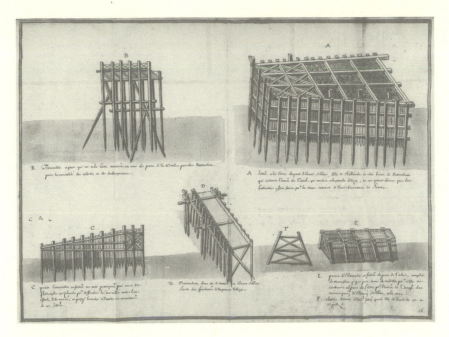

Figure 5.4
Enclosures for cofferdams and jetties, Pons de la Feuille. CCC, no. 448.
Courtesy of the BNF.

building cofferdams and jetties—useful information for perfecting the connection to the Garonne and building the harbor at Sète. He also drew detailed pictures of dredging machines used in the Netherlands where silting
and sand buildup were problems. Clerville was intent on inventing a new
dredging machine that could be used to improve French ports, including
Sète, and de la Feuille's information may have been gathered for him (or
against him).[53]

The book as a whole was a nice primer on canal techniques applicable to
the Canal du Midi, but de la Feuille was cautious in the text about whether
the information would prove useful in France. He emphasized the flatness
of the countryside in the Netherlands, and the clay soil there that reduced
the problems of water seepage and the need for locks on the canals. What
locks the Dutch did build did not have to be very deep. Still, de la Feuille
made notes of what saw, and brought them back to Colbert.

If Colbert had confidence in de la Feuille before he went to the Netherlands, he was thoroughly convinced of his expertise when the engineer
returned with this book. The minister finally had some independent way
to judge Riquet's plans and Clerville's specifications. De la Feuille joined
Colbert at his château at Sceaux as soon as he could, and the two men systematically went over his observations, comparing Dutch engineering to

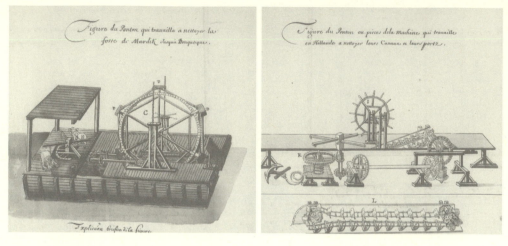

Figure 5.5
Dredging machines, Pons de la Feuille. CCC, no. 448. Courtesy of the BNF.

Riquet's work. Riquet had asked de la Feuille to bring a *plan en relief* that the entrepreneur had made for the minister to study at the same time, but not surprisingly, his new rival failed to do so. Colbert seemed not to mind, and appeared more intent on developing an independent understanding of the problems of the canal, writing to Riquet that he was personally reviewing all the plans for the Canal du Midi with de la Feuille, who would bring the results of their consultations to Languedoc.[54]

On his return to the southwest of France, de la Feuille increasingly participated in engineering decisions for the canal, particularly those concerning Sète. Now there were four major voices in decisions about the Canal du Midi: Riquet, Clerville, Bezons, and de la Feuille. This changed the dynamics significantly, but still assured that the second enterprise as much as the first would be a product of collaborative design.

Subcontracting and Work Structure

Formal authority for the Canal du Midi was once again destabilized by the arrival of de la Feuille in Languedoc. Riquet remained responsible for the canal, but was even less in charge of the daily conduct of the work. With the introduction of the system of subcontracting, Riquet approved the subcontracts, but the waterway was divided up into a set of supervisory regions each with its own inspecteurs and contrôleurs, who oversaw the subcontractors.[55] There were both "ambulant" and stationary controls—ones that reviewed the progress of the subcontractors through frequent verifications, and those who kept the books for a designated region: inspecteurs and contrôleurs généraux as well as *receveurs et payeurs*.[56]

The regional supervisors became the stable points in the system, advising the subcontractors about excavation or construction, and holding them responsible for failures or financial problems. The subcontractors would come and go, and did work of varying quality. The inspectors and financial controllers kept the parts of the project together. Those who worked in areas with the most difficult engineering were the most influential. They included Riquet's early collaborator, Andréossy, who supervised the Toulouse region; two of Clerville's clients, Dominique Gillade and Jean Gasse, sieur de de Contigny, who oversaw the Montagne Noire region at first, and later the Le Somail and Béziers sectors; and finally, de la Feuille, who was put in charge of Sète.[57] Riquet was seemingly disempowered by the arrangement, but in fact he gained some advantages. Responsibility for the canal was placed mainly in the hands of loyal clients: regional supervisors and local subcontractors.[58]

Allocating the work to subcontractors changed the labor process, too. Workers were hired for particular tasks, not in standard-size brigades. In account books, they were now usually listed by skills rather than ateliers— *moilons* (stonecutters), *cartiers* (earth haulers), *niveleurs* (surveyors), *poudres* (explosive experts), *massons* (stonemasons), or laborers with specific duties such as *le contrôle de sables*. Each was paid at a different rate, so they were listed separately in the account books, and were employed in different numbers for each subcontract.

The laborers for the second enterprise were paid less in this period, but were given better working conditions. During the first enterprise, Riquet had paid high wages, offering peasant laborers one and a half times what they normally earned during the crop and grape harvests. By the second enterprise, though, he was broke and reduced the pay scale for the laborers to control the costs of construction. Instead he provided liberal contractual conditions, paying laborers during bad weather and holidays when they could not do their work. The system was effective. A large number of peasants and artisans came to be part of the labor force, working under wage labor conditions in the 1670s that would have seemed generous in the nineteenth century.[59]

The subcontracting system also encouraged distributed problem solving by tailoring the workforce to each task. Each new structure or piece of channel was different, and was addressed with a particular mix of workers and skills. For example, the locks that had to be built from rock and blasted out of the ground were treated differently than those dug from clay soil. The expertise needed for them was radically different, and the people assigned varied appropriately. The locks excavated from stone required stonecutters and explosive experts. Locks dug from soft loam needed laborers who could both dig the trench and stabilize the walls. So construction techniques for each location could be vastly different even for the "same" structure, and both the kinds and numbers of workers employed varied widely from site to site.

The specifications in the devis took on particular importance during the second enterprise because they defined the technical terms of the sub-contracts. Each structure had to be realized with specified materials, and in the size and shape prescribed. But because it was not always clear before-hand what problems would arise on-site, again Clerville left some work methods vague. His phrase "built to normal standards" left plenty of slip-page in the contracts, sometimes freeing up the subcontractors to experi-ment with different approaches, and at other times making it difficult to determine when they had completed the task appropriately. All this made the subcontracting system itself a more flexible and complicated organiza-tional structure than it appeared on paper.

The financial situation also changed for the better for Riquet during this period. In spite of continued problems with tax collection in Roussillon, the États finally agreed to start releasing treasury funds for the subcontracts, and also (after a scuffle) allowed Riquet to sell offices for the canal, provid-ing him with a new source of revenue. They capitulated on the offices both because Colbert was insistent, and because the Canal du Midi clearly needed personnel to guard the waterway and clerks or *greffiers* to supervise traffic. In addition, Colbert agreed to impose taxes on local cabaretiers and others who were getting windfall profits from the enterprise, and he allocated this money to be used for the Canal du Midi.[60]

In this new situation, Riquet was both responsible for the Canal du Midi and not, supported in his enterprise and not. He had the contract to build the canal to the Mediterranean and would pay the penalty if it failed, but he was stripped of authority for daily supervision. He had more funds to spend on the canal, yet at the same time had lost the trust and ear of his pa-tron. De la Feuille had Colbert's confidence, but he was not in charge either. Riquet had to consult him, but since de la Feuille himself was not Riquet's patron, the tax farmer felt no personal obligation to him, only the duty to work with him. De la Feuille also was often in no a position to tell Riquet what to do simply because he was a stranger to the region. He could not say what local subcontractors to hire, where to find construction materials, what they should cost, how and when to find laborers, and what to pay them. And while subcontractors were in principle independent from Riquet's direct supervision, the entrepreneur could and would fire them if he thought they were incompetent, using his legal authority to take over the contract him-self. The canal and the land indemnified for it were his domain, and on his domain, he had the police powers of a noble. This made questions of con-tractual obligation matters of his personal judgment. Still, the entrepreneur did not have the epistemic credibility of a nobleman, so Riquet still needed powerful patrons and had to cooperate with them on technical as well as po-litical issues. In this complicated situation, work on the second enterprise—like on the first—was both contentious and collaborative.

The contradictory supervisory structure became even more complicated when de la Feuille began scheming against Riquet. He wrote a letter saying he wanted to ruin the entrepreneur, and when the letter surfaced, it tarnished de la Feuille's own reputation. Although nominally he kept his position as Clerville's second in command, in fact he was reassigned as supervisor for the region of Sète.[61] He remained Colbert's main interlocutor about the canal and did verifications of the work, but his authority over it was reduced; Père Mourgues, the Jesuit mathematician, then took a greater part in managing the verifications and accounts.

With de la Feuille more marginal, and both supervisors and subcontractors in his sway, Riquet reasserted his authority over the Canal du Midi. Riquet saw moral justification for exercising more power, claiming to be the *pater familias* of the canal. He wrote to Colbert: "My venture is the dearest of my children; I see the glory, your satisfaction, and not the profit. I want to leave honours to my children, and I do not pretend to leave them any big amount of money."[62] He was broke, had neither a powerful ally nor a patron, yet he had a canal that was beginning to be filled and used. Riquet turned away from patrimonial obligations and refashioned his identity as a good steward of the king's lands.

Riquet's words echoed those of mesnagement writers, such as Olivier de Serres, who described stewardship in paternalistic terms."[63] As the canal's progenitor, Riquet was not just a contractor with the state but also a man with a moral and personal commitment to the enterprise. Like any father, he wanted to raise his child well and by definition thought he had the right to do so. Colbert and the nobles of the États could doubt him, but that would not change the situation. He was the one who could dig a canal across the province because he had the experience and dedication to do it. Others might want to father his child, but he was the only one capable of it.

Riquet's desire to treat the canal as his progeny and leave it as his legacy to his (other) children made him all the more insistent about how it should be done. This was nowhere clearer than in debates over the route for the second enterprise. The commission under Clerville had recommended extending the Canal du Midi from Trèbes to Narbonne, the ancient Roman city, and then following the coastline north to Sète. This would have kept the Canal du Midi in the Aude River basin almost until it reached the sea, avoiding the mountains north and south. Clerville was strongly in favor of this, but Riquet wanted to take the canal more directly north to Sète, and he prevailed by eliciting strong support from Mourgues.[64]

In the end, Clerville conceded that the northern route was possible, although he hated to miss his hometown of Narbonne. He admitted that the greatest obstacle to the second enterprise lay just beyond Trèbes. To enter the Aude Valley and stay high enough above the river to avoid flood damage, the waterway had to pass a rock palisade and be excavated into a ledge.[65]

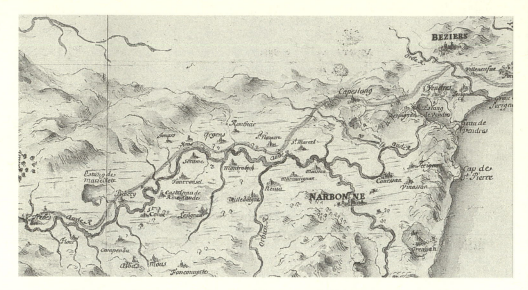

Figure 5.6
Map of the second enterprise to Béziers. Courtesy of the AHMV.

There was nothing beyond the Aude River valley on the northern route more difficult than this. Without more arguments against Riquet's preference, the canal route was redrawn.

The northern route was officially adopted in the "Bail et adjucation des ouvrages à faire pour la continuation du canal et du port de Sète" that was approved by the king in 1669.[66] The Canal du Midi would turn out of the Aude Valley after St. Nazaire, pass by the towns of Capestang and Vendres, and cross the Orb River near Béziers. It was a more direct route to Sète, but it had to pass through the mountains to reach the Orb River valley. The route for the second enterprise posed new and daunting problems of engineering, making the second enterprise a more telling test of the New Romans than anyone in 1669 anticipated.

The Port and the Sea

At the port of Sète, all the shifting political, social, and natural forces that plagued work in other areas of the Canal du Midi seemed to gather in a storm of uncertainty and enduring instability. At Sète, Riquet, Clerville, and de la Feuille again argued about the design of the port, testing their muscles and trying to agree on a way forward. But the powers of earth and sea triumphed against all their efforts. Storms routinely demolished the structures that were erected. Each new technical approach turned out to have flaws just like the preceding ones, making all human efforts to control the coastline seem ineffective. Coastal sand migrated, shifting the balance of power between humans and the environment.

Maps of the Mediterranean coast of Languedoc that contained nice images of Sète sitting quietly beside a placid blue sea might have made the coastline seem stable, but nature denied easy human domination time and again. The king's will as well as the efforts of engineers and the entrepreneur were nothing against natural forces.

Sète was the most troublesome of a number of new ports that Colbert, as minister of the navy, wanted to build to serve the French fleet. It was the only entirely new harbor on the Mediterranean, although Colbert also invested heavily in the existing ports of Antibes and Toulon.[67]

The existing deep harbors in France were small in number and unsettlingly independent from the Crown. Marseille was almost a country of its own, more complexly Mediterranean than specifically French. La Rochelle was a Huguenot center, whose economic success was in many ways more disturbing than satisfying to the Crown. Nantes was more tractable, but its part of France was also thick with adherents to the new religion, and both physically and culturally far from Paris. Colbert wanted ports that he could keep under closer supervision to serve his manufacturing ventures and plans for the colonies. So he established new ports on the Atlantic at Brest, Lorient, Rochefort, and—significantly for the story of the Canal du Midi—on the Mediterranean at Sète.[68]

The new coastal cities on the Atlantic had a strategic logic for their development different than Sète. They were situated close to existing ports—positioned to compete with them or share the profits of overseas trade. Rochefort was built close to La Rochelle. Brest and Lorient were located slightly north of Nantes, and set to capture some of the trade to the Americas. With these new harbors, the minister hoped to bypass rather than tangle with the entrenched interests in seaborne trade. The coast of Languedoc, on the other hand, had no good Mediterranean outlet. This region was rich in wines and textiles, but the French had an ineffective merchant marine in the area, and little infrastructure to support it. French ships in the region were vulnerable to pirates and the Spanish, too.[69]

The new port cities commissioned by Colbert, including Sète, were generally designed in the ideal city tradition to be both tools for and monuments of the New Rome. The towns were typically laid out in geometric patterns, such as pentagons, hexagons, or octagons, and structured internally with streets in measured order.[70]

Because of its location at the base of Mt. Saint Clair, the port at Sète could not be geometric. It had to curve irregularly around the rock outcropping, but it was still designed to have streets in a rough grid and battlements along its exterior like towns in the ideal city tradition.[71] However slight the evocation of Rome, the design of the town and port nonetheless expressed the power of the human hand with its geometric properties, and so became a vivid emblem of human weakness when the sea came to reclaim the harbor.

Figure 5.7
Town and harbor at Sète (Cette) in Nolin, Carte Royal de Languedoc, 1697.
Courtesy of the AHMV.

Figure 5.8
Coastal étangs, detail of a map of the second enterprise. Courtesy of the AHMV.

The problem with building a port at Sète or anywhere in that region was that the coastline in Mediterranean Languedoc tended to fill with sand brought in by a combination of sea currents and silt washed down from inland hills, creating a permanent but shifting bank of sand that cut off estuaries and formed a chain of saltwater ponds (étangs) in Languedoc. These normally shallow areas of water and sand kept the Mediterranean itself far from the mainland, working against the natural development of deepwater ports. Some of these ponds became salt marshes as they desiccated, either filling with reeds and grasses, or being fashioned into salt flats; others were fed with freshwater, becoming nurseries for fish, shellfish, waterfowl, and other creatures; and the largest of them were deep enough to navigate and use as fishing grounds, but were plagued by winds that made them dangerous to cross in bad weather. The abundance of life in the étangs sustained local communities, but did not solve the problem of seaborne navigation.

The rivers that disgorged their waters into the sea sometimes broke through these shoals, creating a grau or channel between an étang and the Mediterranean. If a deep enough estuary opened up after massive outpourings, nearby towns were used as temporary ports while the favorable conditions lasted (ranging from centuries to only decades). But then as river deltas often did, these areas would progressively get shallower. Sand would build up from the sea, eventually isolating the étangs from the Mediterranean again.

During the sixteenth century, the port at Frontignan that had served this region reliably for a long time finally filled in. A small estuary formed at the mouth of the Lez River near Palavas, temporarily providing a harbor for smaller vessels, but it was neither a deep harbor nor an effective shelter to protect ships from the terrible storms of the region. The Crown could not assure the safety of the local inhabitants or French crafts that moved through this part of the Mediterranean.[72] Something had to be done.

The plan Clerville devised for a harbor at Sète in 1664 called for constructing a harbor inside the Étang de Thau, where the small fishing village of Sète was already located. This particular étang was deep and navigable, suggesting that a harbor there could maintain its depth. To connect the new port to the sea, Clerville recommended building a permanent grau or channel that would also serve as a necessary passage point for those wishing to enter the harbor. The result would be a town protected from weather and pirates in large part by the natural buildup of sand along the coast. The existing fishing village of Sète was also situated at the foot of Mont Saint Clair, whose height partly protected the town from Mediterranean storms. It seemed to provide some natural advantage for constructing a safe harbor for French vessels, although in fact the storms on the étang remained dangerous to Sète.

Digging the grau or channel itself was a little tricky because the sea and étang were not at the same level. Clerville brought in the Dutch engineer Regnier Jennse as well as César d'Arcons, "personne très entendue en la nature des mers" (an expert on seas) to look at the problem, and they proposed building a lock in the channel through the sand to raise and lower vessels seeking the harbor or sea. Keeping the channel into the étang free of sand was a more vexing problem given the history of the coastline. Clerville proposed construction of a *mole* or seawall that would protect the channel from storms, and hold back drifting sand from the inlet as well.[73]

There were two main reasons that Sète seemed the best site for a port. One was that Sète was a small fishing village. Its minimal population was in no position to oppose the project. The Mont Saint Clair was also made of pink marble that could be quarried and used for constructing jetties, the harbor, and the port city itself.[74]

In the end, Clerville's proposal for a harbor in the étang was replaced by one locating the harbor on the seaside of the mountain, where vessels could reach it more rapidly in times of danger. A fort was also planned for the top

Figure 5.9
Early plan for the
port of Sète (Cette),
Louis de Froidour.
Courtesy of the New
York Public Library.

of the hill where military men could help identify enemies and call friendly ships to harbor before threatening vessels could come close to land.[75] But the most important structures were the seawall and jetties to enclose the harbor, protecting it from storms and sand.[76]

It seemed at first that building the harbor and port at Sète would be a simple extension of the work being done on the first enterprise. The port needed a set of locks to communicate between the Étang de Thau and the sea that could be modeled on the ones for the Garonne. The seawall and opposing jetty seemed straightforward to build, although hard to design correctly. The Mont Saint Clair would provide construction materials, including the lime for mortar that could be made with chips of marble recovered from quarrying. So nothing seemed to stand in the path of the enterprise except the sea itself and the coastal sands.

Riquet's proposal for the Canal du Midi provided an opportunity to finish the harbor in Languedoc. Connecting the harbor to the canal would make the port and his beloved waterway both more valuable, giving merchants access to the abundant textiles, grains, and wines of Languedoc, and supplying merchant ships an alternative to sailing around Gilbraltar to reach the Atlantic.[77] The opening ceremony for the work site at Sète, on July 1669, echoed these ambitions: "[A] vessel outfitted for the event was placed a hundred meters into the canal between two pavilions with galleries of greenery set on the sand. The first stone for the port was large and black and contained four medals of the king made for the event."

The ceremony itself became a surprising contest of human will and natural forces; the bishop meant to bless the stone could not make it to the sea:

> The stone was carried in a procession to the sea. The bishop could not go to the site where the stone had to be placed because of the length of the walk and the difficulty posed by a relatively tall mountain and a coast of cliffs. . . . One saw the sea and the salt ponds all covered with ships that brought people from all parts and a continual procession of carts and carriages, men on horses and men on

feet, half nude, who walked across the causeway through the salt pond over a good kilometer.[78]

The ceremony marked a new effort to shift the balance of power from the sea to humans; the coast would be transformed into a trading center that would draw strangers and locals alike to Sète. But the ceremony symbolized human dominion over the earth more than the harbor would realize it in the seventeenth century.

Work began slowly on the port as the materials were assembled, workers recruited, and other elements of the infrastructure for this great workshop put together. Riquet was low on funds.[79] Still, furnaces were erected to make lime; forges were built to provide construction materials; ovens were set up to bake bread to feed the workers; houses were constructed for their shelter; and the quarry on Mont Saint Clair started providing the pale pink marble that protected and adorned the nascent port.[80]

Men quarried marble for the seawall from Mont Saint Clair, using gunpowder as well as picks. Women collected the rubble from the stonecutting and blasting that could be used to fill the center of the wall or be heated in nearby furnaces to make lime. Men loaded stones onto carriers. Winches hauled them onto ships and carried them to the seawall. The whole structure formed a massive arm around the southeast side of the new harbor. By 1670, Sète contained a church, stores of food and drink, ovens both for bread and making lime, gunpowder for quarrying, a minimal water supply, four forges, butchers, stables for two hundred horses, and buildings for the workers. By this time, the seawall was also starting to grow rapidly.[81] On January 18, 1670, Riquet wrote to Colbert:

> We are in [Sète] since the day before yesterday, M. la Feuille and myself, where after diverse consultations with the employees that work here under my employ, it was concluded that without difficulty now that the quarries are open and in condition to furnish [adequate building stones], and that I have housing and beds for about one thousand people, I should be able to produce a hundred toises [two hundred meters] of jetties a year, and maybe more than that if the funds are adequate.[82]

By the end of 1670, the seawall was massive, but still no match for the powers of the sea. In December, many dignitaries came to Sète for the holidays. A terrible storm began early in January 1671. Sixty-four ships came to the harbor for shelter. There were four days of high seas. Almost half the seawall was washed away (270 toises remained, and 200 toises had been lost). Only forty-five buildings remained in the town. But many ships survived that would have been lost in the tempest.[83]

Still, Sète was shattered. Successive storms and movements of sand created ongoing problems, too. Different experts were consulted about how to keep the harbor safe and the port deep, and in the end, Clerville, Riquet, de la Feuille, Froidour, Niquet, and Vauban all made proposals. They debated whether to extend or shorten the seawall, giving it a *falaise* or turned tip, and whether to set out a second or different jetties to stop the flow of sand into the port. Riquet also consulted with his workers for ideas, and Froidour interviewed fishermen for their suggestions. Nothing seemed to help.[84] The problem was that when the mouth of the harbor was constricted, the port suffered from silting. Opening up the channel to the Mediterranean helped wash out silt from inland rivers, but also let in waves and sand from the sea—particularly during storms.

Riquet tried to use technology to build a better seawall. He asked an engineer to build a machine to help move rock to fill the jetty. We do not know what happened to it. Colbert sent a Dutch engineer, De Vos, to help de la Feuille with the port, too. It is hard to determine what he did. Still, by August 1673, Sète had started to refill with sand, and another storm in 1676 again damaged the seawall.[85]

Natural Knowledge and Political Territoriality

At Sète, there was a standoff between the powers of the king and the forces of nature. Water became a trickster. Storms threatened not only structures but also the sense of human efficacy and dominion. Technical failures in the face of natural powers triggered new floods of recrimination and blame that flowed down the social hierarchy. There was nothing heroic or superhuman about Riquet's canal at Sète. Dreams of human improvement of the natural world seemed elusive. Stewardship itself was slippery and hard to define.

The conceit of dominating nature was revealed for what it was. Workers on the Canal du Midi were not able to force the land to fit the will of the king. Progress was being made near Toulouse with the design of oval locks that could be both long and deep, but the Mediterranean was less tractable. It did not help solve the problems at Sète that the dam and rigoles in the mountains became usable in 1672, allowing traffic to flow on a permanent basis on the Canal du Midi between Toulouse and Revel. This did not diminish the devastation of storms on the waterway and port. As the harbor started to refill in 1673, Sète remained a reminder of the hubris of the project to redesign Languedoc through engineering. Even as the Canal du Midi started to take on an obdurate appearance near the Garonne, the land in eastern Languedoc failed to substantiate the powers of the king, and realize modern claims of efficacy over the earth and its creatures.

CHAPTER 6

The New Romans

It was clear by 1669 that cutting the canal into eastern Languedoc required new kinds of intelligence—ways of doing hydraulic and structural engineering in an imperfect world. Technical problems in excavating the canal changed as the land became drier as well as both more friable and rocky along the Mediterranean watershed. This part of the province was particularly troubled by water, wind, rock, soil, and shifting coastal sands. The canal had to be blasted out of sheer rock with gunpowder, and at Fonseranes near Béziers, the waterway even had to descend a steep hillside, requiring an eight-lock staircase to be cut from stone. Water was a problem as well as rock. The Canal du Midi had to cross a series of rivers and streams whose currents were dangerous, but whose waters were needed for alimentation. Closer to the coast, the canal had to be carried through the maze of shallow salt ponds and dunes without melting into the water or blowing away with the remnants of topsoil before reaching the harbor at Sète. Meetings of land and water were always tricky, but especially in this dry, rocky, fissured, and sandy land crisscrossed with wild rivers.

Formal knowledge that assumed idealized conditions was of limited value in this part of the world. Engineering the canal called for people who could work in a dynamic environment. Riquet knew that men and women in the towns and villages of the region were accustomed to such conditions, and tapped their indigenous expertise. He did not expect to find New Romans among them, but he did: masons, carpenters, civil engineers, and—most surprisingly and significantly—peasant women from the mountains who maintained classical traditions of hydraulics as commonsense elements of their daily life.

Finishing the Canal du Midi was indeed a job worthy of the New Romans, and was ironically accomplished by workers who—unbeknownst to them—brought classical techniques to the enterprise. Riquet found the intelligence to finish the canal in the common people of Languedoc who came to labor on the waterway. In their towns, they nurtured in modest and unsuspected ways living traditions of hydraulics and architecture with classical provenance. Peasant women from the Pyrenees were the most unlikely among them, but were also perhaps the most important. They apparently brought to the enterprise the most sophisticated tradition of hydraulic engineering extant in France.[1]

Colbert in his own way signaled the need for a change in intellectual strategy for the second enterprise when he sent de la Feuille down to Languedoc to oversee Riquet, hoping that a new dose of formal expertise along with a new system of supervision would improve the speed and quality of the engineering. But Riquet's New Romans were more effective. Inhabitants of the region who took subcontracts or worked for subcontractors knew about soils, weather patterns, topography, and watersheds—the local elements of the environment that were causing most of the problems, and were not easily controlled or taken into account with formal knowledge. The Canal du Midi required workers who could accomplish specific tasks: threading the waterway through mountainous terrain, building retaining walls and overflow drains to minimize local storm damage, taking the waterway over passing streams and rivers, designing larger staircase locks, tunneling along and through rock faces, and following inclines through stretches of hilly land. These were complicated tasks, but ones readily addressed with practices familiar to locals.

If the first enterprise was a work of patronage politics meant to use formal knowledge of the natural world to produce a neoclassical monument to Louis XIV, the second was a work of stewardship, tapping vernacular traditions for ways to etch a new waterway across a wild landscape. Ironically, the latter was the more Roman venture. The Canal du Midi increasingly became a site and occasion for reassembling shards of classical knowledge that had been scattered after the end of the empire, and picked up in local contexts where their use made sense.

The integration of ancient engineering techniques began during the first enterprise in the Montagne Noire as military engineers and artisans worked on the dam, and peasants dug out the rigoles. The dam was a masterpiece of structural engineering, created mainly by stonecutters who chiseled the dam's foundation out of a rock ledge and masons who erected walls above it. They used classical building methods in the work that were familiar in Vitruvius. The laborers digging the rigoles, on the other hand, worked with tacit knowledge of ancient hydraulics, following the contours of the land as Romans had done for their aqueducts. Together, the workers in the Montagne Noire integrated two forms of engineering with classical roots. The result was not only a water supply for the canal but also a new engineering capacity created through collective memory and collaboration. The water system was both one of the most remarkable (and remarked on) features of the Canal du Midi and a fine demonstration of the technical powers of the New Romans.[2]

The most mundane techniques were most important to the engineering and least visible as contributions to the New Rome. Peasant knowledge of contour cutting took the canal through the mountains, and hydraulic cement was used for construction in wet areas. These two methods were never emblems of Rome, and were used without fanfare. But they illustrate better than most techniques the movement of engineering methods with ancient roots into the construction of the Canal du Midi.

Ancient methods wielded by illiterate laborers became more central to the second enterprise partly due to the landscape, but also because of the system of subcontracting. Most subcontractors were local builders from nearby towns. They spoke Occitan, and could understand what the laborers had to say. Women laborers were increasingly used from the 1670s on because they were less expensive, and funding for the second enterprise was constrained by Riquet's loss of his personal fortune. Ironically, this meant that classical engineering gained greater purchase on the Canal du Midi precisely when lower-ranking members of the collaboration gained greater voice in the engineering.

The peasant women laborers from the Pyrenees brought indigenous engineering skills specifically designed to move water through complex topography and put it to multiple uses. In their home communities, they built town and domestic waterworks, tapping high elevation sources (hot as well as cold), using water to improve their meadows, and diverting the rest into channels to reach their towns in the valleys. In some places, they diverted river water in the valleys or streams from the hillsides to use for irrigation, mills, domestic uses, and town supplies. They cultivated their hydraulics where water could never be truly controlled: where spring floods were inevitable, winter storm damage routine, debris in the water system predictable, and repairs an assumed part of the job. Without formal training, they nonetheless understood precisely the methods needed to bring water safely down the Montagne Noire to Naurouze, and carry the canal through the mountains to the sea.

Experience on the first enterprise had already made clear how irrelevant formal knowledge could be for critical problems. The failures of the locks near Toulouse were caused by too many variables to solve by formal methods. They resulted from a wide range of contributing factors that helped define the "forces" of the earth and the techniques to counter them: soil mechanics, the tensile strength of materials, erosion patterns, the water table, and even local weather patterns. Such complexes of specificities were effectively approached with comparable experience rather than methods of calculation.[3]

Just determining the route for the canal became hard to calculate for the second enterprise. Formal tools of survey measurement that had been useful in the broad valleys between Toulouse and Trèbes proved worthless where hills blocked surveyors' views. No wonder the gentlemen of the Académie Royale des Sciences took measures mainly through major valleys or across peaks, leaving the problem of elevation to others.[4]

Trying to thread the canal through the mountains while also keeping the water flowing with a gentle incline was vexed by the lack of reliable instruments and methods of calculation.[5] Experience with the terrain and knowledge of the trails that ran through it provided better clues about how to take a channel around hills, skirting valleys, while slowly descending the Mediterranean watershed.

Calculations also proved unhelpful in designing lock staircases. These locks were supposed to have a series of basins with the same interior volume. If the top basin in a staircase was significantly smaller than the one below it, the water released down would create only a shallow pool in the second lock that would possibly be inadequate to keep boats safely afloat. If the top basin was larger than the lower one, the water from the upper lock would overfill the basin below, floating vessels dangerously up to the top of the lock walls. Even small errors in measurement could lead to large mistakes in volume, and in many instances did. Formal measures were finally abandoned because they had proved faulty so often, resulting in a number of lock staircases being retrofitted or completely rebuilt. Practical tests of volume were more reliable.[6]

> One must again observe that one must correctly place [the locks] as a remedy for the incline of the terrain to find [the] proper level of the canal [in order] to raise and lower it with the same facility. It has been judged appropriate to hold off construction [of the walls] until the excavations have been dug and the levels have been justified with the help of water, which is the only sure proof, and when one can easily remedy any error that has been made in the locks before they are built. And also, one can place them more advantageously when one knows the quality and solidity of the terrain that one could choose for building them.[7]

During the second enterprise, indigenous methods increasingly replaced formal ones as the Canal du Midi was turned over to those who had the practical knowledge to get the work done. These were the New Romans who could finish the canal by employing classical skills and local knowledge.

Classical Artisanal Methods

Both artisans and peasants maintained classical techniques as part of living practices, but they cultivated quite different parts of the Roman tradition and used methods from the past in distinctive ways. Vernacular transmission of classical building methods among artisans was the most obvious conduit of tacit knowledge from the ancients. Arches and Roman corners were everywhere in the landscape of Languedoc embedded in structures that were erected long after the fall of Rome. But there were other classical methods sustained by artisans in the building trades that were not so obvious. One was hydraulic cement. This was a functionally invisible material used on the Canal du Midi. When combined with other classical building methods, it contributed with quiet efficacy to the sophistication of the engineering. It was a method of the New Romans.

The "secret" of Roman hydraulic cement—adding small amounts of volcanic sand called pozzolana to cement mixtures—was supposed to have been "lost" after the fall of Rome and only rediscovered in the eighteenth century. That is why its use on the Canal du Midi in the seventeenth century was so interesting. It was a clear indicator of how much classical knowledge was passed unknowingly, and without the attention of humanist scholars. Vitruvius had specified that pozzolana should be used to make hydraulic cement, but scholars were not sure what this material was, and could not derive the method of mixing hydraulic cement from his writings alone. He did not say where to find pozzolana, how to process it, and in what proportions to combine it with other ingredients. Nonetheless, Riquet knew where to find pozzolana; Clerville understood its applications; and artisans knew how to mix it, drawing on classical knowledge that apparently had been transmitted through practice.[8]

Hydraulic cement was used for two of the most daring engineering projects for the second enterprise: the port at Sète and the first aqueduct-bridge, Le Pont-Canal de Répudre. These were sites where improvisation was required to navigate the dangerous meeting places of land and water. They were also the work of New Romans, bringing classical techniques to modern engineering in collaborative projects that joined artisans and military engineers.

Military engineers, such as Clerville and the regional supervisors he recommended to Riquet, were literate men, acquainted to some extent with the works of Vitruvius, including his words on pozzolana. But their practical knowledge of hydraulic cement likely came from masons who used it in the construction of fortresses, harbors, and shipbuilding facilities. Artisans from the building trades had maintained the tradition of construction with hydraulic cement in the Mediterranean basin. That is why in Languedoc, both supervisors and subcontractors for the second enterprise did not question the value of using hydraulic cement, and knew how to mix it for the Canal du Midi.[9]

The seawall at Sète was the first place that pozzolana was used, and one where underwater building was both necessary and a problem. If the harbor was to be a safe one, the seawall had to be a strong barrier against the sea. It was meant to rise five meters above sea level, extending out from the rock base of Mont Saint Clair into the Mediterranean. It was built like a Roman wall with sloping stone exteriors along with a center filled with stones, sand, and cement. Under most circumstances, cement could not be used under water because it could not harden without exposure to air. Since cofferdams were impossible to use for a structure as large as the seawall, hydraulic cement was the only option for securing the rocks to one another in wet conditions for the harbor at Sète.

The port was originally enclosed by the seawall and a single modest jetty that extended directly out from the beach to stop the flow of sand into the harbor. A second jetty was added later when sand continued to flow past the first. It was designed to keep the sand from the entrance to the canal. The jetty did not have to be as tall as the seawall because it was not intended to stop high seas and the winds from storms but rather to inhibit the flow of sand into the port as it migrated down the coast.

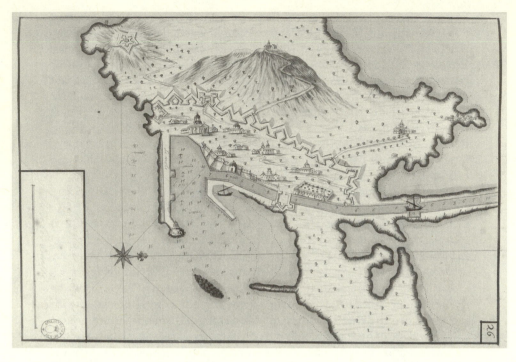

Figure 6.1
Late plan of Sète. Courtesy of the BNF.

The structure consisted of a wooden frame filled with rocks and sand—a form described by Vitruvius. The wooden frame was made with parallel sets of inward-sloping pilings pounded into the seafloor and connected by timbers that were assembled in the water. This scaffolding was filled with rocks and stones fused with hydraulic cement, and then closed on the top. The result was a solid and heavy wall that served as a large silt barrier. Waves could wash over it, but sand flows along the seafloor could not pass through it.[10]

Riquet imported pozzolana from Italy in 1670 for Sète, apparently eager to make use of its unusual properties. Clerville had been urging the entrepreneur to use pozzolana for the harbor, and had even specified its use for the seawall in the devis.[11] Once the purchase was arranged, Riquet seemed particularly optimistic that he could finish the harbor quickly and effectively, apparently assuming he could speed up construction using hydraulic cement. That might explain his suggestion to Colbert that he could provide a usable port within the year.

Colbert was not impressed. Riquet wrote to the minister on December 3, 1670, describing the clay he was obtaining from Italy that when mixed with lime, could make cement solidify under water.[12] Colbert did not reply, and later made it clear that he was dismayed rather than pleased by the news; he was convinced that Riquet's use of cement was simply a way to cut

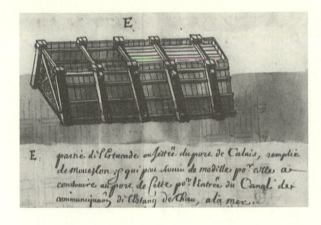

E. partie de l'Estacade ou Jettée du port de Calais, remplie
de mousslon, &c qui peut servir de modelles po' celles a
construire au port de Sette po' l'intrée du Canal des
communiquant de l'Estang de Thau, a la mer...

Figure 6.2
Jettee by Pons de la
Feuille. Courtesy of
the BNF.

the costs of quarrying and stonecutting that would lead to failures—the opposite of a brilliant use of classical engineering. He repeatedly appealed to Riquet to put good masonry on the exterior of the seawall and jetties, and use stones of good size to make them stable.[13] He seemed unconvinced that structures might be strengthened by the use of hydraulic cement, and probably felt vindicated in his skepticism about the method by the destruction of the winter storm in 1670. Riquet's New Romans might have brought classical engineering to the project, but it meant nothing to the sea.

The *pont-canal* de Répudre was the second troubled but important site where classical engineering was used by artisans to solve innovative and complex construction problems for the Canal du Midi. The bridge-aqueduct was designed to carry the canal over a narrow yet deep ravine or gully cut by the Le Répudre River. It was not clear how to design a bridge to carry such a heavy load, and how to make the channel for the canal watertight on top of a bridge. Hydraulic cement was the obvious material to use under these circumstances, particularly since it was understood to be stronger than regular cement.

The pont-canal itself was composed of a central arch, resting on two pilings that reached down into the ravine where the river flowed or sometimes tore through the countryside. The superstructure had a wide base anchored carefully to the arch. It cut across the ravine at a narrow point, and then followed the curve of the adjacent hills to make a horizontal U shape.

An aqueduct wide and deep enough for a navigational canal was well beyond the proportions of Roman aqueducts. Just the channel itself was large for a bridge in the period, extremely heavy when filled with water, and vulnerable to currents that developed in the waterway. In addition to the main channel, the superstructure had to include towpaths for barge horses and exterior railings. At the base of the railings, there were holes for draining excess water brought by floods or simply raised by the splashing of the barges, managing the outward pressures on the exterior walls.[14]

Most of the stonework was done with familiar Roman techniques. The keystone was extended into the superstructure, and parts of the walls used

Figure 6.3
Pont-canal de
Répudre. Photo by
Kenneth Berger.

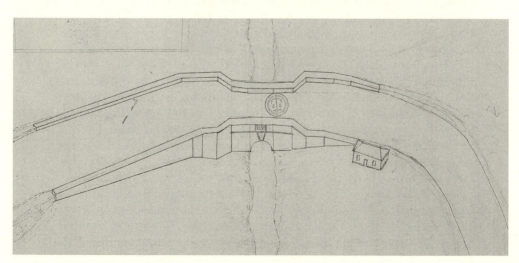

Figure 6.4
Plan for the pont-canal de Répudre. Courtesy of the AHMV.

fingers of stone to link them together—well-used Roman building practices.[15] And in addition, the pont-canal was constructed with hydraulic cement made with pozzolana.

> The masonry . . . of all the walls will be made with good mortar strong against water, in other words, mortar made with lime and pozzolana to assure that the walls of the canal do not deform or bend over the central vault.
>
> Plus the walls over the bridge for carrying the water from the canal across the gorge will be twenty-six yards long from each side to where the structure joins the embankments which form the canal. . . . Just as it is drawn on the plan and profile. The whole exterior facing and most of the interior will be made of cut stone, well shaped, and joined with mortar made with cement and pozzolana, and the center of the pilings of rubble made with lime and sand.[16]

Since the pont-canal de Répudre needed to have footings in a deep, wet ravine, its foundations were made more certain with hydraulic cement. The masonry for the superstructure could in principle dry in the air, but it too was supposed to be built with Roman-style hydraulic cement because of worry about leaks from the canal on top into the structure below.

Clerville set out specifications for the contract, and Riquet put the work up for bid.[17] And the pont-canal de Répudre was subcontracted to a local entrepreneur, Immanuel d'Estan, and his collaborator, André Boyer (*jeune*). D'Estan readily agreed to these specifications and to purchase pozzolana from Riquet at a predetermined price, assuring that Riquet's shipload of pozzolana would find use on the canal and demonstrating artisanal understanding of how to use it.[18] The pont-canal was plagued by problems of weather and financial disputes, but what remained unquestioned was the use of pozzolana, and the effectiveness of the hydraulic cement made with it. It was a taken-for-granted element of artisanal culture that added one clear element of classical engineering to the construction of the Canal du Midi.

The pont-canal de Répudre was clearly a work of New Romans. Its design was based on classical architecture for aqueducts, ways of building bridges with ancient precedents, and the viability of hydraulic cement. The necessary knowledge for the structure was partly documented in books, but the details had to be learned (and understood) as living practice—common sense. The *pont-aqueduc* at Le Répudre was a novelty that relied not only on visible precedents from the classical past but also on an obscure one that artisans used without fanfare—hydraulic cement.

Classical Hydraulics and Indigenous Women Engineers

Women from the Pyrenees were the main source of classical knowledge carried as indigenous technique to the Canal du Midi. Pyrenean knowledge of hydraulics, according the Louis de Froidour, was the best in France, and hydraulics was the Achilles' heel of trained engineers in the period. Peasant women, not gentlemen, knew methods for controlling wild rivers, working with rocky soil, and keeping debris out of a water supply. They could measure and maintain water volumes in a set of locks, control silt, manage drainage, tap water sources, and in other ways negotiate meeting places of water and land. Most important for getting the Canal du Midi to the sea, they could cut contours, making the canal snake through the hills toward the Mediterranean by following the local topography.

Canal contour cutting, or following the topography while maintaining a small but continuous incline, was perhaps the most critical technique of classical provenance needed for the Canal du Midi. It was a skill cultivated by the Romans for their aqueducts to assure a steady flow of fresh water to their cities, and one taken up by Pyrenean peasants and used over centuries to serve their towns.[19]

The Romans did not run their water from mountain sources to lowland settlements only through the arches familiarly called aqueducts. The water flowed mainly through covered channels underground, hugging hills, avoiding river valleys, and tunneling through veins of soft stone or coal. The point was to maintain the incline of the aqueduct to keep the water

Figure 6.5
Open area of the Fréjus aqueduct by the Siagnole source, following the contours above the river. Photo by the author.

flowing continually to its destination by gravity alone. The Romans used arches just to cross valleys where the land dropped precipitously. They were meant to link subterranean channels that mainly followed the natural geomorphology.[20]

Roman colonists, when they settled in Pyrenean valleys, used these methods to bring water from hot and cold springs to the baths in their new towns.[21] After the fall of Rome, the waterworks remained, but some were given new uses. Formal knowledge became vernacular tradition, and baths were replaced with new hydraulic structures. People living in the Pyrenees continued using these ancient techniques in the seventeenth century, applying the water delivery methods mainly to meet domestic needs, and to supply public fountains and laundries. The once-formal techniques disappeared from the written record, so their provenance was lost, but they remained a living tradition of practice. Sophisticated knowledge of hydraulics was useful to mountain peasants, so they did not forget it. That is why they could bring Roman hydraulics to the Canal du Midi, and carry the waterway through the hills of Languedoc to the sea.[22]

The Pyrenean peasants who came to the Canal du Midi were stripped of their identities in the historical record. Women laborers were not named in the account books at all, and men were listed without any indications of their social or geographic origins. Still, Riquet explicitly recruited laborers from Bigorre, a Pyrenean area rich with hydraulic engineering from the Roman period, and Perpignan, on the Mediterranean side of the Pyrenees, where there was a strong tradition of Moorish as well as Roman hydraulics. Most important, Pyrenean techniques of water engineering with Roman provenance were used precisely in those locations along the Canal du Midi where women laborers worked in large numbers.[23]

It was not surprising that mountain women would travel to the Canal du Midi for work in this period. Laborers from the region were known to be itinerant. Women often traveled across the mountains for trade, and also joined men from the Pyrenees who went to both Spain and France to participate in the harvests and grape crush in the fall.[24] Froidour described the patterns with some admiration:

> The thing I want to tell you is that the inhabitants of all these [mountain] valleys and as well those in the plains of Languedoc and Guyenne that are on the border with Spain . . . profit from the laziness and unfitness [*la paresse et de la fêtardise*] of the Spanish for whom they bring in the harvest, collect the grapes for wine, and make oils; they pass into Spain in time for bringing in the grains; they return to France to bring in their harvest that is later; returning to Spain for the wine pressing and returning to do the same in France. Sometimes they have to pass their winters in Spain, and that's the particular way of subsisting for the poor men of this frontier.[25]

Such seasonal, mobile laborers (including women) were obvious candidates for work on the Canal du Midi, and Riquet explicitly sought them out in both Perpignan and Bigorre, timing his labor drives so they would not conflict with the harvests.[26]

If it made sense that women laborers came to work on the Canal du Midi, it was not so obvious why their engineering ideas were taken seriously. It is hard to imagine what made them credible to a gentlemen like de la Feuille, but they were. Elites from Paris and peasants from Languedoc (particularly women) lived in completely different social worlds. Mountain peasants in Languedoc spoke Occitan, not French, and were notoriously suspicious of French as well as Spanish speakers. Gentlemen were skeptical at best of peasant ideas and tended to be derisive of mountain dwellers. In addition, there were traditional hostilities between the people of Languedoc and northern Frenchmen. What those who tried to realize the Canal du Midi shared was the classical tradition. It apparently allowed the participants to recognize each others' ideas and see the value of the practices. Colbert, Riquet, de la Feuille, and Mourgues—men who did not often agree—all spoke approvingly of the women's work. And regional supervisors employed women strategically where water met the mountains.[27]

Reading the Historical Record

There are limited archival records for determining the contribution of laborers who worked on the Canal du Midi, but one way to understand *how* these groups worked together is to determine *where* they were deployed on-site and with whom. Particularly in the area of the second enterprise, when different subcontractors were assigned distinct tasks, one can isolate the engineering strategies used for separate parts of the canal, and associate them with the workers and supervisors (carriers of informal and formal knowledge) who were active on-site. To the extent that women were frequently used together rather than scattered across the labor force, the account books for the canal—at least to some extent—reveal the gender division of labor in the project. Looking at *where* women laborers were used, we can compare the techniques used in the work to peasant hydraulic systems.

One problem with using archival records to look at gender and skill is that the accounting practices were not commensurable over the whole project, or even consistent at any given time. The books did not always specify gender, or what laborers were doing on a project. Worse, labor accounts changed form dramatically between the first and second enterprises. For the first enterprise and early work at Sète, the books often listed workers by gender (*manouvriers, femelles,* or *femalles*), providing relatively reliable counts of the number of men and women at work at a given time and place. But some payments were simply made for brigades or ateliers, that could include either men or women (three women counting as two men). Finally, records for the first enterprise contained little information about what the workers were doing, providing fewer clues about their distinctive contributions (if any) to the enterprise except on the Montagne Noire.[28]

The second enterprise with its subcontracting and new supervisory system required different kinds of accounts. The new books were organized around tasks and projects, since these were the bases of payments to local subcontractors. They listed specific pieces of construction, and lengths of channel that had been dug. The regional contrôles who visited the sites sometimes measured progress by the percent of the structure that was done, or the amounts of dirt that had been displaced in a given period of time. Skilled workers were listed separately since they were paid differently. Each was listed by name and identified as a mason, surveyor, stonecutter, or explosives expert. Their numbers varied in different pay periods, making visible the shifting work requirements for a particular lock, bridge, drain, or stretch of open canal. Ordinary laborers were generally listed by their functions, such as hauling or silt management, but sometimes they were simply described as "employees" of a particular entrepreneur, who was paid for the amount of work done (toises dug). These "employees" could equally well be women or men, making the gender patterns along with the sheer numbers of women used on the Canal du Midi impossible to determine precisely.[29]

Some of this information revealed much about what some work groups were doing at particular times and places, but most did not address the gender of workers except in rare cases.[30] So at the point that de la Feuille and Riquet were hiring more women, it became less clear where these women were working and with whom.[31]

Still, some rough patterns emerge from comparing letters and reports to these financial documents. In 1669, there were a thousand women working on the water system on the Montagne Noire alone, and roughly three thousand who were specified in different sites across the account books.[32] We cannot know how many women returned each season or how often they were replaced by new recruits, but what we do know is that women constituted a major part of the workforce *in key areas*. Moreover, when they were mentioned in the accounts for the second enterprise, they often comprised the bulk of the laborers. For example, in the area of Le Somail on May 14, 1678, the books showed that Estiene Valletter supervised 185 workers, 125 of them "*female*," and Jean Sabarié had 400 "femelles" and 21 "manouvriers."[33] Based on the limited though explicit references to women workers, it is clear that women finished the alimentation system in the Montagne Noire, and worked in the Le Somail and Béziers regions of the second enterprise—precisely where the waterway cut through the hills.

The term in the account books used most frequently to describe women laborers was surprising, even chilling: femelle. According to dictionaries of the late seventeenth and eighteenth centuries, this word was used only to describe animals, never women—even peasant women. The appellation was clearly significant, but hard to understand.[34] It may be that the women who worked on the Canal du Midi were called femelle because they were mainly mountain women who were seen by others (even in the valleys of Languedoc) as essentially wild. Or perhaps the term referenced their backbreaking physical labor, carrying baskets of dirt up and down the mountains like beasts of burden. Most surprising, the term persisted during the second enterprise long after women started to be praised and sought out as workers by Riquet, de la Feuille, and even Colbert.[35] Why this was so remains a mystery.

It is also unclear why gender was mentioned sometimes in account books for the second enterprise, when normally it was not. Subcontractors and supervisors who worked with women in the first enterprise may have simply continued to mention them in the second out of habit. It may also have been that many subcontractors were not sure *how* to keep records, and simply wrote down what they thought was appropriate; that would explain the huge variability in the detail of the accounts. Certainly, some subcontractors may have appreciated the women's work and made a point of mentioning it. Others may have wanted to identify women laborers to explain the low labor costs for the subcontract.

In any case, women laborers working on the Canal du Midi were used in a wide range of places, but were disproportionately employed by particular subcontractors and contrôles. Women were often the majority of laborers or even close to all the workers in the groups supervised by Campmas's colleague, Roux. Probably the largest numbers were overseen by de Contigny, who started out as *receveur et payeur des ouvrages* for the rigole on the Montagne Noire, but soon was appointed *directeur général des travaux*.[36] In both cases, Roux and Contigny started working with the women while designing the water system—where their indigenous skills started to shine.

The original labor force on the Montagne Noire consisted of fifty-three local men, mainly stonecutters and masons, who worked on the foundation for the dam under Campmas and Roux.[37] This group remained more or less constant until 1669, when the mass of new laborers was brought to the project after the War of Devolution. The postwar recruitment brought seven thousand men and one thousand women to the Montagne Noire.[38] These numbers were cut in the 1670s, when the dam was partially filled and started to be used. At that point, the labor force was reduced to four to five hundred women and two to three hundred men—the workers deemed necessary to perfect the system.[39] The men who remained were mainly stonecutters and masons finishing the dam wall or lining leaky parts of the rigole with stone. Women continued to carry dirt up the mountain to fill the dam cavities, but also made changes to the rigoles to assure good water flow.[40] They remained the majority of the workforce on the mountain until the water supply was in good working order. Then many of them followed Roux and Contigny to labor in the mountains between the Aude River valley and Béziers. Clearly, they knew something about mountain hydraulics that was deemed useful.

There is no concrete evidence of why Riquet recruited mountain laborers to the Canal du Midi. He never announced to Colbert that they were good with water systems, but it is likely he had some idea of this. The Roman baths at Bigorre were still in active use in the seventeenth century, being visited by French nobles who came for their curative effects.[41] Riquet knew this region well because it was part of his tax farm, and he had also helped Colbert collect samples of local mineral waters for a survey that the minister was conducting.[42]

In addition, Riquet may have learned about Pyrenean waterworks from one of the men on Clerville's commission, Joseph Avessen, sieur de Tarabel. He came from a town at the foot of the Midi-Pyrenees—a part of the mountains where these systems were prevalent.[43] Alternately, Riquet may have spoken about the mountain waterworks with Froidour. Around the same time that the forestry official came to inspect the Canal du Midi and its water system, he also went into the Midi-Pyrenees, and reported on the indigenous hydraulic systems he found there. He stayed for an extended period in the spa town of Bagnères de Bigorre, recuperating from a devastating illness. While he remained in the area, he came to appreciate the wide uses of water in the region, and may well have talked with or written to Riquet about what he had seen.[44]

Froidour's observations at least documented what Pyrenean waterworks looked like in the seventeenth century. On the baths themselves that he visited in the mountains, he wrote the following:[45]

> What could one not say in praise of the excellent hot baths of [the midi-Pyrenees]? . . . [I]t seems that no matter how degraded and bad a situation nature has devised, here one finds that people devise ways to improve the situation by transforming the waters into a resource so salutary and useful to the public, and one so much more exemplary for the whole kingdom and for our neighbors.[46]

The baths and mineral springs in the Pyrenees were not just remnants from the classical past that had remained intact over the centuries because they were sturdy systems. Frontinus, the aqueduct manager for classical Rome itself, complained that waterworks needed continual attention and broke down all the time. The interior of the conduits would fill with deposits and then the aqueduct would spring leaks.[47] No wonder the water supply of Rome stopped functioning only a short period after the sack of that city.[48] Clearly, the baths in Bigorre were only working in the seventeenth century because mountain peasants knew how repair and rebuild them.

Classical hydraulic techniques were incorporated into local culture and put to new uses in the Pyrenees. Roman-style waterworks existed in the seventeenth century in towns that had not been Roman settlements, suggesting that the techniques had spread geographically as well as temporally from classical times. Also, hydraulic practices that had previously served town fountains and baths now served domestic water systems and public laundries. These applications were familiar in Moorish towns in the eastern Pyrenees. In the Midi-Pyrenees, peasants built these amenities differently— with Roman rather than Moorish techniques. Perhaps interest in hydraulics became so vibrant in these mountains in part because there was a confluence of hydraulic traditions, resulting in waterworks developed with Roman methods and Moorish features that sprang up in towns from the Midi-Pyrenees to the Atlantic.[49]

Froidour's enthusiasm about the waterworks in Pyrenean towns was not matched by a fondness for the people who inhabited them. Like other representatives of the French state on official business for the Crown, he was afraid of locals. People in the Pyrenees for centuries had resisted—sometimes violently—outside authorities like tax collectors, soldiers, and inquisitors. Women as well as men were known to attack them, and run them out of town. In his words:

> The people who live in the Pyrenees from the Mediterranean to the Atlantic Ocean are generally brutish, perfidious, cruel, and nourish among themselves murderers and assassins . . . men who have no more reason than bears. . . . [E]veryone I talked to said that up until now the gentlemen and ordinary people not only in this country but all the mountains had been decidedly difficult to govern, not having recognized any outside authority either judicial or legislative and even the rule of the king was rejected. They never paid taxes of the sort normally paid in other areas of the province, and if there was anyone foolhardy enough to try to collect taxes, there was no difficulty in having that person assassinated. This was from the mouth of the bishop of Couserans. . . . One must agree [with the opinion] that these people are excessively brutal. They have to be disciplined in their domestic lives by necessity, but in return, they are drunken, as I have remarked before. They are also fierce about the conservation of their privileges, and are enemies of all novelties, even the smallest fashions, because since they have little taste, everything gives them umbrage. The women are no less brutal than the men.[50]

Froidour himself witnessed some hostility between local women and tax men. When he arrived in the Luchon valley in Comminge in 1667, there were some women arguing with a salt tax collector. The exchange became increasingly heated until the tax man killed one of the women. The locals thought that Froidour's guards were actually salt tax guards and started to attack them. But Froidour was traveling at the time with a known enemy of the tax men, and explained that he was only an emissary of the king working on forest reform. When the danger passed, Froidour (the good researcher) asked what was wrong with the salt tax. The women explained that they lived by raising flocks of animals, and their animals needed salt, so the women's need for salt was the need for life itself. They said they preferred to lose their lives fighting for their way of life rather than to die of starvation and misery.[51]

This story was instructive in many ways. First, it illustrated the assertiveness of Pyrenean women, and to what extent in these communities of shepherds they took care of the towns in summer while the men were away in the high meadows with their flocks. (Those who had no animals sought out agricultural work in the valleys below the mountains, so even poorer mountain men were away in summer as well.) The confrontation also demonstrated the tenuousness of local existence. The women's complaints made it clear that they were afraid of starving if they were not able to obtain salt. The short growing season in the mountains made agriculture and gardening limited. Flocks were needed to survive.

People of the Pyrenees often fought violently against intruders, as Froidour suggested, also to defend their culture against their powerful neighbors, France and Spain. They traditionally lived in quasi-republics made up of affiliated villages and valleys that practiced extensive collective landownership. The wars of religion had tattered some of this social fabric, but where contiguous valleys still had confederations of towns, they coordinated their uses of natural resources for mutual advantage. They herded animals through commonly held meadows that they improved and used in culturally prescribed manners; they dug ore as they wished from local mines; and (with some restrictions) they cut trees from, fished, and hunted without payment in local forests. Most of the land outside the villages was not privately but collectively held.[52]

The combination of community-shared property and political alliances over vast areas helped to sustain large-scale land-management systems that were maintained cooperatively. Some areas had advanced patterns of forest management and timbering. And towns in valleys along common rivers developed systems of canals, used cooperatively not only for irrigation but also for household purposes. Spring-fed waterworks were tied to their sources with community labor, and were for the use of all those involved.[53]

It was not an ignorant and brutish nature, then, as much as collective land-use patterns that kept this area so hostile to outside authorities, and so effective in staving them off. Confederations of villages could maintain canals that had to span long distances, and put the water to use differently according to the needs of each village; it took group labor and common interests to make such complex waterworks possible, and take advantage of local conditions. Their collective approach to natural resources may explain the ability of mountain communities to maintain classical hydraulics as a living tradition, putting old practices to new purposes.[54]

Froidour described the technical results this way:

The greatest advantage that people of the country here derive from these rivers is that they divert them everywhere they want, and that since their sources are at high elevations and come down steep inclines, they can divert them into canals even in the highest mountains and on high precipices to make meadows there. They also route them around towns to serve as fortifications, and they run them into the majority of private homes for the well-being of the inhabitants; they also disperse water in all parts of the countryside to improve it and to water gardens, fields, meadows, pastures, and to turn mill wheels to grind grain, to tan leather, to cut timber, to forge iron, and to work copper, to full fabrics, to make paper, and in a word, for all sorts of commodities, to such an extent that one could *say that to see all the uses one could make of water, one should see what they do in the Bigorre valley.*[55]

The inhabitants of Bigorre and comparable towns in the mountains used classical methods, but also violated fundamental values of the ancients about water use. These transgressions made it clear that mountain villages might have benefited from Roman hydraulics, but their waterworks were part of a living tradition. Pyrenean peasants, for example, diverted river water for drinking—a practice abhorred by the ancients, who learned to fear the diseases that could be carried in rivers like the Tiber.

On the other hand, many hydraulic techniques used in the Pyrenees had classical precedents: tapping high mountain springs, setting up reservoirs near them, diverting this water to make meadows, and taking supplies into town for collective use. Like the Roman colonists who had inhabited the area, peasants in Pyrenean valleys used contours to control inclines and the flow of water in their canals. They also built settling ponds to reduce the silt content of water entering town supplies, and set out sluice gates to regulate the movement of water into fields and gardens. These were all techniques familiar in Roman times that Pyrenean peasants sustained for their own uses, acting as New Romans well before they were asked to perform that role on the Canal du Midi.

The Powers of Pyrenean Women

Women in the Pyrenees could take on important roles as engineers and defenders of their towns in part because of their power within local family structures. In the Midi-Pyrenees, people lived in clan-based communities in which the oldest sibling—male or female—was the head of the family (*l'ainesse absolut*). This gave women property rights, and allowed them to inherit clan landholdings through the female line. The Basques on the Atlantic side of the Pyrenees took l'ainesse absolut as a firm social rule; others upheld the general principle, favoring men but still allowing inheritance through female descendants when the birth order and social assent supported it.

Culturally, women in the Pyrenees also had significant powers. While in most of Europe the sun was a male figure, in Basque culture the word sun meant grandmother, and a strain of grandmother-goddess-sun worship was sustained. Traditional goddess figures in Pyrenean culture were transformed into the Virgin with Christianity. Mary "appeared" frequently in these mountains even before her most celebrated sighting in nineteenth-century Lourdes.[56] Many pilgrimage towns dedicated their churches to the Virgin. Even in some town churches like the one at Ustou, the local church had an image of the Virgin rather than Christ above the altar.

Stories of fairies and female wood creatures abounded in the Catalan as well as Basque areas. Water sources throughout the Pyrenees were often associated with fertility, cared for by women, and said to be controlled by fairies. A virgin shrine was erected at the intersection of two rivers near the source at Argelès-Gazost, an area rich with both hot and cold springs.[57] The altar was placed on top of a much older rock structure with stones laid vertically and arranged in circles, suggesting a traditional religious site.

The centrality of female imagery in relation to folk hydrology may help explain why women were considered appropriate canal makers, and even

Figure 6.6
Virgin shrine by the source of Argelès-Gazost. Photo by the author.

why the supply systems for community canals were traditionally reconnected to high mountain sources on the day of Our Lady or Notre Dame.[58]

While it might be easier to imagine that men were the major canal builders in the Pyrenees and recruited to the Canal du Midi, that would not explain why so many women went to work on the canal, and why they were employed primarily in mountainous regions.[59] Certainly, both men and women in the Pyrenees worked on the water systems that served their towns, and perhaps both worked together on the Canal du Midi. Nonetheless, Pyrenean town canals were primarily designed for domestic uses, and the women laborers who came to the Canal du Midi seemed to carry with them precisely the skills used for hydraulic engineering in the mountains.[60]

Indigenous Engineering in the Pyrenees

Froidour described in detail the canal systems in the towns of Bigorre. He particularly focused on Bagnères, and the hot, tepid, and cold running water in the baths that typified them as Roman. But he also described the elaborate domestic applications and canals for gardening, pointing toward the hydraulics that served (primarily) the interests of women:

> Bagnères. It is in the mountains encompassed in the territory of this village that the plain of Bigorre begins. It is truly closed on all sides, but it opens up imperceptibly as one descends as the river moves away from its source, and the mountains become smaller. The river along with the water of the rivulets that spring forth in the area, are made into thousands of different canals that the inhabitants of the town use for watering their gardens, their fields, and their lands. The town itself is surrounded, in a fashion used in many locations, with a double and triple *fossé* of running water, and inside the city all the roads are washed by a canal that passes through, and at the same time, there are under the houses some small canals that provide water for all domestic uses; so much so that together with the quantity of fountains in this town of cold water and hot water, one can say in all truth that there is no other spot on earth where nature has been studied to such advantage to make visible these marvelous waterworks.[61]

Some waterworks around Bagnères still exist, and so do water systems in many towns of the Pyrenees from Perpignan to Lourdes, providing physical evidence of how such canals were constructed and used in the seventeenth century. Many of the towns where the canals exist today were mentioned by Froidour in his wanderings through the Midi-Pyrenees, offering some evidence that their waterworks date from this or earlier periods. Others were named on an early seventeenth-century map of the region by Sanson or were mentioned in religious histories of the Pyrenees in this period.

Waterworks located historically by these means supply some evidence of the kind of hydraulic work that laborers from the Pyrenees knew how to do, and the experience they brought to the construction of the Canal du Midi.

In Gerde, a town next to Bagnères, landowners made extensive use of water diverted from the river that was channeled along the streets abutting the city. Sluices along the main channel directed water into smaller conduits, mainly running into gardens and houses, reinforcing Froidour's claim that domestic water systems were ubiquitous in Bigorre. Other waterways split off from the main channels to water gardens, orchards, and small fields that could have been used for grain or grazing. Still other small streams ran into mills.

To conduct water from a single source in all these directions, the canals took advantage of minor differences in elevation, splitting into multiple channels that ran in narrow streams along terraces. One house in Gerde overlooking a small gully had a high channel from the street going into the house, a slightly lower one meandering through a lawn and flower garden, a third just below for another part of the garden, a fourth transecting an orchard, and the lowest one running along a small meadow either watering or draining it. Strategically placed stairs and stones used as bridges allowed people to walk through the complex of waterways. Multiple sluices controlled the movement of water through the various channels.

Other types of waterworks existed in the valleys of the Pyrenees—even well outside the area where Froidour described them. Take, for example, the water system at Ustou in the valley of the Ariège River, the town that Froidour called Houstou. This was a small community in a narrow valley, but its inhabitants nonetheless used weirs in the river to divert water toward two mills—one for grinding grain, and one for tanning. On the surface, this water system did not have the domestic character of the canals at Gerde and elsewhere. It did not seem primarily to serve the public good or provide domestic water for the community, but in fact the stones along the channel to the grain mill constituted a kind of public laundry. The flat rocks were reportedly washing stones, assigned to each family. A woman in Ustou pointed out to me the stone that her mother had used for her washing.[62]

Community water systems of this sort in the Pyrenees were clearly not *just* a product of Roman engineering, although they existed in regions of the mountains influenced by ancient colonists. Ustou was on the road to Aulus-les-Bains, a Roman bath town higher in the valley, smaller yet comparable to Bagnère-de-Bigorre. But the hydraulic engineering here was adapted to the simpler life of the river valley.

Techniques for managing water based on classical hydraulics sometimes led to the construction of elaborate waterworks like those at Mazère-de-Neste. This water system was built where the Neste River flowed through a relatively flat valley, so its waters could be dispersed widely and used for a variety of purposes without elaborate terracing.

A weir consisting of large rocks built diagonally across the Neste River slowed down the currents, and channeled water toward a sluice gate and

Figure 6.7
Two mills and a laundry stone by Oustu. Photo by the author.

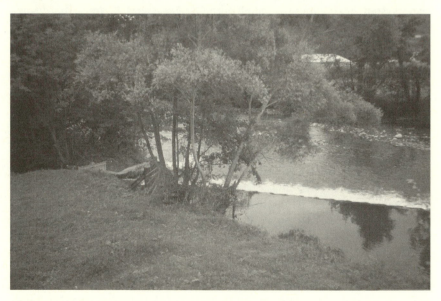

Figure 6.8
Weir on the Neste River. Photo by the author.

Figure 6.9
Sluice gates by Mazère-de-Neste. Photo by the author.

ditch that ran past fields into town. This primary gate could be closed in case of flooding to reduce the threat of high water to the town. The channel was split in two directions before arriving at the village, and met at a public laundry in the center of town. Along the way from the river, the ditch was outfitted with some stone pillars of the sort used by the Romans along their aqueducts.[63] The stones were scored vertically to hold pieces of wood (or perhaps, previously, slate) to close and open channels, sometimes functioning as a drain and at other times as a barrier to stop the flow.

The sluice gates from the Neste River controlled the irrigation and drainage of the adjacent fields. If the top gate was left open, the ditch helped drain the nearby fields because they were slightly higher than the water channel. If the gate into town (right) was blocked, water would back up into the fields and irrigate them. The gate to the fields could also be closed once the area had been irrigated, retaining the water longer to help it seep into the dirt.

The water that continued into town provided domestic water needs for private homes and supplied public fountains. The other branch of the system served as a sewer. The water system met in the central town square and the (recently rebuilt) public laundry. This whole complex constituted an irrigation/drainage system for agriculture, a domestic water supply, a water-based sewage system, and a public laundry.

The public laundries in central squares constituted the most ubiquitous feature of Pyrenean town water systems. They suggested, from both a practical

Figure 6.10
Fréjus aqueduct
detail, showing
grooves for sluices
to drain the
aqueduct near
Siagnole. Photo by
the author.

Figure 6.11
Central laundry at Mazère-de-Neste. Photo by the author.

and symbolic perspective, that women were indeed involved in their design. Town laundries were the center of women's public lives, the places where they would meet and work together, particularly in the summertime when the weather permitted outdoor activity and the bulk of the men were away, reducing domestic pleasures and indoor responsibilities.

Culturally, laundry had surprising importance in the Pyrenees. Fairies were thought to inhabit the inner recesses of the high mountains and were supposed to be obsessed with their nice, white, clean laundry. A periodic spring at Fontesorbes, for example, was said to be turned on and off by fairies doing their laundry. They turned the water on to wash their clothes, and turned it off when they were finished.

Fairies were also described as rich, hiding gold inside the mountains. Men wanting the gold would look for the fairies while they were doing their laundry. These miraculous creatures, being shy of men, would hide if men came forward, and the men would steal the linens and laces hanging up to dry. They could use the laundry as ransom for gold, but they could also hold on to it for good luck. Textiles were handed down over generations in the mountains because of the good luck that the cloth brought to their families. Good wives were even suspected of being fairies, blurring the line between these mystical deities and ordinary women. Legend had it that some men fell in love with the beautiful creatures, and caught the fairies, convincing their captives to become their wives. This was the explanation for particularly happy marriages.

Viewed through the lens of folklore, there was nothing mundane in this culture about public laundries. They were places where fundamental principles of gender and happiness were understood and enacted, and where the powers of women in nature and over water could be publicly displayed.[64] The culture helped make sense of women adapting Roman water engineering to local life. Water sources in the mountains and their use were associated with not only fertility but also the miraculous value of women (as fairies) for making life a pleasure for men and their families.

Women Laborers and the Canal du Midi

Techniques of canal construction from these mountain villages became visible on the Canal du Midi in precisely the areas where a large number of women laborers were employed. These were the mountainous regions for the project: the Montagne Noire and the mountains by Béziers. Work in these areas was clearly facilitated by the indigenous knowledge of the laborers. The canal's water supply system was essentially an extended version of the systems used in the Pyrenees to serve large towns. Water was diverted from rivers to provide alimentation, and managed with sluices and weirs. Some river water was enclosed in a valley by a dam that formed a reservoir, functioning also as a settling pond. From the high mountains, water was carried in trenches over long distances, through fields, along roads, past orchards, and to where it could be stored. Naurouze acted as another settling pond, helping the water shed its sediments before it was channeled to the canal. This was just the kind of water system that made sense to laborers from the Pyrenees.

The laborers who came to the Montagne Noire in large numbers after the war to haul dirt, rocks, and goods in their baskets for the water supply were clearly not local, since Riquet had Colbert require families in the Montagne Noire to house them.[65] From the work they did, it seems that they were from the Pyrenees, bringing indigenous variants of Roman hydraulics to the alimentation system for the Canal du Midi.

After the water supply was complete, women laborers began to be deployed in the area between Le Somail and Béziers where their hydraulic skills could be used to thread the waterway through the mountains toward the sea.[66]

Riquet had hoped that by taking the shorter northern route to the Mediterranean, the Canal du Midi would be easier and cheaper to build.[67] Unfortunately, the rocky terrain and complex topography made the construction more costly and technically difficult than before. In many places, workers had to use gunpowder to blast a canal bed from the rocks. And where the land was hilly, they routed the canal along natural contours to reduce the number of locks. This work called for blasting specialists, miners or former military men, and indigenous experts in hydraulics, mainly women laborers from the mountains.

Women were not listed in the crews building the channel between Trèbes and the Orbiel—the first leg of the second enterprise when the canal was

Figure 6.12
Canal du Midi entering the mountains. Photo by the author.

Figure 6.13
Canal cut into a rock ledge by Poilhes. Photo by the author.

paralleling the Aude River. To avoid flood damage from the river, the canal
had to be elevated and its bed cut from the rocks along the valley wall. Most
of the workers in this region were masons, stonecutters, explosive experts,
and surveyors, and Riquet had to acquire large quantities of gunpowder to
blast through the rock, asking Colbert to get him these supplies at the same
price as the king paid.[68] The Canal du Midi became (in places) a narrow
shelf along the rocky scarps of eastern Languedoc. It ran along ridges as far
as they extended, descending to the next lower ridge with a flight of staircase locks. The main construction problem in the Aude valley was not routing the canal, but cutting it from the hills.

In contrast, in the area of Le Somail toward Béziers, the canal no longer
followed the ridges of the Aude River valley. As it turned north, the canal entered a terrain that was more uneven and hilly, but often less rocky. The work
groups in this subcontracting region often had a large percentage of women
laborers, who apparently could take the canal through the hills for long
distances without locks.[69] In those places where the canal still had to be blasted
from rocky hillsides or taken down hills with locks, such as the region called

Figure 6.14
Canal, following the contours. Photo by the author.

Argeliers, stonecutters, masons, and blasting experts were employed in large numbers.[70] But in most of this region, the main problem was simply following the contours of the landscape, descending the watershed unimpeded while maintaining a gentle incline to run the water toward the Mediterranean.

Between the Cesse and Orb rivers, where many women labored, the canal became "a classic example of contour canal cutting," in the words of L.T.C. Rolt.[71] The waterway followed the inclines with such precision that it surprised visitors in later centuries who understood that the route could not be realized through calculation in the seventeenth century. It was only possible to build in that period with the hydraulic practices based on classical precedents employed by peasants in the Pyrenees to direct spring or river water to their town water supplies.[72]

Women laborers in this region also imported techniques from the Pyrenees to solve silting problems in the main channel when they started to be particularly vexing in the 1670s. Originally, local streams fed directly into the canal, but the storms in 1676—the same ones that devastated Sète—also created havoc in this region. Large quantities of water, silt, and debris rushed into the main channel, doing damage to the levees as well as filling the canal with silt. A number of strategies were used to address the problem. The canal was given a *counter-fossé* or second levee wall to keep exterior floodwaters away from the canal proper. But in 1678 and 1679, large numbers of women

Figure 6.15
Inlet with silt
barrier and
skimmer near Le
Répudre. Photo by
the author.

Figure 6.16
Settling pond with
silt barrier near
Trèbes. Photo by
the author.

were hired to redig the channel, and rebuild some of the connections to the streams and rivulets, where they used forms of water and silt control well-known in the Pyrenees.

They made simple settling ponds by widening the bed of an incoming stream before the water entered the canal, creating a deep pool where the water would lose momentum, and sediments that had been held in suspension could fall to the bottom, helping to purify the water that went into the canal.

The inlets themselves were redesigned, too. Their mouths were furnished with sediment barriers across the bottom to hold back precipitated materials, and some had skimmers on top to prevent large pieces of floating debris from entering the canal. The walls of the canal across from the inlets were also reinforced to prevent erosion by incoming currents. And slightly downstream from the inlets, the canal was given drains to expel any excess water that flowed in. Some of these drains were simple stone-reinforced openings in the levees just above the normal water level. Floodwaters entering the canal would simply flow out over this low barrier. Other drains had sluice gates that could be lifted up, expelling excess water from the floor of the canal and taking silt with it.

These techniques were familiar in Pyrenean towns such as Cheust, a small village on a steep hillside near Bigorre. This was an area where Protestants fled after being chased from Bigorre during the wars of religion. The water system at Cheust resembled the one at Gerde with its terraces and sluices, but it also was designed to manage the force and content of water rushing down the steep mountain into town.

Cheust had perennial problems with the debris and cobbles that inevitably washed down the local stream during storms and the spring thaw. Water was always dangerous, but carrying stones and branches, it could do more than clog up a millstream or domestic water intake. So the water system at Cheust was centered on a settling pond surrounded by stone walls and with a massive overflow drain to protect the village from floods, while making use of the abundant water provided by the mountains.

The settling pond was built near the top of the town, and was large and deep with a wide lip. It functioned partly as a weir for the river, reducing the water's velocity, so it could be dispersed laterally without damaging town

Figure 6.17
Settling pond, Cheust. Photo by the author.

Figure 6.18
Cheust mill. Photo by the author.

canals. The settling pond also held back silt and other debris before the water would move through the remainder of the system. The strong stone walls could stand up to the currents, and to one side of the structure, there was an overflow drain, a lower area of the wall that could discharge excess water from a fast flood into the deep ravine below. The pond walls were also punctuated laterally with sluices and gates to divert water purified of particulates into homes, orchards, gardens, fountains, and mills.

While the main stream of river water that overflowed the pond continued unabated through the town, water from the settling pond meandered through a series of small canals into terraces, gardens, orchards, town fountains, mills, and houses. Terrace walls were pierced with gates so that water from one level could be drained down to the next. And canals that powered one mill would drop down to do more work one terrace below. The central destination of the water supply was (again) the public laundry. In spite of the small size of the town, the maze of canals was elaborate, making wide use of the water offered by the mountains.

These ways of controlling and using water in the mountains were both simple to build and effective. So it was not surprising that mountain peasants remembered them, and brought these techniques to the Canal du Midi. It also was not so surprising that the canal in turn was endowed with a series of public laundries.

Figure 6.19
Cheust laundry. Photo by the author.

Figure 6.20
Laundry by the Argelier lock. Photo by the author.

Peasants at Fonseranes

The most dramatic piece of engineering that relied on peasant techniques and laborers was the staircase lock at Fonseranes. With few exceptions, staircase locks on the canal had only two or three basins in succession. The eight-lock staircase at Fonseranes was dramatically different in scale, and descended a long hill across from the city of Béziers—Riquet's hometown. This was a grand enterprise and no place for a technical failure.

Figure 6.21
Upper half of the Fonseranes staircase. Photo by the author.

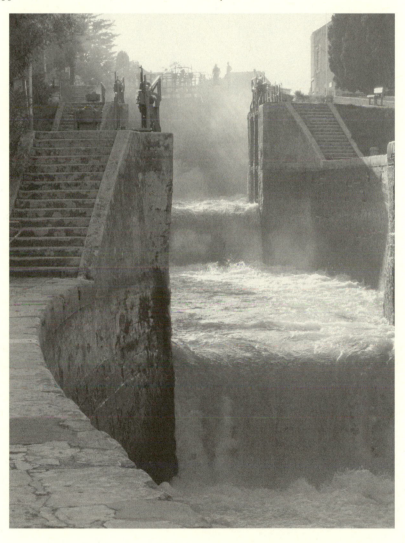

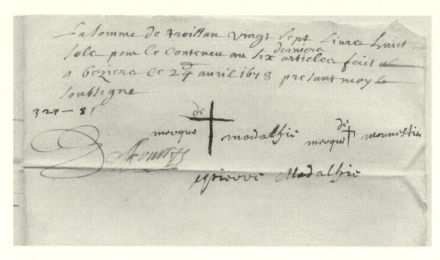

Figure 6.22
Account book for Fonseranes, April 24, 1678. Courtesy of the ACM.

Surprisingly, the entrepreneurs who were awarded the contract for the Fonseranes staircase were illiterate peasants, Michel and Pierre Medailhes. They could not even sign the account books; instead, they put down crosses to verify their work—a large one for the older brother, and a smaller one for the younger brother. Their "signs" were witnessed by père Mourgues.

Why were two illiterate brothers given the contract for such a visible and important piece of engineering? Why did Riquet not take on this task himself—particularly since the staircase at Fonseranes faced his hometown of Béziers? The answer may have been (again) experience with mountain hydraulics.

Most of the other subcontractors who were awarded contracts were described in canal documents as architects or engineers. These men could follow written specifications, sign account books with a fluid hand, and write letters on their own behalf, demonstrating a literacy that was quite the opposite of the brothers who were awarded the subcontract for the staircase locks at Fonseranes.

There was no dearth of educated engineers in the region who worked on the Canal du Midi, and who could have taken the job, many of whom had already been at work on locks or the dam at Saint Ferréol.[73] There were two schools in Languedoc that taught engineering—one in Béziers, and one in Castres. The Académie at Castres had been a major Huguenot school with some notable engineering professors, such as Pierre Borel, who had developed the plan for the canal that Riquet originally copied.[74] Castres was also the town where the great mathematician Pierre de Fermat worked.[75] It was no intellectual backwater, and the Académie in Castres trained engineers in mathematics. Still, no one from this institution was awarded the subcontract for the Fonseranes lock.

The Jesuit college in Béziers was at the other end of Languedoc, close to Fonseranes itself. It was the school where père Mourgues had taught before he started working on the Canal du Midi. He was clearly in a position to find skilled and literate subcontractors from the Jesuit school who could have tried to build the Fonseranes staircase, but he did not. Neither did he oppose awarding the contract to the Medailhes brothers. Why? Again, the best explanation seems to be that the brothers knew something that schooled engineers did not.

The idea that the Medailhes brought peasant skills to the task seems all the more plausible because these brothers employed a workforce composed predominantly of women to build the staircase lock at Fonseranes. This was an odd move in itself, since no women were mentioned in the account books for locks built previously for the second enterprise, and women were also generally not used where the canal had to be cut from stone. Some masons and stonecutters (all men) were hired by the Medailhes, too, since the locks had to be cut from almost solid rock and lined with masonry. But women were there to haul rocks away and form the basins. Why?

The main engineering problem in designing the staircase was that all the locks had to hold the same volume of water even though they were built on a hill that was not a perfect incline plane. The hill curved gently down at first, descended steeply in the middle, and started curving up near the bottom. So although it was assumed to be on a plane, it was not. The basins at the top and bottom could not be the same as those in the middle, and even those in the middle had slightly different depths.

Inequalities of volume had been an ongoing problem in building lock staircases during the second enterprise. The double and triple locks needed in the hills were still not functioning properly when work on Fonseranes was started. Engineers tried repeatedly to compute the water volume in a staircase for a given hillside before they built it, but the locks did not turn out as planned. Some basins required significantly bigger volumes of water than their counterparts, and so the crews had to rework them to compensate for the mistakes. They created special water intakes and side drains (often tambours) for the bigger basins in some staircases to fill and empty them more effectively. Yet these measures were costly and cumbersome

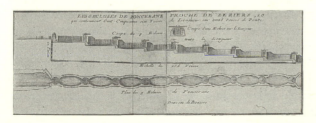

Figure 6.23
Fonseranes staircase
schematic drawing on
the "Carte Royal de
Languedoc, 1697,"
Nolin. Courtesy of
the AHMV.

Figure 6.24
Top lock at Fonseranes, facing Béziers. Photo by the author.

work-arounds, and the problems were repeated in new sites time and again. By 1672, elevation studies for staircase locks were abandoned because inaccuracies in the measures had necessitated so many reconstructions or retrofits.[76]

The problem at Fonseranes was that there was no way to do this kind of retrofitting for a staircase lock with eight basins. So the volumes were tested with indigenous methods. The locks were dug first, and then the cavities were filled with water to check the volumes before the interiors were finished. Only after the final adjustments were made were the locks lined with masonry. Using this method, the women working under the Medailhes brothers made a staircase that was finished without problems.[77]

What did the Medailhes brothers and their women laborers know that made the lock so successful? There was one peasant technique that could explain their abilities. In the high areas of the Pyrenees where forests abutted streams that ran low in the summer and fall, the peasants developed an ingenious technique for floating timbers to larger rivers. They erected a series of temporary dams with room for small reservoirs behind them. When the reservoirs filled, the water became deep enough to float logs. When water in the first reservoir rose to the top, the wall would break, sending logs and water down to the next dam. If each reservoir held the same volume of water, they could be broken in succession, carrying timber in a sequence of waves downstream. The result was a step system for floating timbers in the fall when the sap had stopped running in the trees, snows had not yet impeded access to the forests for timbering, and mountain streams were still too shallow to carry logs down to mills.[78] If in some regions the men cut the trees and the women built the dams, this would explain why the Medailhes brothers would have employed women laborers for the Fonseranes locks.

Indigenous engineers, whether following Roman precedent or not, provided a rich set of skills for constructing the Canal du Midi. The women laborers who carried so many of them proved the efficacy of this kind of expertise. Some of the peasant hydraulic sophistication was derived from classical precedents, but some (like the dams for timbering) was not. Perhaps Pyreneans became so expert in working with water because in the mountains it was plentiful enough to be useful, but difficult enough to manage that it required technical work. And perhaps, too, they benefited from the confluence of Moorish and Roman traditions in the Pyrenees, encouraging them to experiment with dams and laundries as well as settling ponds and aqueducts. Whatever the sources of their skills, women laborers who came to sites like the Fonseranes locks and the Montagne Noire made the Canal du Midi possible, because their indigenous engineering was either more sophisticated or appropriate than its formal equivalent.

Men as well as women who worked on the Canal du Midi brought vernacular skills on-site. The Medailhes brothers were clearly indigenous experts. But it is easier to distinguish what the women laborers could do than the nameless peasant men who shared the project. We can identify what the women built, allowing us to see more clearly the importance of their indigenous traditions to the design of the canal. The staircase lock at Fonseranes was a vivid demonstration of how powerful and useful peasant expertise could be; it was a testimony to the women who maintained the volumes and the illiterate men who hired them, knowing what they could do together.

The staircase locks at Fonseranes and the aqueduct bridge at Le Répudre have come to be celebrated as works of genius, and among the great accomplishments of the second enterprise. Ironically, the indigenous traditions they depended on have remained too lowly to be included in heroic stories about the Canal du Midi. They are heralded as Riquet's works of genius or emblems of the New Rome, not tributes to the New Romans who actually built them.

Much of the work of the New Romans remains obscure. The contour cutting through the mountains, the use of hydraulic cement at Sète, and the insertion of settling ponds and skimmers for purifying water for stream inlets may have been less dramatic bits of engineering, but they were among the most interesting genealogically and technically. They were indigenized forms of classical engineering that peasants and artisans taught to gentlemen, and used to make the canal in the area of the second enterprise a haunting model of impersonal rule.

CHAPTER 7

Thinking Like a King

Riquet was undergoing a personal transformation in 1669—1670 as the canal started to change the province. The outward marks were startling to Colbert; the entrepreneur began to assert a new authority over the project, claiming he had found in himself the knowledge to build it. He admitted his word did not carry the full weight of a gentleman, but his thoughts were nonetheless true because they were products of divine inspiration. Riquet argued that this must be the case because he was clearly no engineer, yet still the canal was succeeding. The only way to explain it was God. With this sleight of hand, Riquet endowed himself with new authority for his enterprise and redefined his relationship to Colbert. He had been given the gift of genius by God. Others might doubt him, but it did not matter; he was an oracle.

According to stewardship principles, land improvement was the exercise of dominion, using the intelligence given man at the Creation to make the world more Edenic. Virtuous men in the service of God could learn how to do His bidding. Riquet found himself bursting with ideas for bringing more prosperity to Languedoc, building more canals, adding new elements to local infrastructure, and turning Sète into a great port city. He saw God as the source of his new creativity, and identified proof of his stewardship in the many inventions on the Canal du Midi that had proved their value. He wrote to Colbert on March 27, 1670:

> God has inspired me with thoughts that I call sacred, since with them my works will be advanced, and I will not lose [my way]: it is in truth, Sir, a philosopher's stone which gives to all and misses no one. . . . [F]rom dreams, I've found such miraculous expedients for the advancement of the canal. My apprenticeship has cost me dearly, and my ruin will follow [without new] funds.[1]

On one level, the letter was a plea for money from a state contactor to the minister of the treasury; on another level, it was an assertion about Riquet's voice. The entrepreneur claimed the moral authority to speak on matters of engineering, having been blessed by God with the ability to transmute land into new wonders.

Long frustrated by the États' calls for verifications, Colbert's insinuations about graft, and his lack of public credibility, Riquet changed the only thing he could: himself. In 1668, he had written Colbert, "I swear to you that it is not endurable to suffer such criticism as I have received until now and [in] these quantities."[2] But now attacks on both his character and ability meant less as he saw himself as God's agent, whose moral virtue gave him epistemic standing. Gentlemen gained their credibility from their high rank, and used their rank to manage the truth of things—a form of hegemonic power. Riquet presented a different epistemic standard and basis for power: evidence on the earth of knowledge from heaven.

Since Riquet was too low in rank to be really trusted, he had always needed until this time men of standing or patrons to vouch for him. But now he promoted a different moral-epistemic compact where ability was dissociated from rank. Good stewards demonstrated their knowledge and connection to God through their "works," their physical power over the earth. Their efficacy on the land was evidence of their moral virtue and capacity to know the truth of things.[3] The success of the canal gave him the authorial status he had always wanted.

These stewardship notions probably made more sense to Riquet than to Colbert in part because the former lived in a part of France where the Huguenot population was both large and powerful.[4] Talk of works was more commonplace there. Local nobles already had made the distinction between moral authority and social rank, with some Huguenots even arguing that they owed no fealty to the king if he lacked moral leadership.[5] Riquet's interests were more epistemic than religious; he wanted and claimed the authority to know, not moral autonomy like his predecessors. The entrepreneur believed that his engineering capacity was essential to territorial politics, so his thoughts should be respected. But more than this, his record of accomplishment seemed to authorize him to redesign France. Riquet, in other words, adopted a modern model of epistemic authority based on things "working," cloaking it in a mantle of spiritual legitimacy.

Efforts by his opponents to find fault in his character reinforced Riquet's drive toward God. He could claim the moral high ground even more easily after de la Feuille and Bezons had scoured his books and the canal, looking for signs of graft and corruption. They had to exonerate Riquet because he had stolen no money from the treasury and was doing what he had promised in the contracts.[6] By 1670, Riquet might not have had the epistemic credibility to enroll Colbert in his way of thinking, but he could claim virtuous motives in his enterprise.

Riquet was also encouraged to see himself as blessed by God by a change in his cognitive powers.[7] Collaborating on the canal had given him (as well as the others) a new ability to think about problems of engineering. If he had started the work assuming either that he would quickly cut the canal with little outside help, he had learned a lesson. The work was only achieved through constant collaboration and the careful recruitment of experts (formal and informal) to help with problems. If Riquet and Clerville knew anything better than all the others on the project, it was this: the Canal du Midi required a wide range of skilled people to debate matters of engineering, including peasants, soldiers, fishermen, artisans, and engineering experts sent by Colbert from the Netherlands and northern France.

By 1670, Riquet was beginning to see his ideas as superhuman, and he was oddly right. They were a social achievement formed in extended discussions. Collaborative work changed Riquet, and his new thoughts—so alien to him—seemed like divine revelations because they were beyond his normal reach. His deepening religious conviction had a Durkheimian character. He gained a sense of the superhuman in the social. But his narcissism propelled him to believe that his thoughts reflected God's will, not his debt to others. *God* had made him an engineer, giving him the technical capacity to build the canal, not the community of practitioners he had so skillfully helped to assemble.

> I am beginning to realize that I am the one who knows more than others. . . .
>
> I am at the moment a kind of experimenter and [designing] new implements is easy beyond my dreams, so that [for the lock doors] a single man without the aide of any mechanical device can open and shut them. A machine [for such work] is simple and ordinary. These [doors] are the best. . . . Even more, I provided for myself, among these implements, novelties that with their aid can fill and empty the locks with incredible speed.[8]

Riquet reimagined himself as engineer and spiritual servant also in opposition to Colbert's presumption that he was not essential to the second enterprise. It was *not* true that others could finish the work; it was *not* true that Riquet could not be trusted; it was *not* true that those with a better education could do the work more effectively than he. Riquet mattered because he had been transformed by God to be His agent.[9]

Without Colbert as a reliable ally, Riquet also needed God as his patron, and couched his spiritual status in patronage terms. He spoke of himself as a recipient of divine gifts and indebted to his Creator for his successes. He now sought to serve the king by listening to God, not Colbert. Of course, this was hubris. The monarch, rather than a tax farmer from Languedoc, was supposed to know and embody God's will on earth.[10]

In James Scott's terms, Riquet started to "see like a state."[11] This was perhaps easier for the entrepreneur because the canal *was* his estate. By mesnagement logic, an estate was the microcosm of the state, and both could and should be administered similarly. On his domain, Riquet already had many powers. He had acquired rights to impose duties and tolls, sell offices, and enforce the law in exchange for the duty of good stewardship. As he dreamed about more means of improving his own domain, he thought of ever more powers of the state he could use to do so. He started to confuse the king's will and power with what he could imagine and do, assuming that if his plans served Languedoc and expressed the will of God, they also served the king.[12] But service to the monarch presupposed a subservience that Riquet did not quite grasp or accept. Riquet did not make the distinction so carefully maintained by Colbert and understood by Machiavelli between thinking from the subject position of the king in order to advise him, and taking on the mantel of monarchical agency.[13] So in imaging himself as a carrier of God's will, dignifying and deifying his own thoughts, and thinking of himself less as a servant of the king because of the *monarch's* role as the executor of God's will on earth, Riquet put himself on even shakier political grounds.

Social Psychology of Territorial Politics

The Canal du Midi was as emotionally disturbing as it was exciting. During the period of the second enterprise, the new scale of the project made it an even greater gamble and a heady thing to contemplate—both good and bad. The project stimulated the imaginations and deep feelings in most of those involved: anger, fear, determination, rage, frustration, and elation. Riquet had become a visionary as Nivelle had predicted.[14] He was Colbert's worst nightmare, and a source of inspiration to Clerville. To have visions, to see phantoms and think of them as real, was to fall into a liminal space between heaven and earth, where flights of fancy were not limited by worldly concerns. A man could dream of a canal through Languedoc, but that did not make it real. And as Riquet learned, a tax man could even build a canal across Languedoc and be ennobled for it, but that would not make him any less a dreamer or any more credible as an engineer.

During the period of the second enterprise, the stresses of the work and the ambiguity of his social position took their toll. By the early 1670s, the combination of successes and failures had become dislocating, disturbing Riquet's sense of self and view of the work before him. On the one hand, the canal started to come together. The oval form for the locks was finally tried and adopted; so was a new profile for the water channel. The canal walls now leaned out farther to make the surface of the waterway even greater compared to its base, providing wider shipping lanes with relatively little more water and effort. The first section of the main canal from Toulouse to Castenet was also flooded in 1670, and the next leg to Naurouze was opened in 1671, connecting Revel to Toulouse as well. The pace of completion quickened again in 1673 as the locks from Naurouze to Castelnaudery were completed, and the reservoir at Saint Ferréol started to be filled and used. Finally, the canal from Toulouse to Castelnaudery—almost the entire first enterprise—was finished and officially opened to navigation in 1674.[15] These were all reasons for Riquet to feel blessed by God and vindicated morally by his accomplishments.

While these successes were heartening, Riquet was chastened by the relentless technical problems that kept arising, the growing depth of his financial distress, and his need for and failure to secure Colbert's unwavering support. Work might be progressing well on some parts of the first enterprise, but the problems on the second were beginning to multiply. Sète was damaged by the devastating storm of January 1671, and Riquet found himself charged excessively by local suppliers for the gunpowder he needed for repairs; Colbert stepped in and assured him gunpowder at the price paid by the king, but not without reprimanding the entrepreneur for building a poor seawall in the beginning.[16] Problems with the lock doors also became evident in 1671 when so many of them were first put into use. Because they had to be large for locks so deep, the doors were hard to open and close. To help redesign them, Colbert sent a Dutch carpenter, the sieur de Vos, to advise.[17] Riquet might have been thanking God for his ideas, but Colbert was frustrated by the entrepreneur's continued inability to solve basic problems.

In 1673, Riquet was again caught in arrears in his payments for the gabelle by three to four hundred thousand livres, leading the minister to begin speaking ill of the entrepreneur to Louis XIV.[18] To add to Riquet's woes, Clerville lost standing with the king in 1673, so his greatest advocate at court was compromised politically. The fortress engineer had been slapped down by the king for his own hubris and assumption of monarchical powers in the pursuit of territorial politics.[19] In the same year, Bezons was replaced as intendant of Languedoc by Henri d'Aguesseau, a man who was an advocate of the Canal du Midi, but not fond of its entrepreneur. This left Riquet without effective supporters in a period that should have been triumphant because the canal began to flow toward both seas, fed reliably by the water supply. To comfort himself perhaps, Riquet treated the success of the first enterprise as the final demonstration of God's grace, regarding the setbacks on the way simply as purifying trials of faith.[20]

Riquet wrote Colbert in January 1671:

> Those who compare the little money that I have received for the great works that I have executed are persuaded that [I have made] miracles, and I myself believe my thoughts are oracular, when I consider what God has made known to me. . . . I am his agent . . . my contribution is that of an oracle.[21]

The entrepreneur had started the Canal du Midi with a more secular approach to power and his political role. Like the miller Domenico Scandella, known as Minocchio, who developed his own cosmography in the sixteenth century, Riquet had developed his own folk theory of power.[22] Some elements were embodied in a statue that Riquet proposed for the basin at Naurouze:

> [We will erect] a statue of the king in a chariot pulled by horses in the sea, having one of their feet on the globe to express the power of your majesty on the earth. . . . The water that will flow out from around the chariot will form a wall around the statue and will fall into hidden conduits and from there will be exuded from large jets

by serpents toward the four corners of the earth to show that *the graces and favors your majesty passes out so graciously pass mainly, sir, through your ministers.*[23]

On the one hand, the statue presumed and represented political administration in patronage terms. Agents of the king like Colbert and to a lesser extent Riquet himself were portrayed as serpents—powerful, menacing, and capable of exercising extraordinary powers. They acted through hidden conduits (like good, self-effacing clients), but were the true distributors of royal "graces and favors." On the other hand, the statue conveyed Riquet's belief in the king's dependence on his agents.[24] Not surprisingly, his plan for the statue and fountain at Naurouze was rejected, and Riquet started on the road to becoming God's client, exercising even more power by speaking as an oracle.

Oracular Visions

Riquet's new faith in his engineering ability was manifested modestly at first in the revival of his ambition to make Pierre Borel's canal through the mountains to Revel. He began by widening the rigole de la plaine enough in 1669 to support navigation. Opponents of the Canal du Midi in the high valley by the Sor provided the pretext. They complained about loss of water from the river, and Riquet proposed to compensate the region with a new inland port. He asked to build it near Revel and to call it Port Louis. It was specified to be "a square basin made of fine cut stone nine pieds deep and twenty-two toises wide," providing a small but usable commercial hub for the region.[25] To "improve" the rigole to make it navigable from Naurouze to Revel, Riquet did not widen it to the proportions of the Canal du Midi but only enough to carry small boats, using a few locks to ascend the mountain. Colbert was at first cynical about the entrepreneur's objectives in making the port, but became convinced that none of this directly benefited the entrepreneur financially, so he approved the work, writing to Bezons that Riquet was inspired by noble motives.[26] That is how in 1669, a small-scale navigational canal began to be built exactly where Riquet had first wanted to construct one. It opened for traffic in 1670—well before the main canal—demonstrating the validity of the plan that Riquet had adopted from Borel that had been discredited by the "experts" of Clerville's commission.[27]

Still, the improved rigole was only half of Borel's plan. The canal to Revel did not connect to the Sor and Agout, or make the rivers navigable to the Garonne. So Riquet persevered. In 1671, he wrote to Colbert, recommending that the Agout River be made navigable from the Garonne to Castres.[28] The tax farmer did not propose the project for himself but rather for his son-in-law and agent, who had "shown interest in the project." He argued that his son-in-law had gained enough experience working with him on the Canal du Midi to take on this new enterprise, and that he had reviewed the offer and plan, and approved them.[29]

In 1670, Riquet gained the confidence to veer away from recommendations of the commission for the Canal du Midi, too. He proposed to develop an inland port out of the basin at Naurouze and create another port at the town of Castelnaudery. Neither project had been suggested or even reviewed by the commission. Significantly, the commission and Colbert had never shown any interest in using the canal to link towns for commercial purposes. The minister wanted the canal for the navy, and the commissioners avoided cities in setting the route—perhaps to make the project less socially disruptive and politically explosive. But Riquet hoped to benefit from trade on the canal and wanted to encourage commercial traffic.

The basin at Naurouze was an ideal site for a port because it lay at the intersection of the Canal du Midi and the rigole de la plaine, where cargo moving to and from Revel had to be transferred from small to large boats, or vice versa. Serving as a quarry for the canal, the basin had become enormous, and was more than adequate for a port. During the quarrying, workers had also hit underground springs, probably draining a catch basin on the Lauragais ridge and giving Naurouze its own water supply. These springs gushed so much water in periods of rain that workers had to construct a large drain to keep the basin from overflowing. Still, the water assured that Naurouze would have the alimentation needed for a functioning port even when demand for water for the canal was high. So Riquet dreamed of erecting an ideal port at Naurouze—a utopian evocation of Rome that was never completed.[30]

The proposed port on the Canal du Midi at Castelnaudery was even more controversial than the one for Naurouze because it required shifting the route of the canal. Central Languedoc was rich in agricultural goods, and full of local manufactures that Riquet hoped would provide traffic for his waterway. The entrepreneur wanted most to take the canal through Carcassone, the second-largest city (next to Toulouse) of western Languedoc where the trading possibilities were greatest, but the city fathers did not want it, fearing that the king's waterway would give royal troops too easy access to their town. Carcassone had a proud heritage of political and religious independence that made locals wary of outside authorities. But nearby Castelnaudery was different, poorer, and in need of the economic opportunities that a port would bring.[31]

The route that the Canal was supposed to take, following the proposal by M. de Riquet and the specification of M. le Chevalier de Clerville, was far from the town of Castelnaudery. . . . Nonetheless, the inhabitants of this town, which has no river, stream nor [other] transport for their merchandise, sensed the advantage they could draw from the passage of the canal near their [city] walls, making the proposition to M. de Riquet, which he communicated to M. de

Colbert, and after the approbation of this minister, a treaty was passed on the 24th of May 1671 . . . by which it was decided that the town of Castelnaudery would give 30,000 livres . . . to make the Canal pass by that town . . . and construct a port commodious for the loading and unloading of merchandise.[32]

At first Colbert opposed the plan, writing to de la Feuille that the detour would make navigation through the canal longer; the work was too advanced on the canal to make any changes; and the desires of a town were less important than the general good served by the canal.[33] But Riquet convinced the minister and king that such a grand port would attract lots of people to this area of Languedoc, and that the detour would be minor. In the end, he was given permission by the Conseil d'État to work with local church and town officials to take the canal into Castelnaudery.[34]

For Riquet, the detour was an attractive way to raise new money. He was, as always, short of funds, and was more likely to get income from Castelnaudery than the treasury. The city fathers put together a handsome

Figure 7.1
Castelnaudery basin. Photo by the author.

package to defray the costs of the detour, and so Riquet happily agreed to develop a substantial port at Castelnaudery.[35]

At Sète, Riquet's dreams also included improvements to the port city. He hoped to make the town into a regional hub as well as a harbor for the Canal du Midi, improving it with infrastructure. The little fishing town was isolated on a long sand spit, cut off from the mainland by the étang. It was hard to reach, had little water, and was far from fertile land for food. Sète could never become a bustling port isolated this way, and the harbor was hard to build without more ready access to materials, so Riquet began to extend a causeway through the étang toward Frontignan. He wrote to Colbert in 1670 that "[the causeway] is 225 toises [480 yards] long, 2 wide, by 1 in height, and so well advanced that in a month carriages, carts, and generally all sorts of vehicles will be able to pass there easily."[36] The new road through the étang facilitated the movement of workers as well as materials to the new port, but it also added to Riquet's costs. To help finance the work, Riquet wrote Colbert asking permission to charge a toll on locals who used the causeway to help him make back the expenses.[37] "What I am able to make does not come without great expense, which is why in justice to give me something for my debts, I am asking for [permission to charge] a small toll."[38] Colbert was sure that Riquet was just lining his pockets with these side ventures, but he approved the toll in any case.[39] The visionary had once again imposed his will on the kingdom by improving his domain.

Riquet also wanted to create a water supply for the town. Since Sète lay so close to the sea, there was not much groundwater to tap. If the town was to have a significant population, it needed a supply of freshwater. Riquet argued that the water could be used to power mills for sawing wood and grinding flour to feed the workers, too. What he did not mention was a secret project he had with Clerville—apparently to use the water to flush the harbor. He wrote to Colbert on October 14, 1671:

> I must find the means to carry water here in quantities sufficient to run some mills. . . . I will take [de la Feuille at the end of the month] to see the places and the means, not simply the sources, but also the conduits [for carrying the water to Sète]. . . . In a word, sire, I will open all my thoughts on this subject to sieur de la Feuille, equally we will make an estimation and calculation of the costs. And [additionally] in this province, there are some experts in these sorts of things, who I would consult without a doubt. . . . And having completed the rigole of the Montagne Noire, and that of the plain of Revel, [I have the] experience to do things of a similar nature.[40]

The aqueduct was apparently not as simple as he had hoped, as it remained one of many projects still under discussion with Clerville in October 1676. Bringing as much water as they wanted to Sète was difficult. So he abandoned hope of doing more than supplying fresh drinking water to the town.[41]

Dredging the harbor remained a problem, so Riquet started talking with experts about buying a dredging machine to use at Sète. In May 1677, Clerville wrote Riquet, disturbed by this change of heart:

> I believe your machinists will leave this morning to execute the noble project of raising the sands from Cette [with a machine] that even the smallest of your workers could do with no other tool than a basket. . . . But [as I understand it] the machine designed by M. Vallois [from Normandy] is nothing more than the one M. de Gillade spoke of in my presence, and which is common in this region, called *la Saume*. . . . One should pay for merit, but . . . it is often that machinists copy each other's inventions, and call them their own.[42]

Whatever happened to the work on this dredging machine, we do not know, and Clerville died in 1677 with his dream of a new technology for maintaining harbors unfulfilled.[43]

In the 1670s, Riquet also began improving the small canals near Aigues-Mortes to connect the Canal du Midi to the Rhône. This pleased Colbert enormously because the work seemed refreshingly sane and necessary. Recovering these canals and connecting them to the Canal du Midi had been recommended by the commission, and could serve trade and the growth of the harbor at Sète. But frustratingly, this was one of the less effective engineering experiments in eastern Languedoc. The coast was a shifting sandscape not hospitable for waterways to flow free of sand and silt. The same problems that plagued Sète confronted this region. Once again, coastal land did not immediately and consistently submit to the will of the king.[44]

The Power and Price of Hubris

One of Riquet's most startling propositions for the Canal du Midi was a tunnel through the mountain at Malpas. In 1677, he proposed this stunningly audacious piece of engineering after a debilitating illness that made it clear he might not live long enough to see the canal completed. Under this new pressure, he began to embrace even more radical engineering schemes to drive the waterway more quickly to the sea.

Taking the canal through the mountain would lead the waterway more directly from the Aude River valley toward Béziers. Malpas lay almost halfway between the two, and created a barrier that was too high to mount with locks and too wide to skirt efficiently. The canal bed in this region already had to be blasted out of rocky hillsides with gunpowder, and the channel shaped by quarrying. Continuing this way around the circumference of a large mountain seemed conceivably as hard as tunneling through it. At first, those considering the problem wondered if an open passageway could be cut through the top of the mountain for the canal, but this seemed even more difficult than boring a tunnel.[45]

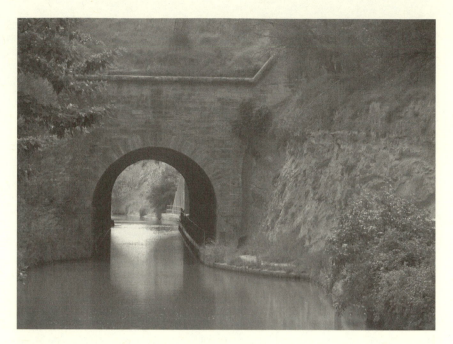

Figure 7.2
Malpas tunnel entrance. Photo by Kenneth Berger.

The passage through the mountain at Malpas did not have to be long. Riquet had kept the canal as high as possible through the mountains so that it would approach Malpas near its peak where it was narrow. The tunnel only had to be carved from soft stone, making excavation easier. But the softness of the stone also made Colbert fear the tunnel walls would crumble, particularly as currents flowed through the passageway and dampness seeped into the walls. The minister was convinced that Malpas would become the greatest of Riquet's expensive failures.[46]

What made Riquet and his collaborators confident it would work? Malpas already had a tunnel through it made by the Visigoths that had lasted for centuries. Under the oppidum of Ensérune, a Greco-Roman settlement next to Malpas, there had once been an étang called Montady. For an unknown reason, local inhabitants had used Roman hydraulic techniques to drain the lake.[47]

From the bottom of the lake, they drilled a subterranean channel that penetrated Malpas, allowing the water from Montady to flow into a lower valley beyond. A series of radiating ditches converging at the bottom of the old lake aided the process by carrying any new precipitation to the central drain, keeping the lake bed available for agriculture. Since Montady was a dramatically visible element of the countryside and everyone in the region knew that the drainage channel through Malpas had lasted for centuries,

Figure 7.3
Drained lake at Montady viewed from the oppidum of Ensérune. Photo by the author.

they had clear evidence that a tunnel through Malpas—even if filled with water—could also endure for centuries.

Confident that a tunnel would be both possible and effective, Riquet proceeded. Local engineers were not afraid of using mining techniques. They could reinforce the passageway for the canal with wooden supports like those used in mines or the underground passages dug by soldiers laying siege to a fortified city.[48]

Mines were plentiful in both the Montagne Noire and Pyrenees, and there was a growing formal literature for military engineers that explained the use of "mines" in war. There was also no lack of workers in Languedoc who could excavate the tunnel, too. Colbert was closing down mines in the Montagne Noire and Pyrenees that had failed to produce silver for the king. Riquet knew of this, having requested control over one of them—a copper mine in the Montagne Noire. The entrepreneur may even have hired some miners during the period in 1669 when the canal labor force expanded so rapidly, since these workers—experienced with digging and perhaps even working with explosives—were being forced to seek new work.[49]

Although the evidence is not conclusive, Clerville, who had wanted to pierce the Cammazes ridge to take water to Saint Ferréol, appears to have opposed the tunnel at Malpas. It does not seem likely he would have done it on technical grounds. The relationship between Riquet and Clerville had

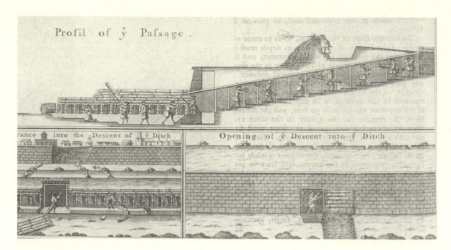

Figure 7.4
Use of mines to attack a fortress, M. Belidor. Courtesy of the University of California.

been strained by problems with their joint water project at Sète as well as conflicts over supervision of the subcontracts. In any case, Clerville did not press the case, and Riquet was not to be stopped.[50]

When Colbert heard about the plan, he was horrified. Sure that the tunnel would collapse into the water, the minister wrote both intendant Henri d'Aguesseau and Pons de la Feuille that if they could not oversee work on the Canal du Midi properly, he would send others to replace them. He also wrote Riquet that he was sending a commission down to inspect the work, and if the channel created so far did not hold up to their standards, he would suspend work on the canal altogether.[51] He was clearly outraged by Riquet's audacity.

Legend has it that the Malpas tunnel was built in the week before the commissioners arrived. It is unlikely that the whole tunnel would have been dug in that time, but quite possible that a small shaft was completed. Riquet apparently took all the laborers in his employ in the Béziers-Agde area to do the job under a trusted local engineer, Pascal de Nissan. What they excavated was probably only the size of a mine shaft, but it was apparently large enough for the commissioners to walk through, and thus exhibit to them the feasibility of the idea. Once again a material demonstration of technical efficacy, much like the rigole d'essai for the water supply, was enough to get the work approved. Riquet, the oracle, prevailed, but lost further political ground for his disobedience.[52]

Delusions and Anomie

Riquet's hubris helped to precipitate a political fall in 1677 that was astonishing in its swiftness, if not predictable in its form. The canal might have been gaining credibility in the 1670s, but the entrepreneur was losing it. It was not just the periodic problems with the gabelles but the erosion of support for the entrepreneur, too.[53] The new intendant, d'Aguesseau, was the main political supervisor to the canal, and he was wary of Riquet. After hearing in 1677 of the entrepreneur's plan for Malpas, he wrote to the minister, confiding in him about Riquet's failings. He received the following famous reply from Colbert:

> I must say [your report] very much disturbed me. . . . You have penetrated more deeply [into] the conduct of M. Riquet and the source of his views on the matter of the vast range which he has given his imagination. Although it might be best to treat him as ill, we must, nevertheless, apply ourselves with care in order that the course and strength of his imaginings does not bring on us a final and grievous end of all his works. . . . This man does as do great liars do who, after telling a story three or four times, persuade themselves it is true. . . . It has been said to him so many times, even in my own presence, that he is the inventor of this great work that in the end he has believed he is in fact the absolute author. And, on the greatness of the work he has founded [the idea of] the grandeur of the service he renders to the state and the greatness of his fortune. It is for this that he has made his son a Master of Requests, and which has given to his spirit, regarding the establishment of his children, a vast career and an inflation which hasn't any proportion or relationship with what he is or with what he has done.[54]

D'Aguesseau showed Colbert's letter to Riquet, possibly to discourage him from trying to build the tunnel, but it had the opposite effect. The entrepreneur responded with a combination of belligerence and wounded despondency. Colbert's sentiments echoed those of Nivelle: "[Riquet] presents fictions, which have more in common with fables than with facts . . . that lack the most important basis, that is, telling the truth."[55] Now not simply Nivelle but a man as great as Colbert was treating Riquet as delusional.

Adding to the strains between Riquet and Colbert, the minister was also finally confronted in 1672 with the real costs of building the canal. By this time, Riquet had already spent his fortune and could not continue to augment the funds for the work. The change was impressive.

Before that time, the treasury paid Riquet roughly 125,000 lives per year. Then in 1672–1673, the *Comptes des Bâtiments du Roy* recorded a supplementary payment to Riquet of over 1.5 million livres in 1672 on top of the base payment of about 125,000. In 1674, Riquet was given over 560,000 livres, and while in 1675, he received only 54,175 livres in regular payments for the canal, this deficit was made up with extraordinary payment in 1673–1675 that amounted to 1.2 million livres. And from 1677 until the end of the 1670s, Riquet was regularly receiving roughly 570,000 livres based on augmentations from the gabelles and the États du Languedoc.[56] For Colbert, the mounting costs of the project were intolerable, and raised questions (again) about where the money was going.

The exasperated Colbert increasingly confided in d'Aguesseau, and asked him for news about the canal, relying on de la Feuille and Mourgues mainly for verifications and financial accounts, not impressions. The minister wrote Riquet an occasional reassuring letter, fearing what would happen if he cut him off, but had little to say other than urging him to work hard and well. There was something deeply disturbing about Riquet that bothered the minister more and more. D'Aguesseau's derogatory observations about the entrepreneur's emotional character mirrored Colbert's own views, and gave him new license to keep the unreliable visionary at a distance.

In fact, Riquet was not so much delusional as suffering from social and technical anomie; he had no firm sense of his social standing, and no way to calibrate what he could or could not do as an engineer.[57] His social standing had been radically elevated, but in indecipherable ways. His family's Italian nobility had been restored and he had a domain, yet he was a functionary. Where did that place him in the social hierarchy? The local nobility rejected his claims of rank, and Colbert rejected many of Riquet's efforts to place his sons in offices appropriate to the nobility. The contradictions between his hopes of social mobility and the reality of it were confusing and deeply disturbing. Was he important or just self-important? Was he a genius or a visionary in Nivelle's negative sense?

In this anomic situation, the entrepreneur had no way to act appropriately to his rank because his social station was not clear. And he had no way to see the limits of his technical expertise because he had failed so often and accomplished so much. Dreaming of an improved Languedoc was the antidote to despair. Riquet focused on finishing and embellishing the canal rather than pleasing Colbert, taking on more debts, embracing more risks, manipulating more money, traveling more frequently, and putting the cheapest labor force possible to work. Someone as irrational as that needed to be handled carefully.

Colbert was wary of men who grew too self-important while in service to the king; it was a temptation he did not allow himself. He understood the danger; the former finance minister, Nicolas Fouquet, had gone to jail for that hubris.[58] State administration under the monarchy was a delicate business. It depended on a sure sense that the king was in charge no matter who was doing the work. To administer France in the name of Louis XIV required assuming the point of view of the king to turn the royal will into reality, but it also required taking the king's will as primary. To do other than this was to violate the principles of the divine right of kings and monarchical authority that Jean Bodin had articulated.[59]

From Colbert's point of view, Riquet failed to acknowledge his dependence on his monarch. He had benefited from the king's power of indemnification to obtain the necessary land for "his" domain, and used massive influxes of treasury money to finance the work. If he was in a position to build a marvelous canal across Languedoc, it was because the king and Colbert made it possible. The administration was meant to be served by grateful clients, not to serve the ambitions of its functionaries.[60]

Colbert was perhaps reassured by d'Aguesseau's view that Riquet was delusional; tunneling through Malpas seemed crazy enough to support his perspective. If Riquet's weaknesses were only technical foolishness and narcissistic self-importance then that would make him dangerous only in ways the minister was ready to handle. But perhaps Colbert understood something more was at stake, and that is why he so strongly cautioned d'Aguesseau and de la Feuille to get Riquet under control. After all, the minister had written: "Although it might be best to treat him as ill, we must, nevertheless, apply ourselves with care in order that the course and strength of his imaginings does not bring on us a final and grievous end of all his works."[61]

Riquet's growing technical efficacy gave him powers over the kingdom that might not have been the product of his own imaginings but nonetheless were real. His insubordination and insistent focus on material politics allowed him to wield a modern political power that was novel and potent enough to be disruptive to the normal flow of patronage relations.[62]

Clerville, who had also been discredited for similar hubris, still defended Riquet to d'Aguesseau, pleading with him to trust the entrepreneur.[63] He wrote to Riquet describing the content of his letter to the intendant in the following terms:

> I have been truly of the advice [to d'Aguesseau] to trust in your zeal and to the passion you have to please the king and M. Colbert and [to give you] the liberty to do what you find appropriate to do, given your facility, the industry of your work, and the good luck you have always had in attracting the number as well as the necessary gifts [bénédictions] of the many laborers that you have required for all your enterprises.[64]

Of course, Riquet did not get free range for his imagination, but he did continue to be insubordinate. When draining and filling locks staircases on the Canal du Midi became problematic because of unequal volumes in the locks or the slow speed of the process, he decided to use tambours or side channels to move the water in new ways. The minister cautioned him that tambours had failed on the Canal du Briare, but Riquet persisted in working on a new design for them, eventually making them a usable part of the system.[65]

Like a modern political functionary, Riquet had mastered an area of authority, and drawn around him a community of practitioners to develop and exercise these powers. He held onto his expertise tightly and touted his abilities widely. He did not try to be modest about his accomplishments or restrain himself from extending his administrative purview; he did the opposite. Riquet continually looked for ways to make his engineering even more consequential. He *was* dangerous, but not as delusional as Colbert had hoped.[66] The entrepreneur had tasted the potency of modern territorial politics and enjoyed the flavor.

New Noble

Perhaps ironically, as Riquet savored the powers of stewardship over the king's lands, he acted increasingly like the nobles of Languedoc who had been his adversaries—independent and resentful of administrative interference. Like the noble Reformers of the region a generation earlier who had distanced themselves from the Crown, he also used his faith in God to justify his autonomy.[67] And like both nobles of the region and peasants like the Miquelets in the mountains, too, he acquired a thirst for independence and a self-righteous disregard for others that was stunning in, yet nonetheless appropriate to, a man so filled with religious rectitude.[68] Languedoc was a land of warriors, particularly Christian soldiers, and he was one of them—especially when pressed.

There were some reasons for the entrepreneur to act more like a nobleman now that he was designated as one. His new standing might not have impressed traditional nobles but it did give him some new powers. Like any noble estate owner, he had a right and duty to improve his properties, and like other nobles of Languedoc, he felt free to ignore demands for obedience from the northern court. He embraced that well-worn local political culture even as he acted as an agent for the state.[69]

Still, the autonomy Riquet sought was different than that of local notables. He did not fight to defend traditional privileges but rather to act as a modern self-regulating subject.[70] He exchanged personal influence for self-management. Like the Canal du Midi itself, the entrepreneur embodied a haunting modernism that seemed both historically out of place and disruptive.

Riquet also became aggressive in the use of legal and police powers. The coldhearted tax collector played doppelgänger to the martyred entrepreneur. As soon as Riquet was faced with difficulties collecting taxes in Roussillon, he proved how self-righteous and aggressive he could be about achieving his ends. He saw the struggles with the rebels in Roussillon as battles of will, and prepared to go to war in the name of the king.

Most of the police powers that Riquet used were relatively benign: edicts for regulating contracts with suppliers, requiring housing for canal workers, and controlling or removing duties on shipping. The minister saw these requests as nagging signs of weakness, but to Riquet, it was a way to use power. He freely called on the state to serve his ends, taking the time to think up new ways to do it. He devised new taxes, tolls, offices to sell, regulations, and contracts. He could not do this alone. Colbert had to help him, sometimes making the minister more Riquet's servant than master.[71] It was striking how often Riquet relied on political imposition rather than political finesse to accomplish his ends. He clearly preferred the iron fist to the velvet glove. But that was the pattern in tax collection.

Riquet also used state powers in his struggles with the États. For example, he asked Colbert for permission to sell offices related to the canal to help finance the project. Once parts of the waterway were in use, there was reason to appoint an economic council (prudhommes) to explore and regulate its uses, and clerks (greffiers) to document the traffic. But doing this usurped the power of the États, the body that normally appointed greffiers.[72] Members of this body, mostly opposed to the project and hostile to Riquet, were loath to give away these powers to the local tax farmer. At first, Colbert was unsure of the propriety of pushing Riquet's cause, postponing any decision until the offices were really needed.[73] The sale of offices was a good way for Riquet to supplement his limited income, though, so Colbert asked the États to consider it.[74] When the États refused, Colbert sent an arrest du conseil d'État, ordering them to approve the sales, ironically finding himself using the power of the king to realize the will of the tax farmer, not vice versa.

Riquet also sought to use the state's military and police powers to violent ends.[75] For example, he had insisted on a clause in the original contract for the Canal du Midi requiring the king to defend the entrepreneur and canal against assaults. This gave him an excuse to request military help to quell the disturbances in Roussillon that disrupted tax collection and work on the canal. When this argument for troops failed, he contended that the opposition from the Miquelets should be treated not as political resistance but rather ongoing warfare, requiring army intervention. Riquet had already used state police power to enforce the gabelle in Languedoc, and was prepared to do the same in Roussillon.[76] He wrote Colbert that local French authorities in the new province recommended that they "seize three or four of the most criminal Miquelets to punish them on the wheel, in order to intimidate their comrades."[77] He went on, "This is fine with me, but it is hard to catch the Miquelets because they are truthfully peasants . . . outside the power of all states."[78]

Riquet preferred another equally harsh option: "I have written death warrants for numerous inhabitants. . . . [F]or the good of peace and service to the king, I wish them captured and brought to death."[79] He justified it this way: "No human power could prevent this kind of men killing each other, and you may therefore infer that the tax collectors are subject to the same fate. In this country, the tax collectors are always on the lookout; they kill just as they are killed; it is the only way to carry on their function."[80]

The tax farmer added that if the king did not want to send four thousand troops to pacify the country, instead the monarch should resolve himself to burning all the villages in Roussillon in order to make the area uninhabitable to the present population. In the meantime, Riquet would try negotiation. He had already made contacts with local people, he wrote, including some Miquelets, and was getting some communities to agree to contracts for salt. If he were successful in calming the villagers, he argued, the king could stop the

Miquelets from further violence by giving them pardons.[81] Death or pardons were equally palatable to him.

Riquet was cold, self-righteous, and calculating in imposing taxes and enforcing rules with sustained vigor—not only seeing like a state, but also thinking with the predatory rationality of a king or a Machiavelli. Even Colbert was shocked by both the death warrants that Riquet wanted and the tax man's call for four thousand troops.[82] Clearly, the functionary was thinking above his station in ways that disturbed the minister. Such bloodlust was appropriate for a monarch, not a tax farmer.

When Riquet was accused by Bezons of high-handedness in dismissing the entrepreneurs at Sète, Colbert did not doubt it. He had seen the hard-hearted side of the entrepreneur. He understood that Riquet could be as aggressive, arbitrary, and self-serving as any recalcitrant nobleman.[83]

The entrepreneur proved himself particularly coldhearted in a legal and labor action against a subcontractor—the man who built the pont-canal de Répudre, Emmanuel d'Estan. The bridge was first contracted out in 1675 to two men, André Boyer and d'Estan, who had apparently worked for Clerville at Rochefort. At first the project seemed to go well. Clerville praised the fine stonework on the bridge in 1675, and it was finished in 1676, according to a dedication on the capstone. But it was damaged by the savage storms that winter and had to be rebuilt.[84]

Meanwhile, Boyer and d'Estan were also hired by Riquet to work on the water supply for Sète, perhaps acting as Clerville's agents there. When Riquet decided to abandon building a water supply adequate for dredging the harbor, d'Estan apparently objected, speaking on Clerville's behalf. He communicated the problem to Clerville, who wrote Riquet in 1676:

> I received that letter that you took the trouble to write me; also the R. de Lestang has conveyed your sentiments about how to build the aqueduct, [taking] water higher than that of the port; but I must tell you sir that I felt some regret that you had not sustained the ideas we conceived [together] and were approved as the best because I presented them to M. de Colbert with the promise that the plan . . . would contribute to the cleaning [removing silt] of the port at the same time, something one had been desperately looking to use in all the ports for harboring warships. . . . I've sent Sr. Boyer to learn from you . . . the reasons you have had for contradicting what you have approved with my advice, concerning the route of the waters, so I can conform to what is most reasonable. . . . I make no game of what is not my business.[85]

Like his other letters about "business," this one was vague about exactly what was at stake, but Clerville was clearly miffed by the entrepreneur's change of heart. Still, that seemed the end of it until tensions started to appear in the partnership of Boyer and d'Estan.[86] In the original contract signed by the two men, the bridge was bundled with the locks at Jouares,

Ons, and Oignon. In 1678, the partners separated their jobs, leaving d'Estan solely in charge of the lock by the Oignon River and the pont-canal over the Répudre River.[87] Boyer was left with the two other locks.

To Riquet, d'Estan was a bad apple, spoiling his projects, and Boyer was the more reasonable man. But a report on the progress of the canal written by Clerville said that Boyer was incompetent and irresponsible ["*mal adroit, negligent*"], and his lack of ability had impeded work on the locks assigned to him. In contrast, Clerville praised the work of d'Estan, calling him skilled ["*fort habile*"].[88] Clearly, the two subcontractors had sided with different patrons and were now being supported by them.

Finally, Riquet brought charges against d'Estan, arguing that the subcontractor had tried to defraud him by including the empty space under the bridge as part of the cubic footage on which payments were based. Riquet refused to pay for a void, and the subcontractor could not pay his bills to his suppliers. The result was that Riquet, who had legal control of the land around the canal (it was his domain, not his property), had d'Estan thrown in jail.[89]

In fact, masons *were* customarily paid in part by the cubic footage of the spaces they enclosed. In a Gothic cathedral, for example, it was the interior space, not the walls, that was the object of the work. The same thing was true of bridges. It was the size of the span across a valley or gorge that was the main building problem. Structures depended on their footings and needed their walls, of course, but the skill in masonry entailed managing the forces in the structure to achieve open space.[90]

With Riquet's refusal to recognize this convention, there began a long series of legal debates about costs, materials, and bill payments. After ten months in prison, d'Estan appealed the charges, and they were examined by experts. Clerville supported d'Estan, asking for his freedom not only on the grounds that the subcontractor used accepted building and billing conventions but also because the work itself showed great care. Yet de la Feuille sided with Riquet, writing:

> Emanuel d'Estan, engineer to the king, contracted for diverse enterprises on the Canal [du Midi], particularly the bridge called de Repude [*sic*], for which he expected considerable sums that he demanded from M. de Riquet, entrepreneur of the aforementioned canal. [The latter] argued that he did not owe him the whole amount demanded by d'Estan, and made an account . . . to le sieur de la Feuille in which they marked the sums legitimately due d'Estan, on the basis of which Riquet claimed that d'Estan was in his debt for roughly five hundred livres, for which he had [d'Estan] imprisoned at the royal prison at Béziers where he has stayed for ten full months during which time he was forced to sell or rent all he had to feed his family.[91]

In spite of the traditional legitimacy of the reasoning, Riquet objected to d'Estan's request, and de la Feuille concurred that there was no reason to pay for empty space. So the subcontractor remained in jail.[92]

Riquet had also argued that d'Estan was overcharging him when he presented the entrepreneur with expenses of 42,998 livres for the bridge, while the expenses from Boyer were only 10,166 livres. Riquet had said that the project had been contracted to them both equally, suggesting that the bills from the two men should have been the same. D'Estan claimed he, not Boyer, carried the responsibility for finishing the stonework, and his former partner concurred. Thus, in a legal document dated May 14, 1678, Boyer did what he could to help. He wrote that they had jointly received the contract for the bridge and locks, but had separated the work, and claimed they had tracked their costs separately, even though they submitted their receipts together. [93] Still, on July 10, 1680, two years later, d'Estan remained in jail, and de la Feuille was arguing that the subcontractor now owed Riquet not the 500 livres originally claimed, but the full 4,282 livres that he had billed the entrepreneur.[94]

D'Estan was caught between two worlds, two standards of practice, and two powerful men. And like many underlings before him, he was crushed by it. Riquet exercised his noble legal powers, and defined accounting practices in ways that suited his pocketbook, walking away from his financial problems by doing so. Riquet was running out of money when he refused to pay d'Estan. The budget was squeezing everyone, and the smallest man the most.

Clearly, Riquet could be as self-righteous as he was a pious steward of the land. His respect for the skills of local men and women was matched by a willingness to claim their inventions as his own; his use of subcontractors was paralleled by his determination to jail those who crossed him; and his willingness to hire some peasants from the Pyrenees was matched by his willingness to kill others for his tax money. He was no madman, but rather a man straddling worlds, New Rome and Old Gaul, using moral rationales from both for extending his secular powers.

Colbert called Riquet dangerous and delusional, but misnamed the real threat he posed. Stewardship politics was meant to reduce noble hegemony, not create a new type of functionary who thought like a king and acted as arbitrarily as a nobleman. Riquet used stewardship principles to pursue power, and recognized the changes he could create through engineering. He had learned from Colbert that building a New Rome was a route to power, and he was mining that ore for all he could get. Seeing himself as God's oracle only allowed him to become even more ruthless and stubborn than before.

Riquet cultivated modern forms of power using tools of traditional authority. He was confused by the anomic status he held, and alternated between the haughty arrogance of a nobleman and the dedicated otherworldliness of an oracle. Most astonishing to Colbert, Riquet in his pursuit of power showed no gratitude for the patronage he received. Instead, the entrepreneur attributed his successes to God and his own imagination. The entrepreneur found himself with uncanny powers that he could not explain without God; he had changed his engineering abilities through collaborative work, and he had changed the habits of life in Languedoc just by altering the landscape. These powers might have been only dimly visible at Versailles, but they were vivid to Riquet, who saw in them his special relationship to God. He had been given the ability and calling to reshape the face of the Creation itself. He suffered from arrogance, bloodlust, and even spiritual greed, but his sense of the power in his hands was hardly delusional.

CHAPTER 8

Monumental Achievement

In November 1672, Riquet became gravely ill. Fearing he would die, the tax farmer wrote to Colbert, presenting his son, Jean Mathias, as his trusted replacement. Colbert wrote back to Riquet with uncharacteristic concern and respect.

> The goodwill that I have for you, the service that you render for the king and state in most of the duties you take on, the vigorous application you have given the great work of the canal de la communication des mers [all] have made me very unhappy about the bad state to which you have been reduced by your illness. But I am comforted by the letters that I received from your son on the 23rd of this month, that it seems you are out of mortal danger and that it is only a matter of you reestablishing yourself and regaining the strength that you need for achieving as grand an enterprise as your zeal for serving the king has encouraged you to engage. Although this news has brought me much joy, I will not be reassured until I receive from your hand assurances of your good health.[1]

Colbert's compassionate letter was an exception in his correspondence with Riquet in the 1670s. The moment seemed transient, but marked a real change. Faced with Riquet's death, the minister realized how much he wanted the entrepreneur alive to assure the canal's completion. The project left unfinished had the potential to become a dirty scar on the province and a blot on Colbert's record. The turn to stewardship politics in Languedoc had burdened the administration with new responsibilities for its material actions as well as new powers over local nobles. The minister could rail against Riquet's insubordination, and the king could reject Riquet's appeals for his son's advancement, but the power of the word was not enough either to build the canal or make it go away. Colbert needed Riquet to make it work. This made Riquet more important than Colbert had ever wanted to admit.

The efficacy of stewardship politics in France had tipped the playing field of power toward works rather than words. God's will could reside not only in the word of the king but also in works—worldly things that were simultaneously profane and divine. In material politics, the king's word was not enough to assure a good outcome; it took intelligence about nature, God's handiwork. The brute fact of things *did* make a difference. And Colbert was now faced with the consequences of his policies.

Riquet had undergone a psychological transformation in 1669–1670; now in 1672–1674, it was Colbert's turn. With trepidation, he started to recognize the political significance of the Canal du Midi, and although loath to accept it, he had to recognize that material politics had veered so far from his control. But imagining the canal left unfinished and treasury funds wasted, Colbert saw things in a new light. The minister could not step in to finish the canal himself. Riquet was no engineer, but he was committed to the project and knew enough about the skills of others to make it work. Without him, Colbert might lose it all, so he wrote, probably in all sincerity, that he would not be reassured until Riquet recovered his health. While this note of compassion was in many ways a momentary impulse that only welled up when Riquet was on his presumed deathbed, nonetheless, it was also part of a permanent awakening. Colbert found working with Riquet difficult, but he did not want the entrepreneur to die.

Clerville's letters in the same period were a study in contrast; they were filled with a deep affection and respect that Colbert could not muster, highlighting the difference in the relationships. The aging soldier and devoted patron truly grieved Riquet's ill health. In February, he wrote endearingly that Riquet's well-being was no less important to him that that of his father.[2] The two men had become partners and confidants, both "visionaries" enjoying the experimental thrill of territorial engineering.[3] They also shared intimacies. Clerville unburdened himself to Riquet when the soldier's wife suffered an extended and painful illness. Later, the soldier confided to Riquet that he was becoming emotionally exhausted by warfare, and did not feel prepared to face another battle.[4] By 1674, Clerville was writing letters with a hand barely able to keep from shaking. Still, he continued to write Riquet frequently, inquiring about how he could help his friend with his enterprises.[5]

In comparison, Colbert's response to Riquet's illness seemed an expression of political more than emotional angst. It was painful for him to admit that the visionary was crucial, just as Riquet had always claimed, but nonetheless, it was true. Colbert had come to rely on the entrepreneur's devotion to the canal, and willingness to sacrifice his family's fortune and personal reputation.

If the entrepreneur passed away, Riquet's son, Mathias, was prepared to take control of the Canal du Midi, but he had no money. Mathias would be held responsible for the contracts for the first and second enterprises, yet he would not inherit Riquet's tax farm, and had only a modest income of his own. In addition, on Riquet's demise, locals whose property had been indemnified for the canal would surely call into question the family's rights to the lands. The whole enterprise could be halted through legal wrangling—just as the Canal de Briare had been stopped on the death of Henri IV. The Canal du Midi had to be completed to maintain its integrity. But it was hard to imagine how to do this if Riquet died, particularly if he was found to be involved in graft when his estate started to be settled.

As Riquet's health remained bad into August 1673, the minister wrote d'Aguesseau to help make preparations. First, Colbert wanted to clarify the contractual and financial commitments of the Riquets for the first and second enterprises, so there would be no confusion about the obligations of Riquet's heirs to the king. The minister also asked d'Aguesseau to make a trip with de la Feuille to visit all the work sites on the canal and study the jetties at Sète, presumably to document the state of the work at the time of Riquet's death in case the entrepreneur did not recover. Most tellingly, he asked the intendant to be prepared to impound money from the gabelle that was still in the tax farmer's hand immediately at Riquet's death to prevent the family from trying to take away those funds before they were deposited in Montpellier.[6]

Riquet recovered his health shortly after that, but Colbert had only momentary reason to celebrate, since the entrepreneur soon announced that he wanted to take the Canal du Midi through a tunnel at Malpas.[7] It was too much for the minister. The palpable outrage he felt at this proposal was probably fueled by his growing anxieties about Riquet's longevity. He did not want a mountain collapsing into the canal any more than he wanted Riquet to fall over dead.

When the minister sent a commission to Languedoc to inspect the work and stop the tunnel, threatening to take the canal away from Riquet if this review did not pass muster, he may well have hoped secretly to get the canal away from the entrepreneur before his health declined again.[8] To his chagrin, Riquet and his colleagues demonstrated the feasibility of the tunnel, and the folie à deux between the minister and entrepreneur continued.

Colbert said repeatedly that Riquet was out of control, but the gnawing truth was that Colbert was out of control; he had lost any ability he had had in the past to manage the canal project, and was not sure why. The reason, in fact, was simple. Work for the Canal du Midi called for experiments, and Colbert was no experimentalist. No matter how many experts the minister sent to supervise Riquet—Mourgues, de la Feuille, and Clerville—and no matter how dutifully the entrepreneur discussed engineering with them, there were failures and disasters as well as triumphs and progress. No one was sure how to build a canal across Languedoc. Riquet claimed the knowledge and acted insubordinately to get his way, but he kept needing help from the minister, and this strained Colbert's patience.[9] As long as Colbert did not trust Riquet to do the work unsupervised, building the Canal du Midi remained a collaborative experiment that was in the end technically fruitful, but left Colbert anxious and Riquet resentful. Both men sought the power of personal rule over the project even as it took form as a model of impersonal power. The minister's growing dependence on Riquet was mortifying to Colbert, and a reality he only began to confront in earnest when he realized what might happen if the entrepreneur died.

Once the entrepreneur had regained his strength, Colbert was impatient for faster progress on the second enterprise, but he understood that the slow pace in the mountains was normal. He also inured himself to the idea that the last efforts would create cost overruns.

> It is maddening that the excavations would be so difficult at the end of the work; it is what happens nearly always in these sorts of large enterprises. The entrepreneurs always do first the easiest work, leaving to the last the most difficult, which almost always leaves a large miscalculation about the costs.[10]

Colbert was correct; the entrepreneur needed more money. In April 1679, d'Aguesseau wrote, "Riquet cannot finish work on the canal without the help of 300,000 livres that he does not presently have due to the diversions and dissipations of the first years [of the project,] the effect of which are felt today."[11] The money from the contract for the second enterprise was spent, and while it might have been satisfying to attribute the cost overruns to dissipations rather than risky engineering, both fiscal problems and the physical limits of the canal remained concerns to the minister. Riquet had indeed used high wages and good working conditions as incentives to increase the labor force in 1669 and the early 1670s. He had hired massive numbers of workers at great cost to make dramatic progress. It was infuriating, but now Colbert was faced with a conundrum. To get the Canal du Midi finished in good time would be equally expensive. Slowing down the pace to allow tax income to cover the costs risked the possibility that Riquet would die before the canal was completed.

In September, Colbert wrote d'Aguesseau making it clear the degree to which his worries about the canal were affecting his ability to concentrate on it:

> It has been a long time since I wrote you on the subject of the canal for the communication of the seas and port of Cette [Sète] even though there have often been letters from you on the subject. I admit that often the embarrassment of work stops one from giving everything equal application, and as I see, from your last letters and frequent ones from the sieur Riquet, the work that he owes is not advancing due to lack of funds although he is almost completely paid [what is owed]. . . . I swear that the end of this affair is annoying me for the reason that I have never seen enough resolve in Riquet's spirit to come out well from such a weighty affair as this.[12]

Riquet's problem was not a lack of character; he had in abundance the "resolve of spirit" to finish the canal. He was running out of money, straining against political impediments, and trying to run the canal through the difficult landscape of eastern Languedoc. His experiments were both dangerous and expensive. So it was comforting to Colbert and d'Aguesseau to blame his character, making it the scapegoat.

Colbert was suffering about the canal in large part because he was suffering politically from its costs.[13] Colbert's own stature in the administration was being diminished by problems at the treasury; the state was in substantial debt, and the canal was an expensive investment that had not yet paid off.[14] Colbert leaned on the États to increase their contributions for the work, and authorized new taxes, but he was caught with a heavy burden that he could not abandon.

Colbert may have imputed financial desperation to Riquet because of his own, and this is why he insistently looked for evidence of graft. He had wanted to punish Riquet for taking the canal through the Malpas tunnel against his orders, but he was worried about cutting him off, sure that it would impel Riquet to siphon off tax funds. Even while he stopped payments for the second enterprise because the state's contractual obligations had been met, he still made other treasury money available to Riquet, hoping this would keep the entrepreneur from taking funds from the gabelle for the canal.

When Clerville died in 1677, Colbert became even more concerned; he had depended on the old soldier to keep some control over Riquet. Their relationship had been reassuring to the minister, and Clerville remained a source of information. De la Feuille had never gained Riquet's confidence, and had lost any chance of it after his indiscreet letter about ruining Riquet had come to light.[15] As much as the entrepreneur was losing his advocates in this period, Colbert was losing his spies and advisers, leaving Riquet freer to be the oracle he wanted to be.

Then in September 1680, Riquet fell ill again, and died quickly on October 1. The canal was not finished. According to legend, he asked about the canal in his last conversation with his son. Learning that there were only a few more meters of the main channel to dig, he died.

In future accounts of the enterprise, this tale was used to present Riquet as a martyr who had given up his life for a canal he would never see. The story was politically valuable; it served to elevate the Canal du Midi by celebrating its "true author" as a good servant to the king. With surprising rapidity, the tax farmer started to emerge like a phoenix from the fire as a great engineer and local hero. It required delicate work on Colbert's part, allowing the minister to attend to the complicated fiscal and political problems that were left by Riquet's death.

Death and Taxes

As he had been instructed by Colbert, the intendant immediately impounded four hundred thousand livres from the États when Riquet died. This assured that the money would not be confused with personal assets. Colbert told him to keep a close eye on how the funds were employed, "because there is still reason to fear that . . . the departed Riquet diverted monies from the work for the establishment of his family."[16]

It soon became clear that the tax farmer had died in debt and there were no hidden assets that Riquet had left to his descendants. He had not diverted state funds or acquired properties with them. The family was faced instead with having all their belongings confiscated to pay their debts to the États and other creditors. The story of Riquet's martyrdom was acquiring some empirical basis.

On hearing of his father's death, the young Mathias left his post in Paris for Languedoc, having prepared himself for this eventuality.[17] He knew what needed to be done to complete the project, but that still left the problem of finances. Mathias studied the books with Dominique Gillade, d'Aguesseau, and Mourgues, who all agreed that he needed extra funds to finish the Canal du Midi, and that to continue with the current arrangements would ruin the family completely.[18]

D'Aguesseau was sympathetic to Riquet's son, and worried about the fate of the canal. On Octber 5 and again on October 19, he raised the question with Colbert of the debts left behind by Riquet. If the funds that he had sequestered from the Riquet family were released soon, he argued, Mathias could take advantage of the rest of the good fall weather to do as much as possible to finish the canal that season. The four hundred thousand livres of treasury money earmarked for the canal had been impounded until Riquet's debts were settled, but that would be a long time. Keeping the funds from being used made little sense if the project was almost done.

The Montpellier provincial authorities, who were trying to liquidate the entrepreneur's property for tax debts, wrote that the family had little more than some silver and dishes to its name. The Riquet children estimated to d'Aguesseau that their father left them more than two million livres of debt. So waiting for the young Mathias to inherit family money did not make any sense either.

Mathias did not and would not have the personal means to continue work on the canal, so to make it work, he needed state funds. Although how it happened was not clear, soon a special account for the project was set up with the monies that d'Aguesseau had sequestered, and large numbers of laborers started arriving at the work sites. The intendant also suggested to Colbert that the treasury forgive some of Riquet's debts. Shortly after, the king forgave most of the tax debts to the Crown.[19] The family still owed money to creditors, but it now could use its assets, getting loans and selling property. Throughout the 1680s, the family "sold" parts of the canal, while maintaining the family rights in perpetuity to the lands themselves. The "buyers" could use the land and profit from the trade on the canal, but the family could still repurchase the estate and hold it as a domain. Only in 1724 were the Riquets finally able to do this.[20]

Soon after the financial arrangements began to take form, Colbert began to repair the reputation of the Riquet family. He wrote d'Aguesseau:

> M. de Bonrepos, *maistre des requestes*, having left for Languedoc for the continuation of the work on the canal for the communication of the seas, I pray you to give him all the assistance that you can to

achieve this grand enterprise and to let it be known in all the province that the king still honors his family with his protection, it being very important and necessary that all the province, convinced of this truth, provide the sieur de Bonrepos and his family, including the sieur Pouget and all those who are involved in their affairs, the credit and help that are necessary to finish the grand enterprise that the deceased sieur Riquet made, and in the execution of which he died.[21]

To save the enterprise, Colbert worked to represent the canal publicly as a success, writing d'Aguesseau to figure out how to publicize it so that people from all over the world would know about and make use of the waterway. He suggested something in the journal *Mercure* perhaps.[22] The "disaster" in Languedoc, the "muddy scar" across the countryside, the continual headache to Colbert, the expensive "boondoggle" by an uppity tax farmer, the impossible piece of engineering, or the stupidity and hubris of the "visionary" was all transmuted into political gold: a modern technical accomplishment of personal genius.[23]

Transmuting Lead into Gold

The social rehabilitation and spiritual elevation of Riquet began, requiring first the completion of the canal. Early in 1681, "the king sent an order to the sieur d'Aguesseau, intendant of Languedoc, to inspect the canal dry, begin to fill it with water, and try the first navigation."[24] Verification of the dry channel began on May 2, 1681. The canal was deemed ready, empty parts of the channel were flooded, and the inauguration of the canal took place starting in Castelnaudery on May 19, 1681.[25]

The voyage itself began at Toulouse, although that part of the canal had been in use for most of a decade. But the official verification had to cover the entire length of the canal, so d'Aguesseau took soundings to determine if the depths required in the devis had been achieved along the entire route. "He had soundings made of the depth every hundred toises, writing a *procez verbal* on the state of the canal."[26]

At Castelnaudery, the navigation turned into a ritual occasion, full of pomp and ceremony. A large entourage joined the intendant as he approached the grand basin. They were local notables, counterparts to the commissioners who had originally verified the plan under Clerville. As always, the official truth of things had to be established by authority, and required both expert measurements and highborn witnesses of the results. The assembled group was a local who's who for the time. Ritual demanded it, and so did politics. This was a moment that could entice notables to find pleasure in the waterway.

> [The] cardinal de Bonzi [president of the États] . . . traveled to Saint-Papoul accompanied by the bishops of Béziers et d'Alet, the

marquis de Villeneuve baron des États, the sieur de Monbel, syndic général of the province, the sieur de Pujols, secretary of the États de Languedoc, and the sieur Mariotte, clerk of the États. The sieur d'Aguessau joined them with his entourage, [and] all of them were magnificently entertained by the bishop of Saint-Papoul.[27]

This opening ceremony for the canal itself was a sacred event with the necessary ritual processions, a mass, blessings, cannon fire, and music.

> On the 19th at seven in the morning, the whole entourage [at Castelnaudery] . . . marched in procession to l'église de Saint-Roch. The bishop of Saint-Papoul . . . led the ceremony . . . dressed in pontifical habits and [was] proceeded by the cardinal Bonzi, the prelates, the sieur d'Aguessau and by other persons of importance from neighboring towns, and by an infinite number of ordinary people from nearby areas. The père Mourgues having celebrated the mass, the procession advanced to the edge of the Canal. . . . [T] he bishop of Saint-Papoul blessed the waters of the Canal, the boats, and all those assembled, and the ceremony finished with the Te Deum and prayers for the king. . . . The boat carrying the cardinal Bonzi . . . [had] a large gallery where there were violins, hautbois and trumpets . . . [and the group] set sail to the noise of these instruments, cannon fire, musket fire, and cries of "Vive le Roy."[28]

The soundings were official business, but also ways to "satisfy the curiosity of those who could not be there to admire the awesome spectacle, and to watch a flotilla pass through places where one had previously had difficulty finding water for ordinary necessities."[29] The measurements determined the rhythm of the journey, keeping the pace slow, and allowing the entourage to enjoy many celebrations at locks and towns along the way.

The success of the voyage had given a new face to the enterprise. The Canal du Midi was firmly located in both the countryside and the public imagination. The uncanny sight of dignitaries floating by, and sounds of their music echoing in the fields and hills, made the project seem as startling and important as the propaganda was starting to say.

Unfortunately, the studies from this first navigation also revealed some imperfections in the construction that required attention, so parts of the waterway had to be drained for repairs. The work was brief, and in March 1683, the canal was again full of water, blessed a second time, and opened (at least in principle) on a permanent basis. Now Riquet, the charlatan and madman who had ruined his family to finance the project, turned out to be a good steward after all, making Languedoc more Edenic by helping ships pass peacefully from sea to sea.

Still, Colbert understood that not all parts of the canal were functioning well, and he struggled to learn the details of what was happening in Languedoc. For this purpose, he sent a new man to act as his agent on-site,

D. Montaigu, *ingénieur originaire du Roi*, who had been in Languedoc as a military engineer.[30]

The minister wrote to d'Aguesseau on October 30, 1683: "[I need] a very exact general map of the Canal de communication des mers with maps of particulars and plans, profiles, and elevations of the locks. . . . I pray you to write the communities in Languedoc . . . [to make] sure the sieur Montegu [gets their help]."[31] Then the minister waited, and waited, and waited.[32] When no general map quickly appeared, Colbert asked again.[33] He soon learned that maps were slow to draw precisely, and Montegu's maps could not keep up with changes on the canal. There was no way to use cartography for administrative action at a distance.[34]

Colbert lived to see the second opening of the Canal du Midi, but was still waiting for maps from Montaigu when he died in November 1683—ironically just before the contract for the canal was fulfilled. Colbert also died without ever seeing the canal.

In 1684, a final inspection determined that the Riquets had officially met their obligations, finally freeing Pierre-Paul Riquet to become a misunderstood genius and the Canal du Midi to become a model of modern hydraulic engineering.

Trustworthy Engineering

Determining when the canal was "really" finished was a matter of definition. It was also fundamentally tied to the shifting meaning of "working" as a measure of efficacy in the period. The Canal du Midi was done in one sense in 1681, when it was first capable of carrying boats from one end to the other. It was complete in another sense when the contract was officially fulfilled and the canal opened in 1684. But it was only made "reliable" after improvements designed by Vauban in 1686 were complete in 1694. Like any construction project, it was easier to say when the canal was started than when it was really finished.

The conceptual problems of defining "finished" reflected the growing importance of working as a measure of empirical accuracy. Colbert used a series of different words to describe completion that shifted in a significant way over time, although not in a simple linear progression. These linguistic migrations traced the minister's shifting understanding of engineering, belying his insistent focus on following the contract.[35] He was clearly learning that incising a canal into the countryside of Languedoc took more than simple obedience to the royal will. Colbert was reluctantly learning the complex and weighty significance of the term "working."

During the 1660s, Colbert spoke mainly of the project being *achevé*, or achieved. This implied fulfilling the terms of the contracts according to the devis. Parts of the canal were sorted into two categories: work to be done (*ouvrages à faire*) and things that were done (*faits*). The canal had a shape, size and length, and set of artifacts detailed by the devis. It defined the "things" that had to be done for the canal to be achevé. In this conceptualization of completing the canal, it was a matter of living up to a set of terms. It required living by the word, the personal word of the entrepreneur, the words of the contracts, and the specifications written in the devis for the first and second enterprise.

By the 1670s, Colbert increasingly used another verb, *parachever*, to describe Riquet's responsibility for the work. The early failures of the locks by Toulouse had made it clear that following the devis in itself did not assure a working canal. The term parachever, set in contrast to achever, implied doing more than fulfilling the terms of the contract; it suggested an obligation to make the waterway workable following the guidelines set down by the devis, but not stopping there if they proved faulty. Only when the canal was parachevé or could "work" would Riquet meet his contractual obligations to the king. This usage implied a significant shift from word to works, paralleling the shift from the system of patrimonial politics based on the authority of the word and patron-client ties to the stewardship system of territorial administration. There was a deeply Christian moral thread implied, too, that went beyond the mere words of the devis. Riquet had a moral obligation to provide a working canal.

Toward the end of the 1670s, Colbert, Riquet, de la Feuille, and Clerville began increasingly to employ another verb, *perfecter*, to describe what was required to finish the second enterprise. The verb implied that the waterway had to be built according to the highest standards, and be a permanent structure with a *durée éternelle*. The specifications of the contract had to be met, of course, but only using the best materials, respected methods, and most diligent techniques. By this period, many of the finished structures of the second enterprise had been destroyed by storms and floods. To fulfill the contract, Riquet had to repair or replace them. But even that was not enough. The canal had to be perfecté or made a material legacy in the classical tradition that would endure like a mountain. This was the highest measure of working because it implied a higher level of engineering: the proper improvement of nature to make it more like Eden itself. It was a test of moral character and political possibility, too. Working at this level implied knowing the Creation as God intended it.

In Colbert's terms, the canal was achevé in 1681, when the canal passed inspection and was filled. It was parachevé in 1684, when the required repairs were made and the contract fulfilled. And although it never reached a durée éternelle, it did get perfecté with Vauban's recommended improvements and the other repairs made by 1694. Finally, the New Rome had gained its monumental form in Languedoc.

The fact that Colbert increasingly credited working as a measure of success on the canal suggests he was more persuaded by Riquet's stewardship logic than he would ever admit. He could champion traditional authority, but he was apparently learning that territorial politics necessarily exceeded the power of "the word" and the authority that he tried to exercise with it. He followed his emergent understandings of infrastructural engineering with his verbs, struggling for terms to bridge words and works. He struggled for vocabulary to try to convey how the works in Languedoc could mirror the word of the king.

The two men fought out with each other the contradictions inherent in territorial politics—the tendencies toward both personal and impersonal rule. They embodied the contradictions, and projected the conflicts onto each other. Even as they ushered into French administration new methods of material governance, each struggled for personal rule over the canal. Colbert and Riquet neither acknowledged the profound significance of their collaborative work to the French state, nor recognized their deep mutual debt in bringing to Languedoc a navigational canal to connect the two seas.

Life after Death

The two men who had nurtured the highest hopes for the waterway, borne the major risks of the project, and grown tired and angry at the slow pace of the work died near its official opening.[36] The Canal du Midi was finally perfected after all of the original leaders of the project had passed away: Clerville in 1677, Riquet in 1680, and Colbert in 1683. As the Canal du Midi emerged from the shadow of the folie à deux between Riquet and Colbert, it was a different kind of project. It was placed in the hands of new collaborators, who did not share the same suspicions of each other or personal ambitions for the waterway through Languedoc. The new generation in charge of the Canal du Midi consisted of more modern men, prepared to collaborate for stewardship purposes, turning the waterway into a success worthy of Rome.

The new leadership consisted of Mathias; the irascible military engineer Niquet; his boss and patron, the famous Vauban; the new director of the gabelles for Languedoc, Pouget; and the new minister of the treasury, Colbert's son, the Marquis de Seignelay. Some early participants remained, and gained new importance. Mourgues, the Jesuit who had been keeping the books, was put in charge of the project under Mathias. And many engineers who had worked for Riquet senior, particularly Gillade, Andréossy, Contigny, and de la Feuille, remained in the core set in this period.[37]

Vauban became involved in the canal only after Colbert's death, when intractable problems of silting and flooding limited the project's usefulness. Louis XIV asked Vauban to study the Canal du Midi and make recommendations about what to do. Vauban had previously been contemptuous of the canal and Clerville. He was miffed that de la Feuille had been sent to Languedoc as Clerville's second in command, never clear why Colbert had not chosen him. He had been through Languedoc before and seen the canal briefly, but he had little knowledge of the waterway and not much good to say about it until the job of perfecting it was placed in his lap.[38] Now that he studied it in earnest, Vauban found reason to praise the Canal du Midi.

> The canal for the junction of the seas is without contradiction the most beautiful and noble work of its type undertaken in our age, and which should become a wonder of the century, if it was made as great as it ought to be. . . . By the greatest bad luck of the world, one never finished this work, and *the entrepreneur who was also the inventor, was neither advised nor aided as he should have been. . . .* There are many places to admire what was able to be realized to make it navigable in such difficult countryside and in a time when so little was known, that there is nothing to do but give [the Canal du Midi] all the improvements [perfections] needed to ensure its endurance.[39]

Still interjecting elements of patrimonial politics into the process, Vauban identified Clerville as the source of the canal's problems, and accused him of providing such poor supervision that the enterprise fell short. Vauban distinguished implicitly between a "trustworthy" canal that real military engineering could produce and the untrustworthy one that had been engineered by Clerville, presenting himself as a savior of the waterway, while actually advocating a large number of technical improvements for the canal that had been previously proposed or designed by Clerville.[40]

The main problem with the canal, Vauban argued, was not silting but that it had not been made big enough to carry warships from one sea to the other. To solve this issue, Vauban advocated widening and deepening the canal at the cost of twenty-three million livres. It was a tempting idea in this period of warfare for France, but too expensive, so Vauban (like Clerville before him) had to abandon the idea.[41]

The silting problems, the military engineer suggested, resulted from the direct connections to rivers. Debris in the water supply—that had even led to the formation of a sandbar in the basin at Naurouze. This resulted, he argued, from diverting water from the rigole de la montagne into the Sor, and from the Sor River directly into the rigole to Naurouze. Vauban advocated digging a tunnel through the Cammazes ridge to bring water from the rigole de la montagne into the Saint Férreol reservoir, just as Clerville had planned it.[42] Vauban reasoned that Saint Férreol could function as a huge settling pond for the whole water supply.[43] The tunnel through the Cammazes became known as the voûte or *percée* de Vauban, sadly dissociating Clerville's name from an idea that he had originally proposed to the commission.

The other main sources of silt, according to Vauban, were the rivers and streams on the Mediterranean side of the Languedoc. They carried massive amounts of suspended materials into the canal. Existing silt barriers and settling ponds were only effective for small streams, and had not proved adequate to stop the influx of silt where the canal connected either to a major

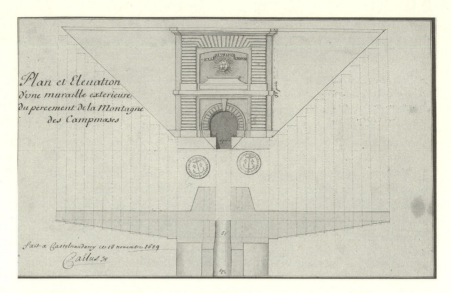

Figure 8.1
Plan and elevation for the percée des Cammazes ou voûte de Vauban.
Courtesy of the AHMV.

Figure 8.2
Water intake ca. 1690 with settling pond and silt barrier, looking more Roman.
Photo by the author.

river or a stream prone to periodic floods. A *procès-verbal* by d'Aguesseau put it this way: "The rivers and streams that enter into the Canal . . . leave sand there daily, and make it impassible even with the greatest care to keep it clean. . . . One cannot blame the entrepreneur, but the sands and gravel that the rivers and streams carry."[44]

The silting was so bad that by 1684, the canal in many areas had become too shallow to use.[45] To address this problem, Vauban advocated building a new set of aqueduct bridges similar to the pont-canal over the Répudre River. This would keep waters from the other rivers from mixing with those of the canal. Like Vauban's other pieces of engineering, these structures were massive (like a fortress) to give them a durée éternelle.[46] The pilings were even reinforced at the base to withstand the forces of floodwaters and attendant debris. No forces on these rivers would move them.

Along with these new bridges, Vauban designed new connections to the rivers themselves so the canal could be replenished with river water. To purify the alimentation before allowing it to enter the Canal du Midi, he used large-scale weirs, settling ponds, and sluice gates to shed the water of most of its debris before entering the canal proper. All the techniques copied the indigenous engineering by peasant women, but made them formal structures that were later attributed to Vauban.

The more original works designed by the military engineer were overflow drains for expelling excess water from the canal capable of discharging silt as well.

His famous multiarched *épanchoir* was a stone retaining wall at a curve of the canal over a gully that had been frequently damaged by floodwaters. He built open arches in the wall just above normal water level to expel any surges. The épanchoir also had sluice gates on the canal floor that when raised, discharged water from the bottom of the channel, sucking silt out with it. It functioned very much like the lower vault of the Saint Férreol dam. A similar system of sluice gates and overflow drains were used by the lock where the Laudot River (coming from Saint Férreol) joined the rigole de la plaine (from Revel). This assured that floodwaters from the region of Saint Férreol did not fill the rigole with silt or devastate it with fast-moving currents.[47]

The lock was on the downhill side of the drain (to the left of the sluice gates in the picture). When its doors were shut, they forced water from the mountain to be discharged through the arches of the drain into the old bed of the Laudot below. Mud that accumulated on the floor of the drain could be discharged by opening the sluice gates. The gates would suck sand along with the water from the bottom of the structure. This design was ingenious, but also was based on principles of the dual discharge system at Saint Ferréol designed by the commission under Clerville.

In spite of Colbert's misgivings, Malpas had surprisingly few problems until the 1690s. When Mourgues and de la Feuille did inventories of the work that was needed to fix the Canal du Midi in 1683, they reported that the tunnel at Malpas seemed to be in good condition.[48] And it was working well

Figure 8.3
Épanchoir de Bagnas or de Vauban. Photo by the author.

when Vauban looked at it. But by 1691, part of the entrance to the tunnel collapsed, dumping materials from the surrounding hillsides into the waterway, and stopping the passage of traffic.

The interior walls of Malpas had originally been reinforced with wood—a style of construction that had been typical in mines and military tunnels. Yet with so much water running through it, and currents and waves weakening the reinforcements, the wood had deteriorated badly. By 1691, it was no longer providing protection from falling rock for boats passing under the mountain. Niquet wanted to erect a stone vault to line the interior of the tunnel, but this was no simple task.[49] Vaults were generally built using wooden scaffolding to lay the stones on. In the tunnel over the canal, though, there was no obvious way to create secure scaffolding. So Niquet had an ingenious idea: he designed a boat for the purpose. A stone vault was finally erected under the part of the tunnel that had suffered from the slides. And a stone entrance to the tunnel was also erected to act as a retaining wall for the crumbling hillside on the other side. This combination of repairs seemed to work.[50]

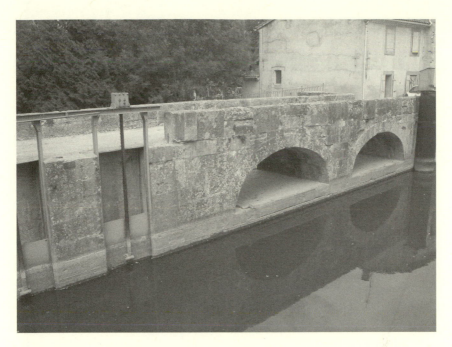

Figure 8.4
Drains on the Laudot by the rigole de la plaine. Photo by the author.

Niquet was a military engineer from Narbonne who was put in charge of realizing Vauban's recommendations. Like Vauban, he developed a rich contempt for Clerville, and helped to taint the older engineer's reputation with stories of graft and incompetence. He was particularly snide about Clerville's dredging project at Sète, presenting it as evidence of Clerville's technical incompetence, financial greed, and personal dishonesty. Ironically, Niquet was driven by a similar combination of motives in his dealings with the Canal du Midi. When the Riquet family had to sell off parts of the canal in order to finance its conclusion, he had bought a section of the waterway, giving him personal financial incentive to improve the canal.[51]

Except for their disdainful comments about Clerville that circulated to their advantage in patrimonial networks, Vauban and Niquet acted as good stewards of the king's lands. They applied natural knowledge to technical problems, and tried to make the natural forces in Languedoc more benign. Their purposes were partly selfish, and Vauban's self-aggrandizement may have been unfair to Clerville, but the work they did on the Canal du Midi was indeed an honest effort to make a peaceful waterway that would run without failure across the province of Languedoc.

By the 1690s, the Canal du Midi was perfecté. It would last for centuries. Vauban became the new hero of the canal who saved Riquet's wonder. The

Figure 8.5
Interior of the Malpas tunnel with new masonry in back. Photo by Kenneth Berger.

waterway across Languedoc became a product of genius—settling into the laps of superhuman men. Gone were the women of the Pyrenees, the artisans of Languedoc, Campmas, Boutheroue, and even Clerville. Absent were the indigenous methods needed for designing staircase locks and cutting contours through the hills.

Vauban gave the Canal du Midi a durée éternelle both on the ground and on paper. He gave formal, inscribed stature to informal techniques, and made the canal appear more Roman as well as more durable that way. As the Canal du Midi was perfecté with his designs under the supervision of Niquet, it was still the product of collaborative work. Different generations of participants in the enterprise might now be separated by time as well as rank and culture, but the lines of influence remained. Ghosts of the past and living actors made the engineering of the Canal du Midi into a living tradition. It was a tradition that finally gained its credibility by working, relying on the efficacy of works rather than words. The hydraulic techniques were scrubbed clean of their disreputable origins as women's work, and formalized and attributed to "great men." But the haunting presence of the canal itself was hard to reduce to thoughts of gentlemen. Impersonal rule in this context took on another layer of meaning, both embodying and masking the unwritten bases of monumental power.

While the canal was being perfected, maps started to represent the Canal du Midi as an engineering marvel in the Roman tradition. Cartography was understood in the period to provide a means of "eye travel," or the virtual witnessing of distant locales inflected visually with their histories.[52]

As Robin Wagner-Pacifici has shown, maps have important roles in symbolic transfers of sovereignty.[53] Early maps of the Canal du Midi, such as Andréossy's 1681 "Carte du Canal Royal de Communication des Mers en Languedoc," fit this pattern—visually transferring the waterway from Riquet and Languedoc to Louis XIV. The monarch was portrayed as a god in Roman garb, bringing the canal into being with his finger and will—the ultimate action at a distance. Viewers of the map saw not only the Canal du Midi embedded in the landscape but also a narrative of power that aligned past and present, the material and the ineffable in visual form. Andréossy used a bird's-eye view of the canal and province in part to emphasize the mountainous topography of the region through which the canal had to pass, and in

Figure 8.6
Carte du Canal Royal de Communication des Mers en Languedoc, François Andréossy. Courtesy of the AHMV.

part to enhance the image's narrative function. Bird's-eye views were typically used for maps of cities, since this perspective made the images partly portraits or "views."[54] They were representations that could depict human action on the earth and facilitate it, too. This perspective worked well for Andréossy's map because it allowed viewers to locate the canal in Languedoc, see how it flowed through the countryside, and imagine how it would look if they approached it from the sea, while also establishing the superhuman scale and significance of the canal as an emblem of the New Rome.

The narrative possibilities of this style were also exploited, with the images of ships hurrying in the direction of the harbor as clouds overhead threatened them like a gathering storm, impelling the vessels to port. In the heavens above the clouds sat Louis XIV, depicted as a young Apollo, exercising the powers of the heavens on earth. From his seat high above Languedoc, the waterway seemed small rather than imposing; it was something for a divine being to draw across the landscape, not something humans blasted or dug from the earth. The king was seated in front of a group of muses, not women of the province, but heavenly creatures whose inspiration stirred the king. On this map, lowly matters were transmuted into heavenly designs.

Andréossy not only presented the canal as a work in the classical tradition by representing the king as Apollo, he also presented it as a meeting place of male and female, a marriage of the ocean and sea.

Figure 8.7
Detail of Carte du
Canal Royal de
Communication
des Mers en
Languedoc,
François Andréossy.
Courtesy of the
AHMV.

Until you, sire, the Ocean and Mediterranean have always kept their liberty entirely against all the works of men. Your majesty has found a way to enchain them, but agreeably . . . by having constructed an immense canal that served as their nuptial [bed].[55]

Described as an act of forced union of the seas, the canal was beyond the work of men. It required Apollo himself. The forced marriage was taken as evidence of the king's divine powers, his ability to command nature to follow his bidding. But it was an act of stewardship, too, perfecting nature by giving it a more pleasurable form.

Other maps of the Canal du Midi offered different viewing pleasures. Some inventoried the things to see, arraying important sights in insets around the border that surrounded the Canal du Midi. The celebratory borders and geographic perspective of these images helped to make the waterway seem worthy of the appellation Canal Royal: a heroic achievement.

A map by Nolin published in 1697 used decorative laurels and technical details to assert the connection between the Canal du Midi and Rome. Inserts of engineering sites framed the main map, and constituted a visual encyclopedia of the techniques used for the work. The ribbons draped with medals around them stood for the notable families of the province all joined around this image of New Rome. The insets along the edge of the map not only inventoried every lock and aqueduct but also underscored the canal as a technical achievement. The dizzying quantity as well as quality of the works added the visual story of power. The larger insets highlighted the most impressive structures: the harbor at Sète, the dam and reservoir at Saint Ferréol, a round lock near Agde, two *pont-canaux* designed by Vauban, and the eight-lock staircase at Fonseranes. These were grand works of engineering, depicted as formal ideas that helped associate the Canal du Midi with the New Rome.

Figure 8.8
Left side of the Canal Royal de Languedoc by Nolin, 1697. Courtesy of the AHMV.

At the top of Nolin's map was a portrait of Louis XIV, represented like an emperor on a Roman-style coin. On this map, the king's powers were presented as secular and historical rather than sacred. Louis XIV was an emperor, not a god, and his powers were embedded in patronage ties and land stewardship, not divine inspiration. Ironically, the God's-eye view of the map removed the heavens from the image and from consideration of the significance of the waterway. This Canal du Midi was a wonder, not a miracle.

The importance of patronage relations to the monarch's power and the province's successes was underscored by the medals with the names and attributes of major noble families of Languedoc that were connected through ribbons and laurels to Louis XIV. They symbolized the power in the ties connecting the sovereign to the nobility, a visual model of patrimonial power. In a surprising twist, the image enrolled the canal's former opponents into the iconography of the regime. They could share in the glory of the Canal du Midi now that they had submitted to the will of the king and helped pay for it.

Virtual witnessing reached its extreme in a game made for the archbishop of Toulouse in that period, Cardinal Bonzi. "Le Jeu du Canal Royal" was an itinerary of the canal (a series of identifying landmarks), spiraling around a small map of the province.

Figure 8.9
Jeu du Canal Royal. Courtesy of the AHMV.

Figure 8.10
Detail of the Carte Royal de la Province de Languedoc by Garipuy, 1774, including the
étang de Montady and the Malpas tunnel. Courtesy of the AHMV.

The game allowed players to cross Languedoc vicariously from the Garonne River to Sète. They advanced from lock to lock, passing key bridges
and overpasses, and even though the Malpas tunnel. As they moved, they
could inventory the engineering features necessary to allow boats to pass.
The game did not try to explain the importance of the canal in terms of gods
and heroes but instead showed it as a marvel of stewardship—nature tamed.

A very different set of principles guided the design of eighteenth-century
maps for the États of Languedoc. They were made in 1774 under the direction of M. Garipuy, ingénieur directeur des Ponts et Chausées, the great engineering school founded in 1747.[56] Garipuy foregrounded the interplay of
the waterworks and the topography—how water moved through the landscape, and how it was captured and used for the canal.

Garipuy's series on the Canal du Midi had four sections and was comprised of fifteen separate leaves. It included massive amounts of geographic
detail about the location and characteristics of the canal, helping to clarify
the character of its construction. Like the engineering taught at the Écoles
de Ponts et Chaussées, this map referred to traditions of practice. The detail
of hills, fields, rivers, and roads was reminiscent of military maps, but emphasized the relationship of works of engineering to their environment. It
also served as a regional map, however, defining the province by its towns
and roads. It contained forestry information, too, indicating the location and
character of woods and orchards. And finally, it was an engineering map,
indicating the location of locks, tunnels, bridges, ports, settling ponds, and
the like, and their relationship to the topography. The map placed the Canal
du Midi in a context that was cartographically layered, suggesting that its
power was multileveled. The image also presented the canal as a completely
impersonal fact, like nature itself.

The technical detail on Garipuy's maps was its own explanation of the canal's significance. It carried a humanist message about intelligence by showing exactly how the canal worked. The section that included Malpas, Enserune, and Montady, for example, showed the old lake bed at Montady and the drainage channel under Malpas near the tunnel, associating the two. Garipuy's cartography also contained enough topographical information to allow viewers to see how the canal was kept above the valley floor approaching Malpas, skirting along the edge of rocky hillsides from the town of Poilhes all the way to the tunnel. The map did not speak of the need for gunpowder to blast a bed for the canal from the hillsides but it did illustrate why this would have been necessary. The Canal du Midi in this representation was a work of modern men, not gods or kings, and the canal was the proof of the efficacy of human thought played out on the ground.

Worthy of Rome

The books on Canal du Midi followed a similar trajectory, starting with accounts that emphasized its Roman attributes, the glory of the king who sanctioned it, and the brilliance of Riquet. As the history was told and retold, it turned into a legend of human genius validated by the efficacy of the result.[57] All the books presented the canal in Languedoc as a modern triumph, surpassing what the ancients had achieved on earth, and a primer for future engineering. In Schutz's terms, it depicted the future of the past.[58]

Bernard Forest de Belidor, from a Catalan area in Spain, was among the most respected of these writers, and an unlikely champion of the Canal du Midi given Riquet's actions in Roussillon. This military engineer was particularly interested in hydraulic engineering in the Roman tradition, and saw the Canal du Midi as a modern instantiation of that tradition. The Canal du Midi was thought to demonstrate that the French could follow the ancients along this path to empire, putting their material and intellectual tools at the disposal of the king.

As a military engineer, Belidor was most interested in the use of hydraulics for war. The value of water in the modern era, Belidor argued, had been proved in the Netherlands—both during the battle for Dutch independence from Spain, and in Dutch battles with France. The Low Countries could and did flood the countryside to keep foreigners from their towns:

> When one reads the history of wars waged in the Netherlands, one cannot help but be astonished by the marvelous use of locks and how the Dutch can divert water to stop the progress of an enemy that is pressing its advantages. This subject merits attention not only by fortress engineers, but also superior officers who want to conduct an important operation. . . . [This treatise] includes the de-

scription of locks for flooding [and methods] of diverting water for defensive and offensive warfare.[59]

But Belidor was even more appreciative of the innovative hydraulics of the Canal du Midi, and saw it as a beneficiary of military engineering.

Belidor justified royal indemnification of lands for the Canal du Midi not by principles of royal sovereignty over the kingdom but rather the efficacy of the result (working). It was an impersonal decision with impersonal positive effects. He recognized that no one wanted to give the land to Riquet for the canal, but without indemnification, he asserted, territorial management could not be carried out properly because people would always lobby against having their land put to state use.[60] The canal was worth the individual losses of land, he suggested. The innovative locks at Agde and near Béziers (Fonseranes), he contended, were beautiful to behold, as was the Malpas tunnel.[61] He maintained that fifty leagues of the canal had to be cut from the rock before piercing the mountain with a tunnel that was twenty-five toises in length. In his words, "This work was something innovative and elegant enough to be worthy of the ancient Romans."[62] Of Saint Ferréol, he said it was "the largest and most magnificent work executed by the moderns."[63] He resolved in his text the conflict between ancients and moderns. The moderns were both worthy of the ancient heritage and capable of surpassing it.

Like a good military engineer, Belidor was particularly admiring of the repairs to the canal called for by Vauban, ignoring the precedents by Clerville and Riquet.[64] In this account, even silt barriers and settling ponds used in connectors to local rivers and streams, and the conduits under the waterway, were attributed to Vauban, and the aqueduct bridge at Répudre was given to him as well. All the collaborative work of earlier participants in the project, while generally acknowledged, became part of Vauban's legend.

It was the astronomer Joseph Jérôme Lalande, however, who established the canal as a work of engineering genius by an ordinary man. He presented Riquet as that genius, using some of the entrepreneur's letters to explain his rightful place as a martyred man of vision. In the late eighteenth century on the verge of the French Revolution, the image of the scientist as a misunderstood genius, struggling against the narrow-mindedness of his own age, made sense. Riquet seemed a perfect example of a modern man in a premodern world who was a victim of petty noble infighting, but through his own perseverance succeeded in providing the people of his province with an extraordinary, useful tool of social betterment.[65] Here was an argument for the primacy of stewardship over patronage as a political philosophy.

Lalande defined the value of the canal in economic terms—ones pleasing to an agronomist, but far from the military interests of Louis XIV. He argued that while haute Languedoc was rich with wheat, bas Languedoc was rich with wine and manufactures from Lyons. The canal allowed them to trade and gain new economic opportunities. Following Riquet's letter to

Colbert, he claimed that foreigners using this route for commerce brought new revenues to the region. Lalande had an example. The Spanish supplied wool to local manufactures, and commercial shippers paid tolls that brought bullion into the kingdom. The canal served the French economy more generally by joining Mediterranean and Altantic markets for the first time. As Riquet had promised, this internal waterway allowed French boats to avoid Gibralter, and it even served the military, permitting French soldiers to get more easily to enemies in Roussillon, Catalonia, Provence, and Italy.[66]

Still, the Canal du Midi was not simply a useful route even to Lalande. The astronomer quoted Pierre Corneille on the brilliance of the reign of Louis XIV.

> The Garonne and Aude, in their deepest caves,
> Sighed for time immemorial to see the linking of their waves
> And to flow thus, happily pooling together in thanks,
> Bringing the treasures of the dawn to the sunset's riverbanks.
> But even with wishes so soft, and a fervor so beautiful,
> Nature, attached to its laws, eternal,
> Sternly placed in opposition invincible blocks,
> A dreadful chain of mountains and rocks.
> France, your great King speaks, and rocks tumble;
> The earth opens its center, and mountains crumble;
> All yields, and water, which by nature follows cavities,
> Reveals him to be all powerful on the earth and on the seas.[67]

Lalande's own praise focused on Riquet's accomplishments and the uncanny power of the Canal du Midi. In his introduction, he quoted the Abbé Maunenet as saying: "This canal is something [for people] from all ends of the earth to see. There has never been in any part of the universe something even approaching the sight of boats, and perhaps one day . . . even warships, [passing] beneath the mountains."[68]

Lalande saw the Canal du Midi as a triumph of the moderns. He compared it to canals in China, Egypt, and Greece, emphasizing the limitations of these earlier traditions. The great improvements in canal design, particularly locks, were started by the Italians and Dutch—and to some extent the Germans.[69] What made the Canal du Midi great, according to Lalande, was the way it used these precedents, and the myriad of novelties it added to the engineering repertoire.[70]

In Lalande's individualistic account of the canal, Andréossy and Campmas (much less indigenous women engineers) disappeared from the picture completely, and Clerville and Boutheroue had just cursory importance as members of the commission, not authors of the canal. Riquet did it all: "*Riquet went* into the Mt. Noire and saw how to divert [water from] the Alzau. . . . At first *Riquet wanted* to run the canal from the divide to the Agout river. . . . *Riquet decided* to avoid these unreliable rivers and connect the Garonne to the Aude with the canal."[71]

Riquet was to Lalande the model of a modern man, who could bring out his own natural gifts with education. He was a rational man of genius for a rational age of science.

> He devoted the best years of his life to this grand and beautiful work that would enrich his land like nothing else; guided by his natural genius, he made up for the knowledge he lacked of geometry and hydraulics; he conceived of and executed the Canal de communication des mers without preliminary studies/drawings. The grandeur of the plan, the way he verified it according to means only suggested by his native genius, the obstacles that he conquered both at Court and in the Province, his constancy, patriotism and success all made it apparent that he was one of the most extraordinary men of the 17th century. What should be included in his legacy is also that [he] was one of the most virtuous, too.[72]

More significantly for subsequent histories, Lalande reprinted as an appendix a series of letters and documents on the canal.[73] They included: the "Arrêt" of October 1666, for the first enterprise; the "Bail and Adjudication" of January 23, 1669, for the second enterprise; and a set of ten of the hundreds of letters exchanged by Colbert and Riquet.[74] These were full of optimistic thoughts about the canal. Only a letter on August 9, 1670, made visible any strain between the two men, when the minister urged the entrepreneur not to divert the Canal du Midi to serve Castelnaudery and Carcassone.[75]

Lalande ended the documents with the report by d'Aguesseau, accepting the canal as complete in 1684.[76] In his selection of documents, Lalande treated the canal as beyond political contention, and the child of a man of genius who could make things work.

Books from Belidor, Lalande, and others made the hydraulics for the Canal du Midi part of a literate tradition. They excluded from their histories all mention of tacit knowledge, knowledge of local materials, silting problems, and soils. Even as they spoke of the canal as a work in the Roman tradition, they denied Roman engineering and architecture as a living tradition. Peasant women's accomplishments as well as the work of *fontainiers* and artisans were formalized and attributed to educated men.

These authors established the frame by which Riquet and the Canal du Midi became known.[77] This view was summarized in English by Charles Vallancey, antiquarian and founder of the Royal Irish Academy:

> Of all the great Works executed by *Lewis* XIV, there is none more useful, more magnificent, nor that does more honour to that Reign than the Canal which joins the two Seas by *Languedoc*. . . .
> M. *Riquet* has not less Merit for putting the Success of [the Canal du Midi] past all doubt. . . . M. *Colbert*, pleased at his great genius, took [Riquet] into Protection, by which he surmounted

those Obstacles which personal Interest too often opposes to the Publick Good.[78]

Vallancey was known as an odd character, but he followed the history set down by Belidor faithfully. He helped to establish in English the importance of the military engineer, Vauban, to the Canal du Midi, making him the hero of the story that he wanted him to be and emphasizing the significance of perfecting the waterway.

> Thus *Vauban* had the honour of bringing this Canal to perfection. A Canal which all the World acknowledges to be the greatest piece of Hydraulic Architecture, that ever was undertaken, and which is of infinite Consequence to the finest Princes in France, thro' which a great Trade is carried on from Sea to Sea.[79]

As the Canal du Midi was raised in cultural stature from a muddy scar to a triumph of engineering, it was stripped of its provenance and practices. Artisans and laborers, who were not thought to be lofty enough to be real actors in the New Rome, were excised from the account. A heritage of greatness could not be a product of tacit knowledge and peasant community life. Success had to be attributed to genius. These rules were ways of cultural genealogy—purifying and masculinizing the past, and individualizing the collective knowledge that made the Canal du Midi possible.[80]

Ironically, the Canal du Midi had become such an asset by the eighteenth century that the États debated whether to claim it for the province, taking it away from the Riquets. They were inhibited from doing this only because by this time Riquet was a local hero, a genius, and the "author" of the enterprise. The enterprise once based on fantasy, not fact, was a model of rationality and object of local pride. The modern material efficacy it represented came to define the province that had fought it tooth and nail.[81]

The wonder of seeing ships pass under the Montagne Noire replaced the wonder of seeing peasants, artisans, engineers, and entrepreneur collaborating, intent on solving problems that academics could not address. Nolin's map with insignia of noble families left the people of the province who had worked on the canal behind. Still, these books and maps were not entirely misguided even in their rewriting of history. What was built at the Canal du Midi was not, as Belidor and Lalande knew full well, just a canal but also an intellectual capacity. The Canal du Midi was the product of an extraordinary intelligence—only it was social and not just located in Riquet, Vauban, or Louis XIV. The impersonal power of this collective intelligence was what made the Canal du Midi so modern. It embodied an anonymous efficacy that was breathtaking—almost as glorious as a king—that spoke to the powers that could be derived from collaborative work and territorial engineering designed for political effect.

CHAPTER 9

Powers of Impersonal Rule

The history of the Canal du Midi is a story of the invention not simply of a canal but also of a political culture of impersonal rule. The waterway became a triumph of territorial governance, embodying both stewardship principles and an anonymous, collective intelligence that characterized and empowered the French state. Its impersonality stood in stark contrast to the personal rule of the king even as it served the regime of Louis XIV. Time and the growth of modernist governments have since naturalized its presence, making the canal seem only a quaint piece of economic infrastructure and obscuring the novel power it embodied. But in its period, the canal was an early experiment in impersonal governance that helped to set in motion a novel dynamic—Bruno Latour's "knot," binding together government, natural objects, public representations, and epistemic communities.[1] Impersonal rule was a method of political transmutation that Colbert sought and Riquet realized with indigenous women engineers, artisans, military men, educated mathematicians, fishermen, salt makers, and the landscape of Languedoc itself. Natural forces were turned into political "gold" that affirmed at once both the glory of the king and the impersonal powers of the administration. The Canal du Midi could serve the government because it was a model of material governance. In working, it manifested the uncanny power of things.

The saga of the construction of the Canal du Midi indicates not only how this piece of impossible engineering was made possible but also how modern technocracy was born in the pursuit of political territoriality, the organized use of material governance to augment the state and disempower the nobility. The administration under Colbert did not and could not circumvent patrimonial politics, but the programs of stewardship it initiated did something more telling and disconcerting: it depersonalized knowledge, land, power, and moral reasoning—the foundations of patrimonial politics. Impersonal rule emerged not in contest with patrimonialism but rather under its feet, changing life around it through material means. It was born of dynamics that were themselves depersonalized: collaborative work, moral measures of legitimacy, material transformation of places, and collective representations of France as the New Rome. As the kingdom was aligned with its territory, impersonal governance both colonized the political imagination and gained a foothold on the landscape.[2]

The history of the Canal du Midi told in this book is at once a story of the pursuit of impersonal powers through territorial governance and also a posthumanist tale of the limits of political domination over the natural world—the failure of the knot to hold things together. The forces of water, wind, rock, and sand, while skillfully enrolled to produce a navigational canal to cross Languedoc, were never fully mastered; dominion as an ideal of stewardship was not realized on the ground. The Canal du Midi became an emblem of governmental efficacy, but only by masking the struggles entailed in the work.[3] Rainwater undermined levees and floods destroyed even Vauban's overbuilt pont-canal at Capestang; the port at Sète filled with sand and part of its seawall collapsed again under storm swells; meanwhile, the rigole de la plaine silted up, becoming too shallow for navigation to Revel, and the basin at Naurouze developed sandbars. Everywhere along the canal the earth and water evaded the control of those who tried to build it, displaying the excess powers that nature could bring to human projects while denying the religious mandate for human dominion.

The opening of the Canal du Midi may have been celebrated with all the pomp of divertissements at Versailles, touting France as the New Rome, but when the trumpets died down and the notables went home, the movement of water along the Canal du Midi, not the will of men, was the medium that floated boats. Nature had to join in the enterprise to take vessels toward the two seas. People could open and close the spillways of the dam at Saint Férreol, and work the locks with their doors and tambours, but without the water, that activity meant nothing.

Water was an independent nonhuman actant in Michel Callon's sense, demanding human attention if people wanted to make the canal work.[4] Water did not need to "speak" to be heard; its mute deviations from normal channels were enough to get attention. Hydraulics remained an ongoing problem of technological imagination that continued to draw people together, acting as "boundary persons" for the water to join rivers, sources, and seas to serve the canal.[5] Humans linked waters before the canal could link humans, reordering the ontology of hydrology in Languedoc.

The Canal du Midi *had* to be an ongoing site of cooperative activity just to keep water flowing in the quantities needed for navigation. The waterway became both an unquestionable fact of local life like a mountain or wild river, laying down conditions of possibility for all those living near it, and the opposite: a product of collective action and an ongoing raison d'être for social coordination. Riquet, Colbert, and Louis XIV might have been variously assigned authorship to the Canal du Midi, but no individual could design, build, or maintain the waterway. Many people over time shared authorial intent, and agency lay in natural forces as well as in the human will.[6]

Powers of Impersonal Rule

There were four constitutive elements of impersonal rule at work on the Canal du Midi: the pursuit of natural knowledge as impersonal truth (defining facts and forces), territorial politics (using land management as a political tool), principles of material improvement (stewardship, dominion, building the New Rome), and material techniques (formal engineering and indigenous land-management practices). All of these processes were themselves impersonalizing forces, at odds with patrimonial politics and the personal exercise of power. They dissociated principles, practices, truth, and land from persons, and defined a social capacity for impersonal rule that became central to both the political ascendancy of modern states and the cultural expansion of scientific thinking into the Enlightenment.

These dynamics of impersonal rule can be summarized in a diagram based on Latour's model of the circulatory system of science.[7] It builds on the four elements described above, but also illustrates the intertwining of natural, moral, intellectual, and political forces that gave impersonal rule its novel efficacy in the seventeenth century:

This linked set of impersonal processes was a dynamic for organizing human relations to the natural world based on both peasant practices and learned thoughts. It was not in itself a social organization of power but rather a logistical and epistemic culture of working that linked material techniques and both ideas and ideals about governing the earth. It could never be entirely contained by the administration, even though it was central to the genesis of a more powerful state. Impersonal rule took administrative

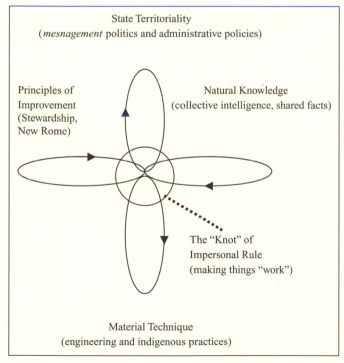

Figure 9.1
The system of impersonal rule.

State Territoriality
(*mesnagement* politics and administrative policies)

Principles of Improvement (Stewardship, New Rome)

Natural Knowledge
(collective intelligence, shared facts)

The "Knot" of Impersonal Rule
(making things "work")

Material Technique
(engineering and indigenous practices)

authority away from officeholders, and assigned it to more impersonal acts and measures: technical expertise, principles of governance, material measures of efficacy, and engineered effects.[8]

Territorial politics both contributed to the pursuit of impersonal modes of governance and benefited from the techniques that were honed for political ends; state administration under Colbert not only gained new infrastructure but also began to stand for and wield an anonymous intelligence that could be called "French." Impersonal rule invested new dignity and significance in interactive and fleeting forms of intelligence that the administration tried to manage for its advantage. It gathered into state political repertoires the indigenous expertise of women peasants and the social acumen of financiers as well as the knowledge of other untrustworthy social groups whose material efficacy now proved essential to the push toward empire. Colbert's plan to contain the authority of nobles with stewardship principles—designed to protect the singularity of the king—led instead to a realignment of nature and power that complicated all the lines of authority, and made singularity itself more problematic, if not immediately less potent.

The political culture of impersonal rule distorted the distinctions between high and low, good and bad, true and false that had traditionally defined the social hierarchy and the authority of the political elite. It both linked and decoupled natural measures and the word of gentlemen. At the same time, it defined a moral basis for government apart from (or as Colbert hoped, in addition to) the divine right of kings. Impersonal rule was powerful and hard to counter precisely because it was born at an intersection of social, moral, and material worlds, a "spreading center" of social possibility where no one lived in traditional ways.[9]

This New Rome was built through collaborative social practices and distributed intelligence, yielding an anonymous form of power. It defined a political environment that lay beyond the negotiations among interest groups or the personal influence of officeholders; it was a place that gentlemen of standing did not know, did not make, and could not easily erase.[10] The normal flows of information and favors that sustained patron-client networks did not carry the technical information for assuring that land would work in the interests of men of power. So engineering projects like the Canal du Midi—precisely because they engaged natural processes like water flows rather than political processes—eluded capture not only by the state but by gentlemen of rank as well.

The knot of the nascent techno-scientific culture nurtured at the Canal du Midi, then, was fragile and elusive because of its collaborative character, but also robust precisely because it was hard to capture and contain.[11] To the extent that the Canal du Midi depended on administrative support, Colbert could have stopped Riquet's project by withholding funds—as the États had tried to do. But the administration had further uses for the material tools of power and abstract measures of legitimacy being honed in Languedoc. The efficacy of territorial politics made it too valuable to let go—even when the king and Colbert began to recognize the dangers that stewardship principles carried. Impossible engineering worked as a mode of governance. It was more effective in constraining nobles than the expense of court life at Versailles so frequently cited as the cause of aristocratic subordination in this period.[12] So Colbert continued to fund Riquet even when he hated dealing with him, and peasant women from the Pyrenees continued to thread the waterway through the hills of Languedoc in the name of a king they were normally loathe to serve. The Canal du Midi shifted the foundations of order in subtle ways that served multiple ends. It crossed southwestern France as a model of human dominion and the opposite: abilities without apparent personal agency.[13]

The political possibilities of impersonal rule gave new meaning and form to knowledge of the earth, elevating its significance with its strategic value. The efficacy of using natural forces for political ends became an impetus for new formalizations of peasant knowledge and artisanal practices. Laboring on the earth—work that had been left to low-status groups in the past—was now integral to territorial governance. Workers turned their logistical "weapons of the weak" into public political assets.[14] Artisans and laborers alike had learned to make things work, and make working a measure of competence. Once this ability from peasant culture became useful to the state, it "mattered," but the communities of practice that had sustained it no longer mattered. "Truths" of nature became anonymous and autonomous.

New interest in hydraulics grew in France during the eighteenth century, stimulated partly by the opening of the Canal du Midi. It led to a proliferation of books on the subject that joined the broader shift in literary practices toward the formalization of tacit knowledge. The *Encyclopédie* was the ultimate compendium of collective intelligence and codified practical knowledge, not only responding to the Baconian tradition in the sciences and utopian hopes about integrating human knowledge, but also reflecting the effectiveness of material governance and impersonal rule that had begun shifting the ground of power in France. Engineering knowledge, Riquet's "philosopher's stone," became too important to circulate freely among peasants so it was captured on paper for gentlemen.

The state may not have been fully able to control nature or local knowledge of it, but the state disproportionately benefited from these material practices of power because of its special access to both land and skilled clients. Using natural forces for political ends, taking logistical control of material life seriously, made the state simultaneously more autonomous of patrimonial power and more vulnerable to natural forces. Political life became "risky" in the modern sense, subject to public criteria of material adequacy that tested the ability of government to maintain the well-being of its people. The effectiveness of a regime was now connected to its response to floods and drought, and not just the control of the social elites that concerned Machiavelli. Politicians needed those with natural knowledge and the ability to toil on the earth to maintain the legitimacy of government. Engineering stable political regimes and landscapes were interdependent practices in the emerging dynamics of impersonal rule.[15]

The Canal du Midi was a piece of impossible engineering that was powerful and dangerous for just this reason; it was contingent, dynamic, and effective—impossible to erase, impossible to manage fully, and potent in its effects. It had no single author, and no "natural" beneficiary. Colbert had to engage in a propaganda campaign to present it as a success of the regime—the work of the Sun King.

The collective authorship of the work was not intended but instead—as we have seen—resulted from a dearth of formal engineering knowledge of the period. No one person knew enough in the 1660s or even 1680s to design the canal alone, so no one had the social stature to claim full personal authority over its development. Given Riquet's low epistemic credibility as a financier, Colbert never would give the entrepreneur full control of the project. He also took Riquet's claims of personal genius and spiritual sanction as evidence of his delusionary tendencies, rather than a reason to trust his judgment.

Still, the waterway could never come under Colbert's control either; it was a product of the minister's government but not his to govern. It was not and could not be a product of "action at a distance."[16] Even if Colbert had been able to get the plans, engineering drawings, and maps he wanted, he would still have been plagued by the sands on the Garonne and the Mediterranean coast along with the floodwaters from the Montagne Noire. Engineering was an interactive process that required close readings of the earth as well as specifications. The land of Languedoc was not an abstraction amenable to abstract planning; it was vital, local, shifting, and not "clay" prepared for human hands.

Accounts of the Canal du Midi that have focused on Riquet's genius, obscuring the collaborative engineering that made it possible and denying the impersonal power it came to embody, have veiled the importance of the canal and even Riquet himself. By substituting Riquet's genius for the contributions of indigenous women engineers and illiterate artisans and subcontractors, they have dissociated modern engineering from its ancient sources. They have hidden the elements of the classical past that remained living traditions in seventeenth-century France, and reduced our ability to understand the variety of engineering heritages in the ancien régime.

Heroic tales of the Canal du Midi have also done Riquet a disservice by erasing his regional significance and political legacy. By making the canal an act of naive genius, they have denied the logistical skills that made him a successful tax farmer, supplier to the French army, and entrepreneur for the Canal du Midi. By treating engineering knowledge as the *only* form of intelligence worthy of note, they have neglected his deep understanding of the intellectual topography of the province as home to men and women of deep natural knowledge as well as strong character. He championed their knowledge even as he exploited it to explore the modern possibility of (impersonal) politics. While later scholars may have ignored the power of Pyrenean women and the local architects and artisans of Languedoc, Riquet did not.

On the other hand, the elevation of Riquet had deep cultural and political purpose, masking the disorienting reality of an anonymous waterway. The story of genius protected the canal from local enemies by making it a personal accomplishment of a client of the king. Elevating Riquet also buried potentially delegitimating elements of the Canal du Midi—the contentious politics about the canal in Languedoc, the clashes between Riquet and Colbert, Riquet's taste for power, and the use of femelles to solve engineering problems. The purified canal became worthy of the New Rome by being cleansed of the New Romans, and remade as the product of a superhuman individual. This prepared Riquet to become a national hero in the nineteenth century as well as an exemplary engineer in the eighteenth, and masked the impersonal powers and political import of the canal itself. Denying Riquet's (personal) genius is important both to properly recognize the sociopolitical importance of the entrepreneur and to reveal the historical significance of the Canal du Midi as a collaborative reconstruction of classical engineering and a model of impersonal rule.

Political Culture and State Territoriality

Rejecting canonical history makes clearer the political as well as technical significance of the Canal du Midi, highlighting its role in a broader program of French territorial improvement. The project provides an unusual vantage point for assessing Colbert's politics, and the role of mesnagement policies in his administration. It also highlights the interconnection of stewardship and patrimonialism in state territoriality, and the material dynamics at the heart of seventeenth-century French state formation.

In spite of the contradictions inherent to political territoriality, the regime under Louis XIV and Colbert was generally successful in extending state powers, yielding a form of government often called "state absolutism" because it wedded administrative and monarchical power so effectively. Whatever one might think of the term absolutism, there was indeed a notable shift of power from the aristocracy to the king in the period that was palpable at Versailles.

The fundamental antagonism between personal and impersonal rule did not ordinarily disturb the flow of power at court in any significant way. The contradictions were easy to ignore as long as Colbert was appropriately subservient to the king. Administrative stewardship seen from Versailles *was* the will of the king, and served his purposes.

James Griesemer, in his studies of the historical coexistence of incommensurable scientific theories, offers (by analogy) a model of how incommensurable political strategies could have coexisted in seventeenth-century

France. He shows that new ideas in science are (and perhaps have to be) first articulated in an older language that shares some background assumptions with the new theory. Revolutionary ideas only come to seem incommensurable over time as new vocabulary is developed to underscore the differences. This was true, he notes, of Mendelian genetics as Mendel's followers explored the radical intellectual potential they saw in the botanist's work.[17]

In a similar way, stewardship emerged from and coexisted with patrimonialism. The two shared a political culture connecting worldly power to otherworldly forces, and this at first obscured their differences. If politics was a way to realize God's will on earth, and God's will was unitary, pursuing territorial improvements implied no contradiction with the will of the king. As long as the legitimacy of the sovereign was taken as unassailable—and this was the case under Louis XIV—then acts of stewardship to serve God could easily be represented by a subservient minister such as Colbert as reflections of the king's will.

The Treaty of Augsburg also provided a positive political rationale for linking territorial politics and personal rule. It defined the territorial integrity of kingdoms in relation to the singularity of princes by giving them the authority to determine what religion(s) could be practiced in their domains. Connecting geographic control to spiritual practices furnished land with a clear role in defining the rights of the sovereign. In return for the integrity of their own domains, princes gave up the right to invade the lands of others to change the faith of the people, reinforcing the importance of territory as a basis for singularity. The point of the Treaty of Augsburg was to keep warring factions apart from one another during the wars of religion, not define state territoriality, but it provided a logic for associating territorial stewardship with personal rule.[18]

If the Treaty of Augsburg reaffirmed the prince's right to rule his lands without intervention, stewardship principles provided guidelines for the virtuous use of this power. As Bernard Palissy suggested, mesnagement practices aligned the countryside with morality as well as politics. His "garden" was a utopian place where the love of God was pursued through tender acts of devout stewardship protected by walls from outside threat. This was a model of a kingdom, fitting the political and moral order defined by the Treaty of Augsburg.

The dream of making France a New Rome was an outgrowth of this moral imaginary, but a less tolerant one.[19] Louis XIV wanted to lead a new Catholic empire to replace the Hapsburg one, eventually repealing the Edict of Nantes, so that his subjects would have to follow his religious beliefs. The architects of the New Rome might have been moderns, using the classical legacy as a guide to the future, but dreams of the New Rome for the king was based on a tradition of moral as well as territorial sovereignty. The "good" in government, as Friedrich Nietzsche put it, underwent a redefinition to serve the New Rome, maintaining the moral foundation of political legitimacy that made simultaneous sense of territorial stewardship and monarchical singularity.[20]

Humanist ideas also provided a foundation for territorial politics that simultaneously made sense of patrimonial and mesnagement ideals. The gardens of Versailles celebrated the king's grandeur with a stunning display of land stewardship. The park was a stage set for literary dramas about human efficacy and godlike powers, and also a showplace of mesnagement gardening technique. Works and words were aligned as expressions of human reason and power cloaked in classical garb.

Territorial politics, then, seemed on the surface thoroughly integrated and effective—harmonized within a political culture that embraced both the will of the king and stewardship land management. The Canal du Midi was easy to portray in these terms, and was in Andréossy's map depicting the king as Apollo, drawing the canal across Languedoc. Cultural associations of the king with his lands masked the contradictions in political territoriality, the rift between personal and impersonal rule that strained relations between Riquet and Colbert.

The Threat of Impersonal Rule

While in principle state territoriality easily encompassed both stewardship practices and the king's exercise of will, the two were harder to hold together in practice. The architects of the New Rome were the ones to feel this first. They began to recognize the distinctiveness and efficacy of logistical politics, and saw virtues in stewardship practices that started to draw them away from the king. Riquet was the extreme case with his oracular sense of self, listening directly to God to follow His will through improvement, but he was not alone. In his embrace of stewardship principles, sense of political self-importance, and alienation from the king and Colbert, Riquet was joined by Clerville and Vauban, who were also integral to French territorial politics.

Clerville and Vauban were clearly more effective engineers than Riquet, and more respected men at court, but they were equally beneficiaries and victims of stewardship politics. They, too, improved French territory for the king and Colbert, had visions of a better France, and ended up being vilified for their hubris. They saw the kingdom change under their supervision, and saw significance to the natural knowledge that they deployed. They could alter the lives of thousands of people by refashioning the landscape and bringing new rationality to the administration of the land. This was the power of governmentality, but not godless rationality.[21] Stewardship was logistical power serving the common good.

In the end, they were both reprimanded and marginalized as individuals for their hubris, but the problem they posed the regime was political, not personal. They had learned to know and manage Creation wisely, exercising dominion, and had changed France with powers that the king did not possess.

If Louis XIV was a good strategist, they were good at logistics, and could administer the kingdom wisely (in his name). They became a new kind of political actor with a liminal status—the technocrat.

Although he trusted them for most of their careers, Louis XIV turned against Clerville and Vauban when they began to exercise independent judgment in logistical matters, and obliquely challenged his singular authority.[22] It is well-known that the aging king attacked previously trusted agents of the Crown, but the motives have traditionally been treated as personal. Louis de Saint-Simon asserted that Louis XIV was a self-centered man who demanded strict adherence to his desires and tastes.[23] Those who showed an independent spirit were rejected from court and alienated from the administration.

This psychological portrait of the king might have been correct, but it does not itself explain the phenomenon. Saint-Simon does not tell us what encouraged men of power, who were already familiar with the king's character and used to placating him, to take risks and exercise their own judgment. Their political faux pas resulted not from personal characteristics but rather stewardship principles that reconfigured their imaginations and reasoning. Surprising acts of political independence—not by traditional nobles but new clients of the king—were unintended consequences of territorial politics.[24]

Clerville's fall from the king's good graces was rumored by Vauban to be a result of incompetence. Clerville, after all, had been the chief engineer for the Château de Trompette when one of the walls of the fortress simply fell to the ground. Apparently, part of the foundation had been poorly constructed while Clerville was away. Colbert worried that all was lost, but as soon as Clerville came to the area and consulted with experts to design a plan to repair the wall, the fortress was restored to good condition and the minister happily forgave him.[25] This moment of incompetence did not ruin Clerville's standing.

Clerville's fall from the king's affections was even less a problem of patronage relations. He handled a delicate situation with Colbert beautifully in Alsace when the minister's cousin became intendant there. The intendant hired a contractor for repairing the fortresses at Brisach and Philippsbourg who was a disaster both technically and politically. Clerville stepped in to fix the situation by supervising work on the fortresses and helping the intendant with his bookkeeping without embarrassingly exposing the incompetence of the beloved contractor.[26]

Clerville's fall came instead from an act of stewardship that was thought at the time to contradict the will of the king. Clerville had become accustomed to overseeing the intendant's bookkeeping, but a new intendant appointed to the region took offense at the practice.[27] He complained to the king, and Clerville was informed by Colbert in June 1673 that he had overstepped his authority:

I am obliged to tell you that all of this [administrative work] is not in any way your authority; since the king has chosen an intendant, his majesty has judged him worthy of his confidence both in the disposition of the taxes he collects and for everything else on which he depends, and that you have no authority to act as his *contrôleur publique*. It is true that if you see that the king is manifestly mistaken [in his trust of the intendant], you must advise the intendant about this, honestly and civilly, so that he can take care [of the problem]; and in case that he does not remedy the abuses that you have discovered, all that you should do is to keep memoirs [of what you have seen] to alert the king, which his majesty could reflect on if he wanted to do so.[28]

Clerville had become a dedicated functionary who saw virtue in making the local administration work. He had started thinking of himself as a skilled supervisor of the province, solving problems successfully with his expertise in fortresses and finances. But he was the king's servant, and Louis XIV made this clear. It was Clerville's job to obey his monarch, and it was the king's job to manage his provinces. With this slap in the face, Clerville lost ground to Vauban, who became the king's main adviser on matters of military engineering.[29]

Vauban in turn suffered an even worse fate some time later. Again, the problem was not a lack of engineering competence or an inability to play patronage politics. Vauban was master of both. The young engineer had even survived acts of treason early in his career, working as a soldier for the Prince de Condé during the Fronde. He was forgiven for even this massive breach of loyalty to the sovereign because the king needed men of military talent. Vauban's fall came later when he developed independent ideas about the well-being of the kingdom.[30]

Vauban had the audacity to propose a tax reform—one that even suggested taxes on the nobility. His interest in developing economic policy for France was not simply a lapse of judgment by a man who should have stuck to fortress engineering. It was a straightforward continuation of stewardship principles of good government, using rational measures to improve the well-being of the kingdom and its people. He felt taxes should be rational, not a way of doing favors to clients. So he proposed a "Dîme royale," a 10 percent tax on nobles as well as nonnobles.[31] But his stewardship reasoning about the kingdom was at odds with the patronage system that was more vital to the king than the military engineer, so he was treated as a madman and exiled from court.[32]

In a way, Clerville and Vauban had become like Riquet—madmen for their times. They made things uncomfortable for others by taking their administrative powers too seriously. In valuing their logistical abilities and stewardship moral reasoning, they were potentially disruptive to the traditional hierarchy that demanded patrimonial subservience to the monarch.

These men manifested a distinctive form of hubris, starting to treat France as though it was their land to imagine and manage. Clearly, this was not tolerable under Louis XIV, and the monarch responded by acting willfully, limiting or destroying their reputations or careers. This political technique demonstrated his final authority, but was also testimony to the threat of administrative stewardship and territorial engineering as a form of power. The architects of the French state who were finally discredited may have been politically compromised by the king's wrath or Colbert's hostility, but the foundations of their power and the potency of their abilities were not erased from historical memory, since territorial stewardship remained part of French public administration.[33] Inadvertently instead, they helped to make visible the dangers of impersonal governance to monarchical singularity.

Strategics, Logistics, and Impersonal Rule

The tensions between personal and impersonal rule were not just matters of political philosophy. The problem was that they exercised fundamentally distinct forms of power, strategics versus logistics, based on entirely different logics: domination of people and dominion over things.

The strategic use of (legitimate) violence for domination is widely recognized in social theory as a form of power (or more precisely, *the* form of power), but there is another form of power illustrated in the history of the Canal du Midi that was explored by Riquet, Clerville, and Vauban: logistics or the mobilization of natural forces for collective purposes. Most social scientists take it for granted that the exercise of power is always an agonistic process, and usually takes the form of struggles among interest groups to dominate one another. But just as war requires both strategics on the battlefield and logistical work to support it, so the exercise of power in other settings also depends on a combination of strategics and logistics. Strategies are efforts to organize human relations while logistics are efforts to organize things.[34] Although they are often used in tandem, both strategics and logistics are important to differentiate because they bring different types of power to social life. Struggles among political actors are clearly central to the political process, but so is land control.[35] Reworking the countryside requires different knowledge and logics of action, and it affects social order in distinctive ways.

The construction of border fortresses, roads, bridges, ports, and shipbuilding facilities for Colbert was obviously logistical work with "strategic" purposes, serving military ends quite explicitly. Fortresses supported the exercise of legitimate violence, but many of them also defined frontiers that were never transgressed, affirming state territoriality without agonistic struggle. They still affected local life, providing construction jobs, taking down forests, and occupying the time of the intendants and contractors who worked on them. The canal that cut through Languedoc was similarly divested of its original agonistic purpose; it was never able to carry warships between the two seas. Its powers lay more in the simple fact that water flowed through lands that had been part of noble estates and now stood for France.

Infrastructure changed the political significance of nature, using the sturdiness of stone, the buoyant qualities of water, the binding properties of cement mixtures, and the flexibility of timber. Ordering things was not in itself, however, an agonistic activity. Enrolling nature in politics was *not* the same as the struggles among interest groups or elites seeking advantages, but the material transformation of places could nonetheless change the playing field on which political contests took place. Noble control in Languedoc was diminished when land began to represent the administration.

Power in the Weberian tradition can be subdivided into six elements that can help differentiate it from logistics.[36] These are summarized in the table below:

Strategics in this model consists of the social processes that we normally associate with power: the forms and workings of systems of domination. Elites have the moral authority by rank to govern; they seek strategic advantages over competitors; they benefit from and shape the social hierarchy; and they wield implicit or explicit forms of violence to maintain or enhance their powers.

Logistics, in contrast, is a form of dominion or regulation of the natural order. Power is exercised over the earth (and its creatures), legitimated by the Christian duty for stewardship, guided by natural knowledge, and realized in a built environment.

Strategic domination has been routinely opposed in social theory not to logistics but rather to political resistance—what James Scott calls weapons of the weak, or what Michel de Certeau calls tactics. To the extent that these patterns of opposition take material form, as when peasants poke holes in irrigation ditches on properties of hated landlords, these tactics are forms of logistical power. Still, many tactics are not logistical; they are forms of agonistic confrontation, such as peasant rebellions. So

Table 9.1

Strategics and logistics as forms of power.

Forms of power	Strategics	Logistics
Form of Control	Domination	Dominion
Legitimacy	Moral Authority by Rank	Authority over Nature
Calculation	Strategic Advantage	Natural Knowledge
Reality	Social Hierarchy	Natural Order
Physical relations	Violence	Built Environment
Social Expression	Patron-Client Relations	Collaborative Work

the distinction between strategics and logistics used in this book cuts across the concept of tactics in Certeau, and encompasses only some of Scott's weapons of the weak.[37]

Logistics per se has not been ignored by social theorists, simply blurred with strategics in both Marxist and Foucauldian ways.[38] Those who acknowledge the importance of logistics generally see its value only in relation to agonistic activity or political will. For Foucault, the panopticon is a strategic instrument of power, but how it is engineered does not matter; its power lies in its design.[39] Material analyses in the Marxist tradition similarly minimize the logistics of material life, highlighting instead strategic uses of material powers. Scholars who have worked in these traditions, using very different methods and theoretical approaches, often claim an interest in materiality, but mainly overlook logistics. Immanuel Wallerstein and Michael Mann write of field systems, and their significance to political regimes, studying land control, but not its engineering. Scott examines the built environment, but sees it as a tool of surveillance used for domination, a form of *technē* linked to praxis, not a form of power in itself. Patrick Joyce similarly connects infrastructure to governmentality, and Patrick Carroll studies the role of "engine science" in "taming" wild nature through surveillance. Wallerstein, Mann, Scott, Joyce, and Carroll write historically grounded studies of materiality, but they still rely on Marx or Foucault for their theory of material power so they dismiss or underestimate the importance of logistical activity.[40]

In contrast, scholars in science studies, such as Latour, Callon, Karin Knorr-Cetina, Geoffrey Bowker, and Susan Leigh Star, have studied the power of infrastructures, pointing to the silent, unrecognized, and routine ways they can define social reality.[41] They explain how keys, trains, computer networks, trading markets, and databases—ubiquitous and uncontested elements of material life—construct the order of things. All implicitly distinguish the logistical power of infrastructure from agonistic processes, but only because it is taken for granted. They treat political struggles over the truth of things as preconditions for infrastructures, leaving a pacified, concretized reality in which the political stakes are black boxed.

Bowker and Star, for example, view infrastructures as constitutive of social reality, but they still treat the design of infrastructures as agonistic and politically prior to their instantiation.[42] So nurses' knowledge is excluded from medical databases as doctors' knowledge is formalized there, politically shaping the medical views that inform local decision making. The formalization and concretization in databases translates power differentials from the hospital into the database; for Bowker and Star, it does not constitute a distinct form of power.

Infrastructures do indeed often black box political relations, but they retain distinct technical qualities that become evident when they fail. Canals overflow; roads shatter in earthquakes; computer systems crash; and aqueducts collapse. Canals and databases are not simply solidifications of power relations, although they are often intended to be and are represented as such; they also mediate relations with the natural world in contingent ways that at moments can evade human control.

Strategics and Logistics as Tools of Personal and Impersonal Rule

To understand the distinction between strategics and logistics better, it is useful to turn to the analysis of the Roman Empire by the French philosopher Michel Serres.[43] His point in describing Rome is not to explain forms of power but rather to articulate a posthumanist philosophy, emphasizing the animallike qualities in humans. Still, to talk of Rome as an empire is to speak of power, and so Serres addresses its dual forms of power: strategics and logistics.

Serres argues that the ancient empire was built on two foundations: the struggle for dominance by Romulus and Remus, the brothers of myth who fought to the death to found Rome; and the material construction of the empire that Serres equates with the collective behavior of social animals like African termites. Both constitute patterns of human control—one, organized domination, and the other, unpremeditated work on the earth.[44]

The concept of power represented by Serres in the story of fratricide (or murder used to establish domination) is familiar in social theory, resembling Max Weber's definition of power as legitimate violence. But the story of Romulus and Remus, as Serres tells it, contains a significant twist. The brothers fought each other to the death not simply for the right to rule but also for the ability to exercise *singular* power, using violence to try to eliminate their competitors.[45] Serres's model of agonistic relations is particularly useful, then, for considering both monarchical singularity and its unruly counterpart: forms of erratic violence reported by Charles Tilly in weak regimes, where the demand for singularity and its impossibility lead to ongoing violence.[46]

The second form of power described by Serres is a social process of shaping the natural world. It could be called collective logistical power, in contrast to the singular strategic power exercised in fratricide. Serres describes this material approach to power in opposition to the discursive analyses emphasized by poststructuralists.

> Rome does not speculate, does not speak, never converses about the latest refinement. Rome fights, Rome prays; it is pious, it humbly accepts the dark sense of a repeated gesture. It builds, extends itself, preserves. It is not the negative—the destructive work that seems to advance things. . . . It gives flesh to the word; it builds. Rome incarnates itself; it is construction.[47]

Material power as Serres describes it has no Weberian counterpart and little foundation in classical social theory. It stands for collective work in the physical world and the power of impersonal rule. Serres focuses on the collaborative processes that make it possible, what he describes as unpremeditative forms of knowing that allow the empire to extend geographically through construction.[48]

Because of its constructivist leanings and cooperative focus, Serres's view of material power is interestingly quite unlike Foucault's notion of embodied power that focuses on strategic uses of the material world for domination, or in Serres's terms, impersonal means of fratricide. Foucault, of course, is as concerned as Serres with the physical aspects of regime power and the impersonality of collectivities, but in seeing buildings and bodies as tools of hierarchical domination, he never considers who builds the wall systems for prisons and mental hospitals, or how these construction techniques developed historically. Architecture designed for imposing political will is reduced to will alone. Gone is the work of masons. So are the cultural forms used for the body discipline that Foucault takes as the other important form of power. We cannot hear in his theory the military music whose rhythms were used to train soldiers for war. The forms of life that make materialized power possible—so clearly necessary for explaining what happened on the Canal du Midi—are easier to recognize in Serres than in Foucault.

What makes Serres's argument distinctive from those of Foucault is its posthumanist stance—taking human beings as part of nature rather than somehow made distinct from it by the power of the human mind (and language). His point is not to deny consciousness, but to see its forms and limits as one might see it in other creatures. He describes human beings getting lost in their signs—first, flying from cities to the countryside to try to understand themselves as physical beings in the natural world, only to return to built environments to distinguish themselves from other animals. All of this, he contends, is part of the nature of human beings as social animals.[49] Creating a built environment is an expression of this nature, not essentially a means of domination, although it can be used to that end. Most of the time, we build houses just like termites do, placing pieces of clay (or bricks or stone) on top of one another, making additions to what others have constructed before, and ending up with built environments that are collective accomplishments—our *termitières*.

Serres avoids attributing power to semiotic or discursive formations because they are grounded precisely in the humanist assumptions about the distinctiveness of consciousness and language that he wants to get beyond. He wants to treat culture, including language, as a product of human collective practices, not its foundation. So for Serres, ideology, hegemonic narratives, discursive practices, and semiotics hold no independent interest. They are aspects of the termitière, accreting over time through collaboration in the effort to carve out viable forms of social life.[50]

Serres uses sociobiological ideas not to design a scientific sociology but to instead push for a philosophy that does not assume that the human mind is particularly capable of dominion, or that human powers are necessarily virtuous. (Dominion has not worked out so well for other creatures, and human powers often look predatory or parasitic.) Serres wants to show how qualities—like making fine works of art or even empires—emerge *less* from individual genius and more from the animallike qualities of our being.[51] He does not deny that we are thinking creatures; he just implies that our ways of thinking come from natural predispositions that lead us to consciousness and culture, and not the other way around.[52]

Serres's termitière in some ways resembles paths in the countryside formed through long-term collective use. Such places represent an anonymous cultural solution to problems of living in the landscape. These socially shaped environments are not the product of an individual imagination, but repeated collective activity that develops as part of group life. Human habitations are similar to Serres. Houses are a response to a problem that is collectively solved over time. One person may seal cracks in a wall; others fix the floor; the result is a more long-standing structure that carries a heritage of intelligent problem solving that cannot be attributed to any one person. Serres suggests that this kind of power, the capacity of social animals to remake the earth to favor their survival, is an important part of what made Rome so powerful, and certainly what made Roman engineering so effective.[53] It was also the kind of power exercised and embodied in the Canal du Midi.

The problem in applying Serres's theory to the Canal du Midi is that when Serres treats the material side of social life as unpremeditated, he writes as though it is instinctual, unthinking action, not a form of thought.[54] But logistical activity makes sense as a form of collective cognition, expressed physically and refined over time. Ed Hutchins and Brian Hazlehurst have built robots without individual memory capacity, but linked through a common computer so they can "think" only through interaction.[55] Hutchins and Hazlehurst have found that these robots could learn to solve problems and even develop language skills through interaction alone—providing an exemplar of the intellectual productiveness of social activity.

The Canal du Midi clearly was less a product of knowledge than productive of it. The engineering had to be invented for the project to move forward. Although Clerville developed (premeditated) specifications for the project, as we have seen, the waterway could not be built simply following these guidelines. In places like Fonseranes, workers drew mainly on experience and each other to address the task at hand, "inventing" the lock through social improvisation that yielded the knowledge of how to do it. The Canal du Midi in this sense was itself a termitière, not a consequence of premeditated planning, but rather of ongoing intellectual productivity.

Impossible Engineering

On the Canal du Midi, logistical improvisations produced a masterpiece of hydraulics. What started as an "impossible" piece of engineering, exceeding the abilities of engineers to plan, ended up as a waterway that defied comprehension because it veered so far from established ways to work.

The waterway eluded easy understanding. The scale of the canal alone was daunting and confusing. It took days to see the whole thing, and it changed its character in the different landscapes it passed, so seeing one section did not explain the rest. Worse, the water supply, so important to the technical system, disappeared deep into the mountains, where visitors did not easily go. The Canal du Midi required work to comprehend.

Although a huge piece of hydraulic engineering, the waterway was also only a tiny ribbon of water on the landscape compared to the natural features such as the Pyrenees or the Garonne. The sheer amount of water flowing through the canal was impressive compared to most local rivers, and it was no small feat that the Canal du Midi climbed the continental divide, descending the Atlantic and Mediterranean watersheds on either side. Still, when it reached Sète, the canal was dwarfed and battered by the sea.

The confusing scale of the Canal du Midi was matched by its disconcerting technical character. Sometimes the waterway seemed just to flow on its own, curving oddly around hills and following ridges in unlikely ways, of course, but still appearing to move naturally in its channel. Much of the technical detail was subtle or hard to see—underpasses for streams, small settling ponds on water intakes, little sluices used to irrigate nearby fields, overflow drains that looked only like small openings at curves on the canal, and modest connections to small rivers. The dozen or so laundries scattered along the banks added to the confusion, too, because when idle, their function was not so easy to identify.

These modest features of the Canal du Midi were the opposite of the locks—particularly the staircase locks—which were large, noisy, and impressive. To fill and drain the basins quickly, water rushed in and out in quantity, sometimes echoing through the lock walls in hidden tambours. Mostly, it came crashing down with deafening noise from openings in the locks doors above to the great stones below them that held the gap intact. Currents formed around boats as the locks filled or emptied, so the vessels had to be carefully tethered to keep from hitting their hulls on the stones or battering the doors. Muscle, water, stone, wood, and iron all strained as topography and water levels were reconciled in cacophonous performances of hydraulic finesse.

The intelligence that held the Canal du Midi together was logistically brilliant in disconcerting ways. Often techniques were familiar peasant practices that seemed inconceivable once they were blown up to the scale of a navigational canal. A round lock near Agde was a case in point. Located where the Canal du Midi approached the Hérault River, the round basin was large enough that canal boats could turn around inside it to pass between the waterways through the lock. Since the river and canal were at different elevations, the lock was filled or emptied to shift barges to the desired level.

As a lock, the round structure was extraordinary, but it had some precedents in both Pyrenean towns and Roman aqueduct systems. It was similar in some ways to the settling pond at Cheust that captured and diverted river water laterally through sluice gates to different parts of town. The round lock was even more like a large, open-air *castellum* (or round town reservoir) usually set by the Romans at the high point of a town for dispersing water through the neighborhoods. The round lock by Agde was like a castellum simply made large enough to hold barges and designed to use changes in the water level to move boats from the canal to the river and vice versa. The result was not without precedent, but it was disconcertingly counterintuitive and impressive as a lock along the Canal du Midi.

In this case and others, the Canal du Midi turned the world of engineering upside down. Peasant practices (often with hidden Roman roots) were used to devise structures grand enough to stand for Louis XIV and the New Rome. This inversion gave new nobility to the weapons of the weak, but also turned them into tools of the king. The irony was masked by the anonymous intelligence of the waterway that dissociated techniques from the social worlds that sustained them. The result was a hydraulic sophistication that was startling because it seemed to come from nowhere and was not supposed to exist. Nevertheless, if in the 1660s, "to see all the uses one could make of water, one should see what they do in the Bigorre valley," in the 1680s, anyone hoping to understand hydraulics studied the Canal du Midi.[56] The lessons were surprisingly similar, but the effects at the Canal du Midi were grander because the techniques were finally assessed in relation to the dearth of formal knowledge of hydraulics.

A Working Canal

The Canal du Midi became a model of impersonal rule by working, and by demonstrating the power of working as an epistemic culture. As it was built, the waterway developed its own intellectual foundation—one that had not existed, and was not expected, entirely understood, or easily mastered. In general, the canal worked better than anyone expected, but at any moment, parts of it could fail as the seawall did at Sète. Working, then, seemed important because it was hard to achieve and sustain, and also because in the absence of clear authority, it provided a way to measure its intellectual efficacy. Cutting a navigational canal across southwest France was an experiment in engineering that integrated technē, praxis, and *epistemē*, simultaneously changing the world, and transforming the way the earth was known.[57] The Canal du Midi occupied and contributed to the knot of impersonal rule that joined politics, knowledge, morality, and things.

The epistemic culture of working, so opposed by Colbert when the canal began, was ironically itself part of the legacy of the ancients. Classical authors from the late empire, many of them of high rank, wrote tomes for estate owners on the logistics of farming, architecture, gardening, and wine making that were at once compendiums of knowledge and "how-to" books. These Roman authors identified the effectiveness of their methods as the measure of their knowledge, and as men of rank, their names also gave authority to their words.[58]

Working was more problematic in the seventeenth century, when logistical power was wielded mainly by those of lesser rank and had to be legitimated by gentlemen. In the 1660s, Nivelle could easily dismiss Riquet's plan for the canal into Toulouse because he was the client of the city's Capitouls, who had epistemic authority because of their social standing. The authority of gentlemen remained powerful in disputes over the Canal du Midi into the 1670s even as segments of the waterway started to function. But there was no point in arguing about the truth of things when the canal was finally filled from end to end and water was running where it was supposed to go. The Canal du Midi provided its own measure of efficacy—one that was both impersonal and incontrovertible in situ. The canal did not have to debate the adequacy of peasant techniques or formal measures as long as the latter failed and the former did not. Peasants who were "generally brutish, perfidious, cruel and . . . have no more reason than bears" might not have had the rank to authorize the truth, but they could make things work—even femelles.[59]

Working as an epistemic culture was dangerously modern and impersonal, but it still had purchase in the seventeenth century because, in stewardship terms, to know nature well enough to improve it was evidence of knowing God. Studying the earth was a sacred act, a bond with the Creator. Working revealed truth, demonstrating that men had come to understand God's works. God had made men in His image, capable of knowing His design. So the capacity for knowledge of nature was a gift from God that could not be denied, and was manifest in works.

Palissy, for one, saw how the equation between works of God and man informed his search for better techniques of working the land.[60] He was driven to write down his mesnagement ideas out of respect for God, conveying his (sacred) knowledge to other people. Working was a religious duty and measure of spiritual worth. "God did not create these things to leave them idle, therefore each performs its duty according to the commandment it received from God. . . . [T]he earth likewise is never idle."[61] Palissy placed works at the heart of stewardship politics, affirming the moral logic of working as a measure of truth, and Riquet employed this logic to explain his achievements on the Canal du Midi.

For Riquet, this epistemic culture suited both his stewardship principles and lack of social authority to know. As the entrepreneur for the Canal du Midi, he had become an experimentalist, ready to embrace risk to test ideas through works. He did not know whether Borel's plan to take a canal into the mountains would be effective, but it had promise, so he proposed it. He did not know whether anyone could bore a tunnel through Malpas large enough for a canal, but the idea seemed feasible, and it worked. He did not "know" engineering, nor could he ever acquire the knowledge he needed by studying it. He wanted to do impossible works, learning the qualities and limits of nature in the process.

The Canal du Midi, as a laboratory of experimentation, became productive of knowledge rather than its product, and working became its measure of success. This new epistemic culture detached knowledge from elites before their eyes and against their protests. What nature could do was impersonal, anonymous, and hard to resist. Experts, nameless doers and knowers—even femelles—could find truths by trying things out. Groups could know; things could judge them. It was a radical shift in power relations over knowledge.

The importance of the Canal du Midi now lay in its mute facticity; it was the judge of men. The waterway held sway over parts of life in Languedoc, not by telling local inhabitants of the region what to do or determining how they had to live. They simply had to adapt to the canal and its needs— even though how they did this was not precisely prescribed. The canal divided roads, so locals demanded bridges. When it reduced the flow of local rivers, they required reparations. When its walls broke and flooded towns, they called for help to make the needed repairs. They also pierced the walls to irrigate their fields, and used the water to wash their laundry. In the end, the Canal du Midi was both less under human control and more powerful than anyone had imagined because water had no strategy, but simply logistical methods of its own.

The Canal du Midi was ultimately a triumph of stewardship politics and territorial government, a tranquil waterway in a land of wild rivers and a miraculous piece of engineering connecting the two seas. The technical setbacks involved in building the canal through Languedoc were embarrassing to Colbert and Riquet, but they did not diminish the value of the project in stewardship terms. Toil for virtuous ends was a duty of good Christians as recompense for the sins of Adam and Eve. Struggling to thread a ribbon of water through Languedoc fit this cultural topos, and Riquet exploited it, calling his problems with the canal forms of mortification, moral lessons in humility. Colbert may have balked at the entrepreneur's hubris in speaking of his spiritual life this way, but Riquet's logic fit the reasoning that the minister had cultivated for empowering his government. The Canal du Midi merely gave stewardship discourse new efficacy as a linguistic operator of power by demonstrating the ineffable potency of natural knowledge put to logistical use.

Depersonalization, Authorship, and Genealogies of Greatness

Given its role in depersonalizing power in seventeenth-century France, it seems more than ironic that the Canal du Midi was hailed at its completion as a work of personal genius. There was good reason, though. The genealogy of greatness nurtured to describe the canal and its "author" stabilized the canal on paper as well as in the political imagination. Colbert understood this clearly, quickly making Riquet after his death author of the enterprise and a good servant to the monarch. Just a little ideological work could turn the canal into what Louis XIV expected: a tribute to (his) personal rule.[62] The elevation of Riquet stripped the canal of his collaborators—the peasants, artisans, subcontractors, and supervisors who dug the ditches, built the walls, applied the specifications, and did the books. The waterway became a model of personal rule and human accomplishment, the brainchild of a genius and a tribute to the great monarch who he served.

The legacy of the New Rome was purified to elevate French standing. The Canal du Midi was cleansed of polluting elements—dirty labor, collaborative thought, shifting sands, communities of practice, tax farming, and even unruly waters. The canal was reduced to the human mind instead. It was genius incarnate. The ancient sources of technical practices were not just forgotten but also washed away in a process of simultaneous personalization (authorship) and depersonalization (genius) that replaced New Romans with symbolic evocations of Rome as a distant Olympus, not a living tradition. This political repair work did what Colbert wanted. It stabilized the canal as a fact of life in Languedoc too important to disturb. It affirmed the political value of the "muddy scar" by making it less muddy.

Gone from the mythical New Rome was the untrustworthy financier from Languedoc who Colbert loathed. Riquet emerged as a new man. The entrepreneur's character was washed clean of all defects. His engineering ability was polished to a shine. His visionary nature was now an asset. He was no longer a dreamer who could not recognize his limits but rather a man of clear sight and insight.[63] This mythical Riquet was even given epistemic credibility based on his works, named as author of the enterprise, and defined as a man of natural virtue—just what the historical Riquet had wanted.

Making France the New Rome was a strategic program of collective memory and political genealogy, used to assert France's primacy over Italy as heir to the ancients.[64] As the hope of displacing the Hapsburg Empire faded, France still pursued its neoclassical retrofitting. Rome itself remained the central pilgrimage site for humanists in the period, but travelers interested in classical culture also sometimes came to southern France and could be lured to the Canal du Midi. There visitors would see logistical brilliance, the engineering on a classical scale that stood for France's genius now. The political deployment of the Canal du Midi did not make France an empire, even though the project made the state more powerful in Languedoc, but it did help focus European eyes on the court of Louis XIV and make the reign seem more splendid.

The waterway in Languedoc lived in poetry as well as on the ground. Fiction and fact were blended around it in more ways than even Nivelle feared. The two seas were "married" by the French king. The noble families of Languedoc were joined in celebration of the wonder. Everything that did not and could not happen while the canal was being built could now be symbolically affirmed. This was the power of empire that the New Rome supplied.

Ironically, the hidden story of the Canal du Midi was even more fantastic than anything the propagandists fabricated. It was an unbelievable tale of a landscape in the southwest of France that had served as a memory palace for classical thought, preserving ancient techniques over centuries. When Plato, Cicero, and Aristotle spoke of memory palaces, they spoke of (fictional or real) rooms where orators could "place" different parts of a speech they wanted to memorize. The orator would then simply "walk" through the rooms in his or her imagination to remember all the points that needed to be covered.[65]

The countryside of Languedoc was such a "palace" of classical memories—a space where bits of Roman engineering had been physically dispersed and located in the land where they made sense. Some "rooms" were towns with ruins of ancient buildings, aqueducts, bridges and roads that provided templates of classical construction methods reproduced by artisans for new structures. Others were villages in the mountains where ancient hydraulic techniques were used to improve fields, mills, laundries, and domestic water systems. One room was at Montady where the Visigoths had drained a lake by tunneling through a mountain, leaving on the ground a vivid demonstration of large-scale hydraulic methods and a fertile valley for cultivation that continued to be used.[66]

The geographic dispersion of these memories in the countryside helped make them robust by breaking down complex information into smaller bits and employing the landscape itself as a reservoir of collective memory. No ancient orator placed evidence of Roman engineering strategically around Languedoc, but bits of ancient technique were nonetheless there to walk through. It was easier to remember Roman construction methods in the sites where they made local sense. An old aqueduct that tapped a high source could be turned to new uses, providing a water supply for a mill or a new laundry. A Roman-style sluice gate could be used as a clean-out valve for a domestic water supply. These techniques could be reproduced over time because the logistical methods of Roman life were not only cataloged in books by Roman gentlemen but also embedded in the countryside by slaves and soldiers for peasants to inherit.

Small and locally useful, shards of ancient knowledge stayed intact where they were picked up. The wisdom nurtured this way—particularly in mountain towns—remained unnoticed by outsiders, and lost its provenance. But it was there. Until a tax farmer who traveled through this memory palace—himself unable to "read" the truths tucked into the landscape— thought of cutting a canal across Languedoc. He hired those who knew secrets of the memory palace to do the work. They brought their traditions to the project, and reinvigorated, integrated, and extended them, becoming New Romans and inventive engineers.

The Canal du Midi, in Colbert's narrative, became a monument to greatness, the emblem of the New Rome, and a triumph of the king. Authors in books on engineering continued to locate it in a genealogy of brilliance from ancient Gaul into the ancien régime. The canonical story of Riquet's native genius was flexible and inspiring, useful for elevating French engineering over the centuries. Tales of the entrepreneur's martyred life added to the dignity of a profession highly regarded in France. And after the French Revolution, Riquet was depicted as the innocent victim of noble hostility and arbitrary power, standing for the generations of Frenchmen who had been underestimated by men of rank.[67] The price of elaborating this heritage, of course, was to deny other ways of remembering both ancient engineers and the canal. To make Rome represent the highest levels of political and cultural achievement, the classical past had to stay in literate circles and define authority.

The authorship of the canal could be mythologized, but the Canal du Midi itself still remained a brute fact in the countryside, a place to do laundry or carry the mail. It was something to work with and work around like a mountain, not something to debate or query about its history. This kind of logistical power—so much like the power of the Creation itself—seemed godlike in its anonymous effectiveness or unholy in its failings. It appeared

Figure 9.2
Canal bridge with circle reflection. Photo by the author.

beyond anything normal people could imagine or build. It demonstrated how unpredictably powerful a state could be. It suggested, too, what being French might mean or what moderns might do if they became Roman.

The Canal du Midi also quietly became a new and more compact memory palace for Languedoc, a place full of forgotten knowledge and a harbor for unimaginable stories—like those of peasant women coming down from the Pyrenees, teaching Roman hydraulics to gentlemen to join the two seas.

NOTES

CHAPTER 1
Impossible Engineering

1 L.T.C. Rolt, *From Sea to Sea: The Canal du Midi* (London: Allen Lane, 1973), 4. For the basic history of the canal, see also André Maistre, *Le Canal des Deux-Mers: Canal Royal du Languedoc, 1666–1810* (Toulouse: E. Privat, 1998); Jean-Denis Bergasse, ed., *Le Canal du Midi*, 4 vols. (Cessenon: J.-D. Bergasse, 1982–1985).

2 Perry Anderson, *Lineages of the Absolutist State* (London: New Left Books, 1974); John Brewer, *The Sinews of Power: War, Money, and the English State, 1688–1783* (New York: Knopf, 1989); Norbert Elias, *The Civilizing Process* (1939; repr., New York: Urizen Books, 1978); Norbert Elias, *The Court Society* (New York: Pantheon, 1998); Charles Tilly, *Coercion, Capital, and European States, 1000–1999* (Oxford: Blackwell, 1990).

3 Philip Gorski correctly argues the importance of religion to state formation, but he associates it exclusively with social discipline and governmental practices, overlooking the importance of conceptions of dominion to political territoriality. See Philip S. Gorski, *The Disciplinary Revolution: Calvinism and the Rise of the State in Early Modern Europe* (Chicago: University of Chicago Press, 2003). Compare with Chandra Mukerji, "Material Practices of Domination and Techniques of Western Power," *Theory and Society* 31 (2002): 1–31; Chandra Mukerji, "Intelligent Uses of Engineering and the Legitimacy of State Power," *Technology and Culture* 44, no. 4 (2003): 655–76.

4 Alice Stroup, *A Company of Scientists: Botany, Patronage, and Community at the Seventeeth-century Parisian Royal Academy of Sciences* (Berkeley: University of California Press, 1990), chapter 1; Henri Enjalbert, "Les Hardinesses de Riquet: Données Géomorphologiques de la Région que Traverse le Canal du Midi," in *Le Canal du Midi*, ed. Jean-Denis Bergasse (Cessenon: J.-D. Bergasse, 1985), 4:127–42, particularly 130.

5 Giorgius Agricola, *De re Metallica; Translated from the First Latin Edition of 1556, with Biographical Introduction, Annotation, and Appendices upon the Development of Mining, Metallurgical Processes, Geology, Mineralogy, and Mining Law from the Earliest Times to the 16th Century*, trans. and ed. Herbert Clark Hoover and Lou Henry Hoover (New York: Dover, 1950); Joseph de Lalande, *Des canaux de navigation, et spécialement du canal de Languedoc* (Paris: Veuve Desaint, 1778), 26–29; Bernard Forest de Belidor, "De l'Usage des Eaux a la Guerre," in *Architecture Hydraulique seconde partie qui comprend l'Art de diriger les eaux de la Mer & des Rivieres à l'avantage de la défense des Places, du Commerce & de l'Agriculture. par M. Belidor, Colonel d'Infanterie* (Paris: Jombert, 1753), vol. 2.

6 Agricola, *De re Metallica*; Lalande, *Des canaux de navigation*, 26–29; Belidor, "De l'Usage des Eaux a la Guerre," vol. 2.

7 On Dutch flooding, see Belidor, "De l'Usage des Eaux a la Guerre," 2:233, 242–43; on Dutch locks, see ibid., 2: 325–28; on canals prior to the Canal du Midi, see ibid., 4:357–59; Rolt, *From Sea to Sea*, 14–27.

8 Belidor, "De l'Usage des Eaux a la Guerre," 4:357–59.

9 Pierre Pinsseau, *Le Canal Henri IV, ou Canal de Briare, 1604–1943* (Orléans: R. Houzé, 1944).

10 For a description of tax farming, the contracts involved, and Colbert's changes in the procedures for assigning tax farms, see Arthur J. Sargent, *The Economic Policy of Colbert* (London: Longmans, Green, and Co., 1899), 26–29.

11 Bertrand Gabolde, "Riquet à Versailles vu par le conteur Charles Perrault," in *Le Canal du Midi*, ed. Jean-Denis Bergasse (Cessenon: J.-D. Bergasse, 1982), 1:185–89.

12 Inès Murat, "Les rapports de Colbert et de Riquet: Méfiance pour un homme ou pour un système? in *Le Canal du Midi*, ed. Jean-Denis Bergasse (Cessenon: J.-D. Bergasse, 1984), 3:105–21; Pierre Burlats-Brun and Jean-Denis Bergasse, "L'oligarchie gabelière, soutien financier de Riquet," in *Le Canal du Midi*, ed. Jean-Denis Bergasse (Cessenon: J.-D. Bergasse, 1984), 3:123–40; G. Bernet, "Riquet intime dans ses résidences de Revel, Bonrepos, et Toulouse avant 1668," in *Le Canal du Midi*, ed. Jean-Denis Bergasse (Cessenon: J.-D. Bergasse, 1984), 3:50–62, particularly 52. For a description of the contracts for suppliers of the army, see John A. Lynn, *Giant of the Grand Siècle: The French Army, 1610–1715* (Cambridge: Cambridge University Press, 1997), 114–18.

13 For the technical problems along with the groups that could or did solve them, see François Gazelle, "Riquet et les eaux de la Montagne Noire: L'idée géniale de l' alimentation du canal," in *Le Canal du Midi*, ed. Jean-Denis Bergasse (Cessenon: J.-D. Bergasse, 1985), 4:143–67. For a discussion of the amount of geographic learning that went into the project and, more important, was derived from it, see Robert Ferras, "Le Canal Royal et ses cartes: des images d'une province à la connaissance d'un espace géographique," in *Le Canal du Midi*, ed. Jean-Denis Bergasse (Cessenon: J.-D. Bergasse, 1983), 2:271–308, particularly 273–74. And for the work of the commission that verified the project, see L. Malavialle, "Une Excursion dans la Montagne Noire," *Bulletin de la Société Languedocienne de Géographie* 4 (1891): 240–91, 5 (1892): 112–49, 283–530.

14 Murat, "Les rapports de Colbert et de Riquet"; Rolt, *From Sea to Sea*, 42–47, 76; ACM 11-07 on *rigole d'essai* and experts; Chandra Mukerji, "French *mesnagement* Politics and the Canal du Midi: Labor and Economy in Seventeenth-century France," in *The Mindful Hand: Inquiry and Invention from the Late Renaissance to Industrialization*, ed. Lissa Roberts, Simon Schaffer, and Peter Dear (Amsterdam: Edita—The Publishing House of the Royal Netherlands Academy of Arts and Sciences and University of Chicago Press, 2007), 169–88.

15 Compare this to the stewardship language used by the British for colonial expansion. See Richard Drayton, *Nature's Government: Science, Imperial Britain, and the "Improvement" of the World* (New Haven, CT: Yale University Press, 2000). For consideration of language and materiality in history, see Michel de Certeau, *The Practice of Everyday Life* (Berkeley: University of California Press, 1988), 22–28; Michel de Certeau, *The Mystic Fable* (Berkeley: University of Chicago Press, 1992), 126–29. And for a discussion of language in historical sociology and particularly for practice theory, see William H. Sewell Jr., "Language and Practice in Cultural History: Backing Away from the Edge of the Cliff," *French Historical Studies* 21, no. 2 (1998): 241–54.

16 See Alex Preda, "Where Do Analysts Come from? The Case of Financial Chartism," in *Market Devices*, ed. Michel Callon, Fabian Muniesa, and Yullo Millo
(Chicago: University of Chicago Press, 2007), chapter 2.

17 Belidor, quoted in Charles Vallancey, *A Treatise on Inland Navigation or the
Art of Making Rivers Navigable: Of Making Canals in All Sorts of Soils, and of
Constructing Locks and Sluices* (Dublin: George and Alexander Ewing, 1763),
100 (translation by Vallancey).

18 W. W. Powell and Paul DiMaggio, *The New Institutionalism in Organizational
Analysis* (Chicago: University of Chicago Press, 1991).

19 Charles Perrault, *Parallèle des anciens et des modernes en ce qui regarde les arts
et les sciences* (Munich: Eidos Verlag, 1964).

20 P. N. Miller, *Peiresc's Europe: Learning and Virtue in the Seventeenth Century*
(New Haven, CT: Yale University Press, 2000).

21 Peter Burke, *The Fabrication of Louis XIV* (New Haven, CT: Yale University
Press, 1992).

22 The roots for the word technocrat are "techne" and "kratos," meaning skill and
power—the power of skill. For some discussion of technocratic governance
and its distopian effects, see James Scott, *Seeing Like a State* (New Haven, CT:
Yale University Press, 1998). For an analysis of differences in the technologies
of power and the value of empirically studying technologies of power, see
Nikolas Rose, *Powers of Freedom: Reframing Political Thought* (Cambridge:
Cambridge University Press, 1999), 51–60.

23 See, for example, William Beik, *Absolutism and Society in Seventeenth-century
France: State Power and Provincial Aristocracy in Languedoc* (Cambridge:
Cambridge University Press, 1985); Riquet à Colbert 27-03-70, ACM 23-16; Jean
Bodin, *Six Books of the Commonwealth* (New York: Barnes and Noble, 1967).

24 Beik, *Absolutism and Society*, 15–16. The patronage system was grounded in both
the legal writings about monarchical authority, particularly the precepts of Bodin, and the traditional exercise of legal powers by nobles, using Roman legal
precedents. See Bodin, *Six Books*; Roland Mousnier, *The Institutions of France
under the Absolute Monarchy, 1598–1789: Society and the State*, trans. Brian
Pearce (Chicago: University of Chicago Press, 1979), 645–720. For the role of religion in this dual legal system, see Gorski, *The Disciplinary Revolution*, 142–44.
The lack of clear coherence between the two systems of law helped to make this
system fragile, and made patrimonialism and personal rule more important.
See Sharon Kettering, *Patrons, Brokers, and Clients in Seventeenth-century
France* (New York: Oxford University Press, 1986); Julia Adams, "The Rule of
the Father: Patriarchy and Patrimonialism in Early Modern Europe," in *Max
Weber's Economy and Society: A Critical Companion*, ed. Charles Camic, Philip
Gorski, and David Trubek (Stanford, CA: Stanford University Press, 2005). The
techniques of impersonal rule developed in the seventeenth century constituted
tools of political power outside the patrimonial system that disproportionately
favored government, and served the growth of a French state more independent
of noble political networks.

25 Mukerji, "French *mesnagement* Politics"; Gorski, *The Disciplinary Revolution*,
166–67; George Lawson, *Politica Sacra et Civilis*, ed. Conal Condren (1678;
repr., Cambridge: Cambridge University Press, 1992), 68–74; David Hume, *Political Essays*, ed. Knud Haakonssen, (Cambridge: Cambridge University Press,
1994), 94–101.

26 Beik, *Absolutism and Society*; Kettering, *Patrons, Brokers, and Clients.*

27 Charles Tilly, *The Contentious French* (Cambridge, MA: Harvard University Press, 1986). While Tilly emphasizes popular politics in France, studying the clashes between patronage and stewardship politics illustrates the extent to which elite politics were contentious for reasons explained by Niccoló Machiavelli and Beik.

28 Rolt, *From Sea to Sea*, 24–26; Arret de 1666, ACM 3-10; Lettre de Riquet à Colbert, 15 Novembre 1662, ACM 20-2.

29 Jean-Denis Bergasse, "Le Cult de Riquet en Languedoc au XIXe Siècle," in *Le Canal du Midi*, ed. Jean-Denis Bergasse (Cessenon: J.-D. Bergasse, 1982), 1:217–51. For the role of the hero in hagiography, see Michel de Certeau, *The Writing of History*, trans. Tom Conley (New York: Columbia University Press, 1988), 276–77.

30 Lettre de Riquet à Colbert 27-03-70, ACM 23-16; Lalande, introduction to *Des canaux de navigation.*

31 Chandra Mukerji, "Women Engineers and the Culture of the Pyrenees: Indigenous Knowledge and Engineering in 17th-century France," in *Knowledge and Its Making in Europe, 1500–1800*, ed. Pamela Smith and Benjamin Schmidt (Chicago: University of Chicago Press, 2008), 19–44.

32 Compare to Chandra Mukerji, "The Collective Construction of Scientific Genius," in *Cognition and Communication at Work*, ed. Yrjö Engeström and David Middleton (Cambridge: Cambridge University Press, 1997), 257–78.

33 Mukerji, "Women Engineers."

34 Edwin Hutchins, *Cognition in the Wild* (Cambridge, MA: MIT Press, 1995).

35 For a different approach to social cognition, one organized around philosophical debates about the social study of science, see Helen Longino, *The Fate of Knowledge* (Princeton, NJ: Princeton University Press, 2002). And for the significance of cognition for the sociology of culture, see Paul Dimaggio, "Culture and Cognition." Annual Review of Sociology 23 (1997) 263–88.

36 Edwin Hutchins and Tove Klausen, "Distributed Cognition in an Airline Cockpit," in *Cognition and Communication at Work*, ed. Yrjö Engeström and David Middleton (Cambridge: Cambridge University Press, 1996), 15–34.

37 Karin Knorr-Cetina, *Epistemic Cultures: How the Sciences Make Knowledge* (Cambridge, MA: Harvard University Press, 1999).

38 David Lewis, *Counterfactuals* (Malden, MA: Blackwell, 1973), as explained in Nancy Cartwright, "Evidence-Based Policy: What's Evidence" (lecture, Colloquium in Science Studies, University of California at San Diego, February 11, 2008).

39 Michael Cole, *Cultural Psychology: A Once and Future Discipline* (Cambridge, MA: Harvard University Press, 1996); Michael Cole and Sylvia Scribner, *Culture and Thought* (New York: Wiley, 1974).

40 Michael Cole and James Wertsch, *Contemporary Implications of Vygotsky and Luria* (Worcester, MA: Clark University Press, 1996).

41 A. Luria, *The Making of Mind*, ed. Michael Cole and Sheila Cole (Cambridge, MA: Harvard University Press, 1979); Cole, *Cultural Psychology*; Cole and Scribner, *Culture and Thought.*

42 H. Gallet de Santerre, "L'empreinte Romaine," in *Histoire du Languedoc*, ed. Philippe Wolff (Paris: Privat, 2000); Julie Roux, *Les Chemins de Saint-Jacques de Compostelle* (Vic-en-Bigorre: MSM, 1999).

43 Murat, "Les rapports de Colbert et de Riquet"; Rolt, *From Sea to Sea*, 83–86.

44 See, for example, Riquet à Colbert 27-03-70, ACM 23-16.

45 Burke, *The Fabrication of Louis XIV*; Chandra Mukerji, "Unspoken Assumptions: Voice and Absolutism in the Court of Louis XIV," *Journal of Historical Sociology* 2 (1998): 228–315.

46 Donna Haraway, *Primate Visions: Gender, Race, and Nature in the World of Modern Science* (New York: Routledge, 1989).

47 Bruno Latour, *We Have Never Been Modern* (Cambridge, MA: Harvard University Press, 1993). For the local data, see Carolyn Lee and Mary Walshok, *A County Level Analysis of California's R&D Activity, 1993–1999* (San Diego: University of California at San Diego Extension, 2002).

48 The patterns of complexity in forms of life on the Mediterranean side of Languedoc are described in Fernand Braudel, *The Mediterranean and the Mediterranean World in the Age of Philip II* (New York: Harper and Row, 1975). For the complexity of French culture and identity more generally, see Fernand Braudel, *L'identité de la France* (Paris: Arthaud, 1986); Tilly, *The Contentious French*. For the political character of the province, see Beik, *Absolutism and Society*. For Pyrenean culture in this period, see Roux, *Les Chemins de Saint-Jacques de Compostelle*; Isaure Gratacos, *Femmes pyrénéennes, un statut social exceptionel en Europe* (Toulouse: Éditions Privat, 1987), 105–15; Louis de Froidour, *Les Pyrenees Centrales au XVIIe Siècle. Lettres ecrites par M. de Froidour . . . a M. de Haericourt . . . et A M. de Medon . . . publiées avec des notes par Paul de Casteran* (Auch: G. Foix, 1899), 96–97; Peter Sahlins, Boundaries: *The Making of France and Spain in the Pyrenees* (Berkeley: University of California Press, 1989).

49 Sahlins, *The Making of France and Spain*; Beik, *Absolutism and Society*.

CHAPTER 2
Territorial Politics

1 Louis de Saint-Simon, *The Age of Magnificence*, ed. Ted Morgan (New York: Paragon, 1990), 248.

2 In Michael Mann's theory of history, the nation-state as a social system does not exist in this period because the dynamics of power do not cohere in particular places and people. See Michael Mann, *The Sources of Social Power*, vol. 1, *A History of Power from the Beginning to A.D. 1760* (Cambridge: Cambridge University Press, 1986). This lack of clear coherence explains concretely the importance of territoriality to politics in this period, and Colbert's use of stewardship logics to try to constrain the exercise of power through economic, military, and political networks. Territorial politics nominally served the existing social infrastructures, but contradictions between patrimonial politics and mesnagement-based territoriality spurred social change. For an understanding of precedents for state territoriality in the church's use of territorial power, see Philip Gorski, *The Disciplinary Revolution: Calvinism and the Rise of the State in Early Modern Europe* (Chicago: University of Chicago Press, 2003), particularly 17–18. For a comparative case of territorial management and sociopolitical order in Bali, see John Lansing, *Priests and Programmers: Technologies of Power in the Engineered Landscape of Bali* (Princeton, NJ: Princeton University Press, 1991).

3 Bennett Berger, *The Survival of a Counterculture* (Berkeley: University of California Press, 1981).

4 The centrality of performance to politics and the importance of this connection for cultural sociology is illustrated nicely in Jeffrey Alexander, *The Meanings of Social Life* (Oxford: Oxford University Press, 2003); Robin Wagner-Pacifici, *The Art of Surrender: Decomposing Sovereignty at Conflict's End* (Chicago: University of Chicago Press, 2005); Lyn Spillman, *Nation and Commemoration: Creating National Identities in the United States and Australia* (New York: Cambridge University Press, 1997). For material demonstrations as part of performative political culture in France, see Chandra Mukerji, *Territorial Ambitions and the Gardens of Versailles* (Cambridge: Cambridge University Press, 1997); Jean-Marie Apostolidès, *Le Roi-Machine: spectacle et politique au temps de Louis XIV.* (Paris: Éditions de Minuit, 1981).

5 André Corvisier, "Colbert et la guerre," in *Colloque pour le tricentenaire de la mort de Colbert, sous la direction de Roland Mousnier, Un nouveau Colbert* (Paris: Éditions Sedes, 1985), 287–308, particularly 288–92.

6 Elizabeth Clark, *History, Theory, Text: Historians and the Linguistic Turn* (Cambridge, MA: Harvard University Press, 2004); Wagner-Pacifici, *The Art of Surrender*; Spillman, *Nation and Commemoration*; William H. Sewell Jr., *Logics of History: Social Theory and Social Transformation* (Chicago: University of Chicago Press, 2005), particularly 152–224; and Sewell Jr., "The Political Unconscious of Social and Cultural History, or, Confessions of a Former Quantitative Historian," in *The Politics of Methods in the Human Sciences: Positivism and Its Epistemological Others*, ed. George Steinmetz (Durham, NC: Duke University Press, 2005), 173–206.

7 Lettre de Riquet à Colbert à Bonrepos, le 15 Novembre 1662, ACM 20-2. Because Colbert refers to Anglure de Bourlemont, La Feuille, and Anguesseau as d'Anglure de Bourlemont, de la Feuille, and d'Aguggesseau, I use that convention in this book to stay true to period usage.

8 For the importance of high-minded conversation to intellectual authority in France, see Mary Vidal, *Watteau's Painted Conversations: Art, Literature, and Talk in 17th- and 18th-century France* (New Haven, CT: Yale University Press, 1992). For the enduring importance of high-minded conversation as a display of intellectual credibility, see Michèle Lamont, *How Professors Think: Inside the Curious World of Academic Judgment* (Cambridge, MA: Harvard University Press, 2009).

9 William Beik, *Absolutism and Society in Seventeenth-century France: State Power and Provincial Aristocracy in Languedoc* (Cambridge: Cambridge University Press, 1985), 15–16, 223–28. For this the historical formation of clientalism in this period, see Sharon Kettering, "The Historical Development of Political Clientalism," *Journal of Interdisciplinary History* 18 (1988): 419–47.

10 Beik, *Absolutism and Society*, 297–301. For Colbert's use of patronage networks, particularly his use of the clergy to solidify them in Languedoc, see Sharon Kettering, *Patrons, Brokers, and Clients in Seventeenth-century France* (New York: Oxford University Press, 1986), 167–69.

11 For stewardship and Colbert's administration even at its inception, see Chandra Mukerji, "The Great Forest Survey of 1669–1671," *Social Studies of Science* 37, no. 2 (2007): 227–53.

12 P. Burlats-Brun and Jean-Denis Bergasse, "L'Oligarchie gabelière, soutien financier de Riquet," in *Le Canal du Midi*, ed. Jean-Denis Bergasse (Cessenon: J.-D. Bergasse, 1984), 3:123–41.

13 Julian Dent, *Crisis in Finance: Crown Financiers and Society in Seventeenth-century France* (New York: St. Martin's Press, 1973).

14 Kettering, *Patrons, Brokers, and Clients*, 40–67.

15 Lettre de Riquet à Colbert à Bonrepos, le 15 Novembre 1662, ACM 20-2.

16 This work exemplified the kind of struggles involved in giving cultural coherence to political regimes. Sewell, "Political Unconscious of Social and Cultural History," 172–74.

17 Lettre de Riquet à Colbert à Bonrepos, le 15 Novembre 1662, ACM 20-2.

18 Ibid.

19 Kettering, *Patrons, Brokers, and Clients*, 22–26, 167–75, 208–18, 234–37; Mukerji, "The Great Forest Survey."

20 Niccolò Machiavelli, *The Prince*, ed. and trans. D. Donno (New York: Bantam Dell, 1966), 26–27.

21 Kettering, *Patrons, Brokers, and Clients*, 141–83.

22 Ibid., 86–97.

23 Machiavelli, *The Prince*, 26–27.

24 Kettering, *Patrons, Brokers, and Clients*, 190–91; Mukerji, *Territorial Ambitions*, 65–66.

25 Michel de Certeau writes interestingly about the contingent thinking of Machiavelli—his thinking "as if" he were a king. This was the fine art of ministers, thinking like kings, but avoiding any display of monarchical power. Michel de Certeau, *The Writing of History*, trans. Tom Conley (New York: Columbia University Press, 1988), 8–11.

26 Machiavelli, *The Prince*, 44–45.

27 Ibid., 44–46.

28 Mack P. Holt, *The French Wars of Religion, 1562–1629* (Cambridge: Cambridge University Press, 2005), 99–101; Mukerji, *Territorial Ambitions*, 66.

29 Charles W. Cole, *Colbert and a Century of French Mercantilism* (Hamden, CT: Archon Books, 1964); Perry Anderson, *Lineages of the Absolutist State* (London: New Left Books, 1974); Dent, *Crisis in Finance*; David Parker, *The Making of French Absolutism* (London: Edward Arnold, 1983); Beik, *Absolutism and Society*; Roger Mettam, *Power and Faction in Louis XIV's France* (London: Basil Blackwell, 1988).

30 See Samuel Clark, *State and Status: The Rise of the State and Aristocratic Power in Western Europe* (Cardiff: University of Wales Press, 1995), 49–52. For older literature stressing the effectiveness of Colbert as a bureaucrat and celebrating his reforms, see, for example, H. A. de Colyar, "Jean-Baptiste Colbert and the Codifying Ordinances of Louis XIV," *Journal of the Society of Comparative Legislation*, new series 13, no. 1 (1912): 56–86.

31 Mettam, *Power and Faction*, chapter 5.

32 Beik, *Absolutism and Society*; J. A. Lynn, *Giant of the Grand Siècle: The French Army, 1610–1715* (Cambridge: Cambridge University Press, 1997).

33 Mukerji, *Territorial Ambitions*, 116–18. For the long-term development of a parallel set of offices, see Philippe Minard, *La fortune du colbertisme: État et industrie dans la France des Lumières* (Paris: Fayard, 1998), 15–74. For courtesy, favors, and court society, see Norbert Elias, *The Court Society* (New York:

Pantheon, 1998). For the *honnête homme* in court society and the *politesse* of patron-client ties, see Mukerji, *Territorial Ambitions*, 222–23.

34 Stewardship reasoning was being used in other parts of Europe as well, importantly used as a justification for colonial expansion. See Richard Drayton, *Nature's Government: Science, Imperial Britain, and the "Improvement" of the World* (New Haven, CT: Yale University Press, 2000).

35 Mukerji, "The Great Forest Survey." For the other side of territoriality, its relationship to sovereignty, see Stephen D. Krasner, *Sovereignty* (Princeton, NJ: Princeton University Press, 1999), particularly 20–21.

36 Henry Heller, *Labour, Science, and Technology in France, 1500–1620* (Cambridge: Cambridge University Press, 1996), 57–96.

37 For a sense of the violence and politics of the period, see Frederic Baumgartner, *France in the Sixteenth Century* (New York: St. Martin's Press, 1995), 211–37. Humanism in southern France offered new ideas about tolerance, foregrounding human efficacy and creativity in the face of religious strife. See P. N. Miller, *Peiresc's Europe: Learning and Virtue in the Seventeenth Century* (New Haven, CT: Yale University Press, 2000).

38 Anne Blanchard, *Les ingénieurs du roy de Louis XIV à Louis XVI: Étude du corps des fortifications* (Montpellier: Université Paul-Valéry, 1979), 50–55.

39 L.J.M. Columella, *Columella on Agriculture 1-Iv*, ed. and trans. Harrison Boyd Ash (1941; repr., Cambridge, MA: Harvard University Press, 2001). From a 1794 edition, see E. Heitland, "Agriculture," in *The Legacy of Rome*, ed. Cyril Bailey (1923; repr., Oxford: Clarendon Press, 1940), 475–512, particularly 501–7, and also Pamela Long, *Openness, Secrecy and Authorship: Technical Arts and the Culture of Knowledge from Antiquity to the Renaissance* (Baltimore: Johns Hopkins University Press, 2001), 29-45.

40 Charles Estienne and Jean Liebault, *L'agriculture et la maison rustique de Charles Estienne . . . ; paracheuee premierement, puis augmentee par M. Iean Liebaut* (Paris: Chez Iacques Du-Puys, 1570). An English translation appeared shortly after: Charles Estienne and Jean Liebault, *La Maison Rustique of the Country Farme, Compiled in the French Tongue by Charles Stevens and Jean Liebault, Doctors of Physicke, and Translated into English by Richard Surflet, Practitioner in Physicke* (London: Arnold Hatsfield for John Norton and John Bill, 1606).

41 François de Dainville, *Géographie des humanistes* (Geneva: Slatkin Reprints, 1969), 80, 85, 88–93. For a sense of his science and its practice, see also Marguarite Boudon-Duaner, *Bernard Palissy: le potier du roi* (Carrières-sous-Poissy: La Cause, 1989), 44–48.

42 Bernard Palissy, *A Delectable Garden, by Bernard Palissy*, trans. and ed. Helen Morganthau Fox (Peekskill, NY: Watch Hill Press, 1931), 2.

43 Ibid., xxv–xxvi.

44 Bernard Palissy, *Recepte véritable* (Geneva: Froz, 1988); Bernard Rivet, "Aspects Économiques de l'Oeuvre de B. Palissy" et Frank Lestringant, "L'Eden et les Ténèbres Extérieures," in *Bernard Palissy, 1510–1590, l'écrivain, le réforme, le céramiste*, by Frank Lestringant (Saint-Pierre-du-Mont: Coédition Assocation Internationale des Amis d'Agrippa d'Aubigné- Éditions SPEC, 1990), 167–80. For a discussion of techné in this period, see James Scott, *Seeing Like a State* (New Haven, CT: Yale University Press, 1998; Pamela Smith, *The Body of the Artisan* (Chicago: University of Chicago Press, 2004). See also Palissy, *A Delectable Garden*, 7.

45 Henry Morely, *Palissy the Potter: The Life of Bernard Palissy, of Saintes, His Labours and Discoveries in Art and Science* (London: Chapman and Hall, 1852), 2:241–42.

46 Preface letter to Henri I de Montmorency, quoted in ibid., 2:337.

47 Olivier de Serres, *Théâtre d'agriculture et mesnages des champs* (Geneva: Mat Hiev Berjon, 1611), 217; Olivier de Serres, *The Perfect Vse of Silk-wormes* (New York: Da Capo, 1971); Chandra Mukerji, "The Political Mobilization of Nature in French Formal Gardens," *Theory and Society* 23 (1994): 651–77.

48 Serres, *Théâtre d'agriculture*; William Cronon, *Nature's Metropolis* (New York: W. W. Norton, 1991).

49 Serres, *Théâtre d'agriculture*; Yvette Quenot, "Bernard Palissy et Oliver de Serres," in Bernard Palissy, 1510–1590, l'écrivain, le réforme, le céramiste, by Frank Lestringant (Saint-Pierre-du-Mont: Coédition Association Internationale des Amis d'Agrippa d'Aubigné- Éditions SPEC, 1990), 93–103.

50 Serres, *Théâtre d'agriculture*, 28–29.

51 Ibid.; Morely, *Palissy the Potter*; Palissy, *A Delectable Garden*.

52 Roland Mousnier, *The Institutions of France under the Absolute Monarchy, 1598–1789*, 17.

53 Mukerji, The Great Forest Survey."

54 Inès Murat, *Colbert*, trans. Robert Francis Cook and Jeannie Van Asselt (Charlottesville: University Press of Virginia, 1984), chapter 3; Orest Ranum, *Paris in the Age of Absolutism: An Essay* (Bloomington: Indiana University Press, 1979), 259–72; Jean Meyer, "Louis XIV et Colbert: les relations entre un roi et un ministre au XVIIe siècle," in *Colloque pour le tricentenaire de la mort de Colbert, sous la direction de Roland Mousnier, Un nouveau Colbert*, ed. Jean Favier and Roland Mousnier (Paris: Éditions Sedes, 1985), 71–84, particularly 71–75.

55 Machiavelli, *The Prince*, 45.

56 Compare to British views of officeholders. See Conel Condren, "Liberty of Office and Its Defence in Seventeenth-century Political Argument," *History of Thought* 18, no. 3 (1997): 460–82.

57 Charles Engrand, "Clients du roi. Colbert et l'Etat (1661–1715)," in *Colloque pour le tricentenaire de la mort de Colbert, sous la direction de Roland Mousnier, Un nouveau Colbert*, edited by Jean Favier and Roland Mousnier (Paris: Éditions Sedes, 1985), 85–97; Charles Engrand, "Colbert et les commissaires du roi. Grands jours et intendants. La reduction des officiers a l'obéissance," in *Colloque pour le tricentenaire de la mort de Colbert, sous la direction de Roland Mousnier, Un nouveau Colbert*, edited by Jean Favier and Roland Mousnier (Paris: Éditions Sedes, 1985), 133–44; Sharon Kettering, *Patrons, Brokers, and Clients in Seventeenth-century France* (New York: Oxford University Press, 1986); Mukerji, "The Great Forest Survey."

58 Max Weber, *The Theory of Social and Economic Organization*, ed. A. M. Henderson and Talcott Parsons (New York: Free Press, 1947), 33–34.

59 Colbert's lack of reform of the bureaucracy is discussed in Mettam, *Power and Faction*, chapter 5. For the special investments in royal industries, see Cole, *Colbert and a Century of French Mercantilism*, 2:239–86.

60 Mukerji, "The Great Forest Survey;" Conel Condren, "The Politics of the Liberal Archive." *History of the Human Sciences* 12 (1999): 35–49.

61 ACM 11-07.

62 Mukerji, *Territorial Ambitions*.

63 L.T.C. Rolt, *From Sea to Sea: The Canal du Midi* (London: Allen Lance, 1973), 38.

64 Mukerji, *Territorial Ambitions*, 297–99; Steve Shapin and Simon Schaffer, *Leviathan and the Air-Pump: Hobbes, Boyle, and the Experimental Life* (Princeton, NJ: Princeton University Press, 1985).

65 Shapin and Schaffer, *Leviathan and the Air-Pump*.

66 Mukerji, *Territorial Ambitions*, 259–65; Roland Mousnier, *The Institutions of France under the Absolute Monarchy, 1598–1789: Society and the State*, trans. Brian Pearce (Chicago: University of Chicago Press, 1979); Raymond Frank Paskvan, "The Jardin du Roi: The Growth of Its Plant Collection, 1715–1750" (PhD diss., University of Minnesota, 1971); Chandra Mukerji, "Dominion, Demonstration, and Domination: Religious Doctrine, Territorial Politics, and French Plant Collection," in *Colonial Botany: Science, Commerce, and Politics in the Early Modern World*, ed. Londa Schiebinger and Claudia Swan (Philadelphia: University of Pennsylvania Press, 2005), 19–33; Alice Stroup, *A Company of Scientists: Botany, Patronage, and Community at the Seventeenth-century Parisian Royal Academy of Sciences* (Berkeley: University of California Press, 1990), 185–97; Davy de Virville, *Histoire de la Botanique en France* (Paris: Société d'Édition d'Enseignment Supérieure, 1954); Marguerite Duvall, *The King's Garden*, trans. Annette Tomarken and Claudine Cowen (Charlottesville: University Press of Virginia, 1982).

67 Chandra Mukerji, "Printing, Cartography, and Conceptions of Place," *Media, Culture, and Society* 28, no. 5 (2006): 651–69; David Woodward, *Maps as Prints in the Italian Renaissance: Makers, Distributors, and Consumers* (London: British Library, 1996); Denis Cosgrove, "Mapping New Worlds: Culture and Cartography in Sixteenth-century Venice," *Imago Mundi* 44 (1992): 65–89; Emmanuela Casti, "State, Cartography, and Territoriality in Renaissance Veneto and Lombardy." *Imago Mundi* 3 (2001):1–73.

68 For administrative uses of geography, see Jospeh W. Konvitz, *Cartography in France, 1660–1848* (Chicago: University of Chicago Press, 1987); David Buisseret, ed., *Monarchs, Ministers, and Maps* (Chicago: University of Chicago Press, 1992); Mukerji, *Territorial Ambitions*, 259–65. And for a period description of survey methods and their uses, see Sébastian Le Clerc, *Practique de la geometrie sur le papier and sur le terrain* (Paris: Chez Thomas Jolly, 1669).

69 Woodward, *Maps as Prints*. Humanists were involved in elaborating the new forms of geography as they became interested in depicting land as a site of human activity as well as physical features. They emphasized historical and social representations of places while also promoting survey mathematics. For cadastral maps and other new uses of maps in the period, see Dainville, *Géographie des humanistes*.

70 With his patronage, Colbert built up a reserve of experts in survey techniques and geographic information that could be called on for favors. For the political value of scientific patronage, see Chandra Mukerji, *A Fragile Power: Scientists and the State* (Princeton, NJ: Princeton University Press, 1990).

71 Haraway, *Primate Visions*; Longino, *The Fate of Knowledge*.

72 See, for example, Susan Leigh Star, ed., introduction to *Cultures of Computing* (Cambridge, MA: Blackwell, 1995); Susan Leigh Star, ed., *Ecologies of Knowledge: Work and Politics in Science and Technology* (Albany: State University of New York Press, 1995).

73 This new state was simultaneously conceptual and actual: a quasi-object in Bruno Latour's terms. See Bruno Latour, *We Have Never Been Modern* (Cambridge, MA: Harvard University Press, 1993). See also Mukerji, *Territorial Ambitions*, chapter 7, particularly 324.

74 Chandra Mukerji, "Entrepreneurialism, Land Management, and Cartography during the Age of Louis XIV," in *Merchants and Marvels*, ed. Paula Findlen and Pamela Smith (New York: Routledge, 2002), 248–76.

75 Ellis Veryard, *An Account of Divers Choice Remarks as well as Geographical, Historical, Political, Mathematical, and Moral; Taken in a Journey through the Low Countries, France, Italy, and Part of Spain with the Isles of Sicily and Malta* (London: S. Smith and B. Walford, 1701).

76 Ibid.

77 Blanchard, *Les ingénieurs du roy de Louis XIV à Louis XVI*; Janice Langins, *Conserving the Enlightenment: French Military Engineering from Vauban to the Revolution* (Cambridge, MA: MIT Press, 2004).

78 Allain Manesson Mallet, *Les Travaux de Mars ou l'Art de la Guerre* (Amsterdam: Jan et Gillis Janson à Waesbergue, 1684); Mukerji, *Territorial Ambitions*, chapter 2. The importance of surveyors and the military engineers to the politics of the period was symbolic as well as practical. Military engineering had been a hallmark of ancient Rome. "Monuments to the greatness of Rome" were celebrated by Louis XIV and his contemporaries, and stirred them to use French soldiers for comparable work. The army engineers were first employed for the obvious jobs of the building of ports, garrisons, and arsenals, but they were also used to improve the water supplies for Versailles and Paris, and in setting out drainage and flood-control ditches around rivers and swamps, and laying out canals that flowed where rivers did not. All these projects required some survey work, and all of them showed up again on maps. Tantalizingly, these efforts reshaped precisely what was recorded on maps—the shape of the shoreline, the course of rivers, the topography, and the roads crossing the landscape. See Chandra Mukerji, "Unspoken Assumptions: Voice and Absolutism in the Court of Louis XIV," *Journal of Historical Sociology* 2 (1998): 228–315; Lynn, *Giant of the Grand Siècle*; Josef W. Konvitz, *Cities and the Sea: Port City Planning in Early Modern Europe* (Baltimore: Johns Hopkins University Press, 1978).

79 Blanchard, *Les ingénieurs du roy de Louis XIV à Louis XVI*, 42–54. See also Chandra Mukerji, "Engineering and French Formal Gardens in the Reign of Louis XIV," in *New Directions in Garden History*, ed. John Dixon Hunt (Philadelphia: University of Pennsylvania Press, 2002); P.D.A. Harvey, *The History of Topographical Maps* (London: Thames and Hudson, 1980). For a discussion of conventions in maps, see Mukerji, "Printing, Cartography, and Conceptions of Place."

80 Hélen Vérin, *La Gloire des Ingéneurs: L'intelligence techniques du XVIe au XVIIIe siècle* (Paris: Albin Michel, 1993), 188; Mousnier, *The Institutions of France*, 575–76.

81 Jacques Cassini, *De La Grandeur et de la Figure de la Terre. Suite des Memoires de l'Academie Royale des Sciences, Annee MDCCXVIII* (Paris: Imprimerie Royale, 1720); John Noble Wilford, *The Mapmakers* (New York: Vintage, 1981); Monique Pelletier, *La Carte de Cassini: l'Extraordinaire Aventure de la Carte de France* (Paris: Presses de l'École nationale des ponts et chaussées, 1990); Dainville, *Géographie des humanistes*, 445–47; M. Picard, *Traité du Nivellement* (Paris: Estienne Michallet, 1684).

82 Mukerji, *Territorial Ambitions*, 259–65.

83 Picard, *Traité du Nivellement*; Philippe de La Hire, *L'ecole des arpenteurs: ou l'on enseigne toutes les pratiques de geometrie, qui sont necessaires a un arpenteur: On y a ajoute un abrege du nivellement, & les proprietez des eaux, & les manieres de les juger* (Paris: Thomas Moette, 1689). See Peter Sahlins, *Boundaries: The Making of France and Spain in the Pyrenees* (Berkeley: University of California Press, 1989).

84 Henri Enjalbert, "Les Hardinesses de Riquet: Données Géomophologiques de la Region que Traverse le Canal du Midi," in Le *Canal du Midi*, ed. Jean-Denis Bergasse (Cessenon: J.-D. Bergasse, 1985), 4:127–42, particularly 129.

85 Palissy, for example, was asked to survey the local salt marshes for the government, which was trying to redefine the gabelle or taxes on salt. For his maps, Palissy made note not just of the location of the marshes but also of the complex web of canals, bridges, and constructed salt flats that constituted the material basis for salt production in the region. For the problem of elevation measures in the period, see François de Dainville, *Cartes Anciennes du Languedoc XVIe–XVIIIe S* (Montpellier: Société Languedocienne de Géographie, 1961), 55.

86 Louis de Froidour, *Les Pyrenées centrales au XVIIe siècle: lettres par M. de Froidour . . . à M. de Haericourt . . . et à M. de Medon . . . publiées avec des notes par Paul de Casteran* (Auch: G. Foix, 1899); Andrée Corvol, *L'Homme et l'Arbre sous l'Ancien Régime* (Paris: Economica, 1984); M. Devèze, "Une Admirable Réforme Administrative: La Grande Réformation des Forêts Royales sous Colbert (1662–1680)," in *Annales de L'École Nationale des Eaux et Forêts et de la Station de Recherches et Expériences* (Nancy: École Nationale des Eaux et Forêts, 1962); Mukerji, *Territorial Ambitions*, chapter 2; Mukerji, "The Great Forest Survey."

87 For a discussion of databases, see Geoffrey Bowker and Susan Leigh Star, *Sorting Things Out* (Cambridge, MA: MIT Press, 1999).

88 Kathryn Henderson, *On Line and on Paper: Visual Representations, Visual Culture, and Computer Graphics in Design Engineering* (Cambridge, MA: MIT Press, 1999), 25–58. Henderson describes the importance of sketches to the engineering process, and their centrality to collaborative work. She argues that formal drawings with their detail have increasing value as engineering decisions are made, but the rougher sketches are more useful and more frequently used in the conversations that go into the early stages of collaborative design.

89 They constituted what would become in later periods an "elite reserve labor force." See Chandra Mukerji, *From Graven Images: Patterns of Modern Materialism* (New York: Columbia University Press, 1983).

CHAPTER 3
Epistemic Credibility

1 Pierre Bourdieu, *Distinction: A Social Critique of the Judgment of Taste*, trans. Richard Nice (London: Routledge and Kegan Paul, 1984).

2 Lettre de Riquet à Colbert, 27 Novembre 1664, ACM 20-5.

3 Theodore Porter, *Trust in Numbers: The Pursuit of Objectivity in Science* (Princeton, NJ: Princeton University Press, 1995); Marie-Noëlle Bourguet, *Déchiffrer le France* (Paris: Éditions des Archives Contemporaines, 1988).

4 Steve Shapin and Simon Schaffer, *Leviathan and the Air-Pump: Hobbes, Boyle, and the Experimental Life* (Princeton, NJ: Princeton University Press, 1985); Steve Shapin, *A Social History of Truth: Civility and Science in Seventeenth-century England* (Chicago: University of Chicago Press, 1994); François de Dainville, *Cartes Anciennes du Languedoc XVIe–XVIIIe S* (Montpellier, Société Languedocienne de Géographie, 1961), 47. The idea of using the Sor River as a source of the water supply for a canal between the Garonne and the Mediterranean had been discussed by a *géographe du roi*, Pierre Petit, in 1663. See L. Malavialle, "Une Excursion dans la Montagne Noire," part 1, *Société Languedocienne de Géographie Bulletin* (1891): 14:273 ff; lettre de Riquet à Colbert, 27 Novembre 1664, ACM 20-5.

5 Lettre de Riquet à Colbert, 27 Novembre 1664, ACM 20-5.

6 See Pierre Saliès, "De l'Isthme Gaulois au Canal des Deux Mers: Histoire des Canaux et Voies Fluviales du Midi," in *Le Canal du Midi*, ed. Jean-Denis Bergasse (Cessenon: J.-D. Bergasse, 1985), 4:55–97; Louis de Froidour, *Lettre a M. Barrillon Damoncourt contenant la relation & la description des travaux qui se font en Languedoc pour la communication des deux mers* (Toulouse, 1672), 3; André Maistre, *Le Canal des Deux-Mers: Canal Royal du Languedoc, 1666–1810* (Toulouse: E. Privat, 1968), 30–33; Inès Murat, "Les Rapports de Colbert et de Riquet: Méfiance pour un homme ou pour un système?" in *Le Canal du Midi*, ed. Jean-Denis Bergasse, 3:113 (Cessenon: J.-D. Bergasse, 1984).

7 Bruno Latour, "Why Has Critique Run out of Steam?" *Critical Inquiry* 30 (2004): 238–39.

8 Compare to Sheila Jasanoff's work on contemporary public policy, and the need for and problems of scientific knowledge. Sheila Jasanoff, *The Fifth Branch* (Cambridge, MA: Harvard University Press, 1990); Sheila Jasanoff, *Science at the Bar* (Cambridge, MA: Harvard University Press, 1995).

9 L.T.C. Rolt (*From Sea to Sea: The Canal du Midi* [London: Allen Lance, 1973], 29) speaks with an authoritative voice about Riquet's education, but evidence of his education is sketchy at best. He himself claimed little education.

10 For a brief overview of the system of États in France, see Samuel Clark, *State and Status: The Rise of the State and Aristocratic Power in Western Europe* (Cardiff: University of Wales Press, 1995), 40–41.

11 Malavialle, "Une Excursion dans la Montagne Noire," part 1, 8:135n1.

12 Ibid., 8:142. About the view, Riquet wrote to Colbert (Lettre de Riquet à Colbert à Bonrepos, le 15 Novembre 1662, ACM 20-2) that he had shown d'Anglure de Bourlemont where the canal could be built. It is possible that they went to St. Félix or somewhere near it. See Rolt, *From Sea to Sea*, 38.

13 Froidour, *Lettre a M. Barrillon Damoncourt*, 5.

14 Rolt, *From Sea to Sea*, 13–16; François Gazelle, "Riquet et les eaux de la Montagne Noire: L'idée géniale de l'alimentation du canal," in *Le Canal du Midi*, ed. Jean-Denis Bergasse (Cessenon: J.-D. Bergasse, 1985), 4:143–70.

15 Rolt, *From Sea to Sea*, 16; Gazelle, "Riquet et les eaux de la Montagne Noire," 143–70, particularly 147.

16 Rolt, *From Sea to Sea*; Gazelle, "Riquet et les eaux de la Montagne Noire," 146–47. See also Henry Heller, *Labour, Science, and Technology in France, 1500–1620* (Cambridge: Cambridge University Press, 1996), 91.

17 Hubert Pinsseau, "Du Canal de Briare au Canal des Deux Mers: Origines et consequences d'un système inédit de navigation artificielle," in *Le Canal du Midi*, ed. Jean-Denis Bergasse (Cessenon: J.-D. Bergasse, 1985), 4:27–54.

18 Rolt, *From Sea to Sea*, 24–26.

19 Pierre Borel went on to Paris when he left Languedoc and became quite well-known. He entered the Académie Royale des Sciences as a chemist and wrote on Cartesian science. See Joseph Frederick Scott, *The Scientific Work of René Descartes, 1596–1650* (Ann Arbor: University of Michigan Press, 2006), 84.

20 Rolt, *From Sea to Sea*, 24–26.

21 Gazelle, "Riquet et les eaux de la Montagne Noire," 145–47.

22 Malavialle, "Une Excursion dans la Montagne Noire," part 1, 7:287, 8:135.

23 Maistre, *Le Canal des Deux-Mers*, 43.

24 Gazelle, "Riquet et les eaux de la Montagne Noire," 146–47; Bertrand Gabolde, "Revel: Des eaux du Sor à la rigole de la plaine," in *Le Canal du Midi*, ed. Jean-Denis Bergasse (Cessenon: J.-D. Bergasse, 1985), 4:241–44.

25 Gazelle, "Riquet et les eaux de la Montagne Noire," 150–51, 163; Gabolde, "Revel," 243.

26 Rolt, *From Sea to Sea*, 35.

27 For Andréossy's family and education, see Rolt, *From Sea to Sea*, 35; Malavialle, "Une Excursion dans la Montagne Noire," part 1, 8:120–21, 137–41. For the figures about the rigoles as well as their sizes and their water captures as they were understood in the period, see ibid., 7:288.

28 Antoine Andréossy, *Histoire du canal du Midi: connu précédement sous le nom de Languedoc* (Paris: F. Buisson, 1800). See also François Fabre, "La lutte des descendents Andréossy et Riquet pour la paternité du canal," in *Le Canal du Midi*, Jean-Denis Bergasse (Cessenon: J.-D. Bergasse, 1985), 4:343–66.

29 Malavialle, "Une Excursion dans la Montagne Noire," part 1, 8:135–41; Rolt, *From Sea to Sea*, 34–37. Rolt suggests that Riquet presented the idea of the mountain waterway to d'Anglure de Bourlemont, but there is no evidence that the line of influence did not go the other way around.

30 Riquet's dream was in a letter from Riquet to Colbert, July 31, 1665, translated and quoted in Rolt, *From Sea to Sea*, 47–48.

31 A letter from Colbert to Riquet in 12/12/64 (ACM 94-14) illustrates Riquet's lack of credibility with the minister. Colbert tells him that he is pleased with his zeal in developing the project, but he must wait for Clerville, and in his company, not by himself, come up to Paris to talk about the project.

32 Rolt, *From Sea to Sea*, 46.

33 Malavialle, "Une Excursion dans la Montagne Noire," part 1, 7:280–84; Rolt, *From Sea to Sea*, 45–46.

34 Malavialle, "Une Excursion dans la Montagne Noire," part 1, 9:284.

35 Ibid., 9:284–85.

36 For a discussion of Cavalier's work, see Dainville, *Cartes Anciennes du Languedoc*, 38–44; Jean Robert and Jean-Denis Bergasse, "L'Étrange Destin des Andréossy," in *Le Canal du Midi*, ed. Jean-Denis Bergasse (Cessenon: J.-D. Bergasse, 1984), 3:195-232, particularly 203.

37 Rolt, *From Sea to Sea*, 31; Dainville, *Cartes Anciennes du Languedoc*, 55, 60–61; L. Malavialle, "Une Excursion dans la Montagne Noire," part 3, *Bulletin de la Société Languedocienne de Géographie* (1892), 15:283–314. The first map published of the canal plan and its water supply was actually made by the géographe du roi P du Val. See Malavialle, "Une Excursion dans la Montagne Noire," part 1, 8:436–39, 475–76.

38 Malavialle, "Une Excursion dans la Montagne Noire," part 1, 9:286–89.

39 ACM 11-07.

40 Rolt, *From Sea to Sea*, 42–44.

41 See ibid., 42–44; for details of the work, see Malavialle, "Une Excursion dans la Montagne Noire," part 1, 8:146–49.

42 Michel Adgé, "Chronologie des Principaux Évènements de la Construction du Canal, 1662–1694," in *Le Canal du Midi*, ed. Jean-Denis Bergasse (Cessenon: J.-D. Bergasse, 1985), 4:174; Rolt, *From Sea to Sea*, 42–46. Surprisingly, Boutheroue agreed at this time to using the Aude as part of the navigational system—perhaps because Colbert did not want a larger project—but he was probably later instrumental in convincing the other commissioners to change this plan.

43 ACM 29-6.

44 Adgé, "Chronologie des Principaux Évènements," 175.

45 Ibid., 175.

46 Ibid.

47 Ibid.; Rolt, *From Sea to Sea*, 45.

48 It is unlikely that Boutheroue inspired it because it used the Lampy riverbed as part of the alimentation channel. This suggests that it was drawn by one of the lesser-known surveyors for Clerville, but there is no positive evidence to confirm this.

49 December 20, 1664, ACM 20-10.

50 Adgé, "Chronologie des Principaux Évènements," 176; Rolt, *From Sea to Sea*, 46.

51 Froidour, *Lettre a M. Barrillon Damoncourt*, 16–17.

52 Malavialle, "Une Excursion dans la Montagne Noire," part 1, 259–61.

53 Adgé, "Chronologie des Principaux Évènements," 176.

54 Malavialle, "Une Excursion dans la Montagne Noire," part 1, 311.

55 "Arrest de 1666," ACM 3-10.

56 Adgé, "Chronologie des Principaux Évènements," 176.

57 Ibid., 176–77.

58 ACM 11-07.

59 "Relation particuliere de la rigolle dessay," ACM 02-14; Gazelle, "Riquet et les eaux de la Montagne Noire," 143–70. For details of the workers employed and what they did, see lettre de Riquet à Colbert, 18 Août 1665, ACM 02-13.

60 Lettre de Riquet à Colbert, 3 Septembre 1665, ACM 20-19. A toise is 1.949 meters, or six French feet or 6.396 U.S. feet.

61 "Relation particuliere de la rigolle dessay," ACM 02-14; Rolt, *From Sea to Sea*, 35–37; Froidour, *Lettre a M. Barrillon Damoncourt*, 9–10.

62 Lettre de Riquet à Colbert, 28 Septembre 1665, ACM 20-21.

63 "Relation particuliere de la rigolle dessay," ACM 02-14; Rolt, *From Sea to Sea*, 35–37; Froidour, *Lettre a M. Barrillon Damoncourt*, 9–10. For more details about the politics of developing the water system, see Chandra Mukerji, "Entrepreneurialism, Land Management, and Cartography during the Age of Louis XIV," in *Merchants and Marvels*, ed. Paula Findlen and Pamela Smith (New York: Routledge, 2002), 248–76.

64 "Arrest 1666," ACM 3-10, 18.63.

1 Peter Burke, *Fabrication of Louis XIV* (New Haven, CT: Yale University Press, 1992); André Félibien, *Receuil de descriptions de peintures et d'autres ouvrages fait pour le roy* (Paris: Veuve de S. Mabre-Cramoisy, 1689); *Relation de la Feste de Versailles du 18e Juillet 1668* (Paris: Pierre le Petit) (BN Lb37. 360); Chandra Mukerji, *Territorial Ambitions and the Gardens of Versailles* (Cambridge: Cambridge University Press, 1997), 72, 101, 294.

2 J. P. Neraudau, *L'Olympe du Roi Soleil* (Paris: Société des Belles-Lettres, 1986), 234–37; Jean-Marie Apostolidès, *Le Roi-Machine: spectacle et politique au temps de Louis XIV* (Paris: Éditions de Minuit, 1981).

3 Monique Mosser and Georges Teyssot, *The Architecture of Western Gardens* (Cambridge, MA: MIT Press, 1991); L. A. Barbet, *Les Grandes Eaux de Versailles: Installations Méchaniques et Éstangs Artificiels, Descriptions des Fontaines et de leurs Origines* (Paris: H. Dunod et E. Pinat, 1907).

4 See the map by Nicolas de Fer, "Partie de France, et d'Italie par raport a la Route de Cesar pendant la premiere campagne dans les Gaules" (Paris: l'Auteur, ca. 1705).

5 See, for example, Louis de Froidour, *Lettre a M. Barrillon Damoncourt contenant la relation & la description des travaux qui se font en Languedoc pour la communication des deux mers* (Toulouse, 1672), 2–3. The forester refers to this kind of work as not new. Such a canal was considered in the past, he said, and the Romans were in fact the authors of this plan. But while the idea had existed for years, the execution of it in the happy reign of Louis XIV was due to the care of Colbert and genius of Riquet.

6 Charles Vallancey discusses the ancients and Strabo in particular in pointing to France's rich array of rivers that could be put to use. See *A Treatise on Inland Navigation or the Art of Making Rivers Navigable: Of Making Canals in All Sorts of Soils, and of Constructing Locks and Sluices* (Dublin: George and Alexander Ewing, 1763), 102–3.

7 See, for example, Chérine Gébara and Jean-Marie Michel, "L'Aqueduc romain de Fréjus: sa description, son histoire et son environnement," in *Revue Archéologique de Narbonnaise*, supplement 33 (Montpellier: Éditions de l'Association de la Revue Achéologique de Narbonnaise, 2002).

8 P. Burlats-Brun and Jean-Denis Bergasse, "L'oligarchie gabelière, soutien financier de Riquet," in *Le Canal du Midi*, ed. Jean-Denis Bergasse (Cessenon: J.-D. Bergasse, 1984), 3:123–40.

9 Pierre Bourdieu, *Distinction: A Social Critique of the Judgment of Taste*, trans. Richard Nice (London: Routledge and Kegan Paul, 1984).

10 Local city maps in the south often depicted the ancient structures in their towns. Chandra Mukerji, "Printing, Cartography, and Conceptions of Place in Renaissance Europe," *Media, Culture, and Society* 28, no. 5 (2006): 651–69. These cities were also shown from a bird's eye perspective associated with Italian print culture. For the tradition of city mapping and its Renaissance roots in Italy, see Thomas Frangenberg, "Choreographies of Florence: The Use of City Views and City Plans in the Sixteenth Century," *Imago Mundi* 46 (1994): 41–64.

11 Elting E. Morison, *From Know-how to Nowhere* (New York: Basic Books, 1974), 7–9.

12 Vitruvius, *The Ten Books on Architecture*, trans. M. H. Morgan (1914; repr., New York: Dover, 1960).

13 Jacques Heyman, *The Stone Skeleton: Structural Engineering of Masonry Architecture* (Cambridge: Cambridge University Press, 1995).

14 Gustavo Giovannoni, "Building and Engineering," in *The Legacy of Rome*, ed. Cyril Bailey (1923; repr., Oxford: Clarendon Press, 1940), 429–74.

15 For an analysis of prototypes in craft modes of construction, see David Turnbull, *Masons, Tricksters, and Cartographers* (Amsterdam: Harwood, 2000); Simon Schaffer, "'The Charter'd Thames': Naval Architecture and Experimental Spaces in Georgian Britain," in *The Mindful Hand: Inquiry and Invention from the Late Renaissance to Early Industrialisation*, ed. Lissa L. Roberts, Simon Schaffer, and Peter Dear (Amsterdam: Edita— and University of Chicago Press, 2007), 279–307.

16 See Françoise Moreil, *L'Intendance de Languedoc al la fin du XVIIe Siècle* (Paris: C.T.H.S., 1985), edited edition of Nicolas de Lamoignon de Basville, *Mémoire sur la province de Languedoc de 1697.* Basville's section on the great works done in the province and those still needing to be done begins, predictably, with the Romans (263–67).

17 In other words, the canal was 31.7 miles long from Toulouse to Naurouze, and another 37.3 miles from Naurouze to Trèbes. The trench was 6 feet deep and 30 feet wide on the bottom of the channel, expanding to 56 feet wide across the top. The canal had to rise 179 feet from Toulouse to Naurouze, and drop 371 feet from Naurouze to Trèbes.

18 Rolt, *From Sea to Sea: The Canal du Midi* (London: Allen Lance, 1973), 72, derives these figures from Joseph de Lalande, *Des canaux de navigation, et spécialement du canal de Languedoc* (Paris: Veuve Desaint, 1778.) Rolt says that these were the early proportions, and that the width of the surface of the canal was increased, but he also argues that the first part of the canal dug near Toulouse was even narrower than what Lalande says. For a more reliable discussion of these figures, see Michel Adgé, "L'Art de l'Hydraulique," in *Conseil d'Architecture, d'Urbanisme et de l'Environment de la Haute-Garonne, Canal royal de Languedoc: le partage des eaux* (Caue: Loubatières, 1992), 186.

19 André Maistre, *Le Canal des Deux-Mers: Canal Royal du Languedoc, 1666–1810* (Toulouse: E. Privat, 1968), 68.

20 "l'Arrest d'adjudication des ouvrages à faire pour le canal de communication des Mers en Languedoc est promulgué. Ce même jour, le Roi 'fait bail et délivrance à M. de Riquet des ouvrages contenues au Devis,'" 14 Octobre 1666, préalablement défini sous l'autorité du Chevalier de Clerville, ACM 3-10.

21 Arrest of 1666, ACM 3-10, 20.

22 Ibid., 20–21.

23 See Maistre, *Le Canal des Deux-Mers*, 55–57. In a letter from Bordeaux in May 1669, Clerville offered to come look at whatever Riquet wanted him to see and asked the entrepreneur to pay his expenses (ACM 31-25). One of the verifications he did make after M. d'Aguesseau became intendant for Languedoc in 1673 is a detailed inventory of all the major work sites, with commentary on their status given to M. d'Aguesseau by Riquet, and Clerville's review of the work completed in these areas (ACM 13-12). For an example of a different sort of verification, less like an account book, there is a 1672 review of work on the water supply system. New and deeper rigoles were being dug then, and some of

the connections to rivulets and rivers were being fixed. Drains were being installed to get rid of excess water, and return it for use by locals (ACM 13-07). There is also another verification done by the sieur de Montbel in 1674 for payments to property holders near the canal (see ACM 96-13). And the conditions under which the États would approve the project were outlined in "Estraitz du cayer des deliberations prises par les Gens des Trois Etats de la Provence de Languedoc assembles par mande(ment) du Roy à Carcassone . . . Du mercredy deuxième du d. mois de mars Président Monseigneur l'Archevêque de Toulouse," Mars 1667, ACM 96-3.

24 Arrest of 1666, ACM 3-10, 15.

25 Lettre de Clerville à Riquet, 26 Avril 1669, ACM 31-26; Lettre de Clerville to Riquet, 1 Julliet 1669, ACM 31-27. In this letter, he tells Riquet that he has found a contractor who was less expensive than the one recommended by Jean Gasse, sieur de Contigny for the job. He is sending M. la Mar to see Riquet.

26 On Clerville and the devis for the first enterprise, see Maistre, *Le Canal des Deux-Mers*. For the devis of 1665 and 1668 for Sète, and Clerville's role in designing the devis for the first enterprise, see Alain Degage, "Un Nouveau port en Languedoc, de la fin du XVIe siècle au début du XVIIIe," in *Histoire de Sète*, ed. Jean Sagnes (Toulouse: Privat, 1987), 50–51. For the new devis for the locks, see ACM 07-01. On Clerville's fall from power vis-à-vis Vauban, see J. A. Lynn, *Giant of the Grand Siècle: The French Army, 1610–1715* (Cambridge: Cambridge University Press, 1997), 558–59, 588–89. Lynn explains that Vauban worked for and learned from Clerville until 1667, when they were both asked to provide plans for refortification of Lille. Clerville's plan was much less good than the one submitted by Vauban, and the younger man took over his job as engineer in charge of fortifications. (558–59). In 1661–1662 Clerville had been appointed to a new position as commissaire générale des fortifications, working both for Colbert and the then-minister of the army, Michel Le Tellier, but he had never been completely accepted by the latter. His work in fortress engineering was mainly under Colbert. Le Tellier's son, Louvois, also did not respect or use Clerville very much, but readily took on Vauban as a military engineer, transferring responsibilities for fortresses into the army (588–89). Clerville continued to work for Colbert, particularly on the Canal du Midi, bringing his training in military engineering to this work. See Degage, "Un Nouveau port en Languedoc," 265–306.

27 Arrest of 1666, ACM 3-10, 15–16.

28 For Vitruvius on wood and its uses, see *The Ten Books on Architecture*, 58–65, 162–64. For fortress design in this period, see Allain Manesson Mallet, *Les Travaux de Mars ou l'Art de la Guerre* (Amsterdam: Jan et Gillis Janson à Waesbergue, 1684), 68–72, 143–45. For an example of this type of design in the Italian tradition of military engineering, see Antonio Sangallo, *The Architectural Drawings of Antonio da Sangallo the Younger and His Circle* (Cambridge, MA: MIT Press, 1994), 197–98, showing the network of timbers as well as the use of *cateras* or twigs for reinforcements in fortress construction.

29 "Memoire donne a M. de Bezons et Verification fait par M. de Besons, Fev. 1669," ACM 13-3, 2. Bezons notes that digging down to the layer of clay was a Roman technique, and one used for the Robine de Narbonne. The result was that Roman canals did not have a uniform depth. Riquet's workers started following classical precedent, but Riquet was being asked to make the depth what

was proposed in the original devis. This made it necessary to use indigenous practices from the salt flats.

30 For the specifications for the second enterprise, describing the techniques that had been codified during the first enterprise and that continued to be used along the canal, see ACM 07-01.

31 For Roman aqueducts along with their disrepair and repair, see Katherine W. Rinne, "Between Precedent and Experiment: Restoring the Acqua Vergine in Rome, 1560–70," in *The Mindful Hand: Inquiry and Invention from the Late Renaissance to Early Industrialisation*, ed. Lissa L. Roberts, Simon Schaffer, and Peter Dear (Amsterdam: Edita and University of Chicago Press, 2007), 95–116. For images of Moorish dams, see Claude Louis Hiacinthe de la Boulaye, "Proces Verbal de visites des Bois du Pays de l'Abour [Adour]," AHMV, SH no. 278, 3, 8. These maps from the Pyrenees (the area of the Adour River) show a dam in an area where Moorish hydraulic systems were common. Compare these drawings to the specifications for the Roman dams at Subiaco. See Norman Smith, "The Roman Dams of Subiaco," *Technology and Culture* 11, no. 1 (1970): 58–68.

32 Froidour, *Lettre a M. Barrillon Damoncourt*, 16–18.

33 In period terms, it was 500 toises in length across the valley, 25 toises high, and 61 toises thick.

34 Froidour, *Lettre a M. Barrillon Damoncourt*, 17.

35 Ibid., 17–20. The voûte was 61 toises x nine *pieds* x 12 *pieds*.

36 Ibid., 17–19.

37 Ibid., 17–19.

38 Ibid., 17–18.

39 Ibid., 20–22.

40 Michel Adgé, "Les Premiers États du Barrage de Saint-Ferréol," *Les cahiers d'histoire de Revel* 7 (2001): 5–22, particularly 5–12.

41 Ibid., 11–12.

42 Helen Roseneau, *The Ideal City: Its Architectural Evolution* (Boston, MA: Boston Book and Art Shop, 1959); A.E.J. Morris, *History of Urban Form* (London: George Godwin Ltd., 1979).

43 Ideal city planning was brought into France during the sixteenth and early seventeenth centuries when Italian engineers were hired to fortify border cities, mainly along the northeast. Clerville, like other French military engineers of the late seventeenth century, was schooled in this tradition. Anne Blanchard, *Les Ingénieurs du Roy de Louis XIV à Louis XVI: Étude du corps des fortifications* (Montpellier: Université Paul-Valéry, 1979), 47–49; Carlo Cipolla, *Guns, Sails, and Empires* (New York: Pantheon, 1965), chapter 1; Lynn, *Giant of the Grand Siècle*, 547–93.

44 ACM 13-05.

45 Froidour, *Lettre a M. Barrillon Damoncourt*, 29. See also a letter from Riquet to Colbert (13 Mars 1669, ACM 22-8) in which Riquet says his son is bringing this *plan en relief* for the basin at Naurouze to Versailles.

46 According to Riquet, these books would be updated by the hour. Maistre, *Le Canal des Deux-Mers*, 73. But the archival records don't show this. (These are uncataloged papers in the archives.) They simply list workers for a day, not by the hour. For comment on the military style of organization for this canal, see Lalande, *Des canaux de navigation*, 14. He also said that the intendant of the region had problems of lodging not unlike those faced by the military, and used

the same solution: requiring local people to house them, and giving them one sol per worker to pay for food and housing. For the labor organization, see also Bertrand Gabolde, "Les Ouvriers du Chantier," in *Le Canal du Midi*, ed. Jean-Denis Bergasse (Cessenon: J.-D. Bergasse, 1984), 3:233–38, particularly 235–36.

47 For the chains of command being developed in the military in this period, see John Lynn, *Giant of the Grand Siècle*, 67–106, 453–593. For the organization of work on the Montagne Noire as it was described in 1669, see ACM 2-12. For the chain of command more generally in the area of the first enterprise, see "Ordre . . . pour la communication des mers," 1 Janvier 1667, ACM 17-1.

48 Letter from a M. Madron criticizing the old work structure, and signed by Dominique Gillade and other supervisors, ACM 17-7; Froidour (*Lettre a M. Barrillon Damoncourt*, 26) described the piecework labor practices he saw at the dam. For the work groups, and the shifts in wages and work conditions, see Gabolde, "Les Ouvriers du Chantier," 233–38. And for the real character of the work groups in the Montagne Noire in spite of their formal attributes, see Adgé, "Les Premiers États du Barrage de Saint-Ferréol," 7–12.

49 Rituals of land transfer are described in Robin Wagner-Pacifici, *Art of Surrender: Decomposing Sovereignty at Conflict's End* (Chicago: University of Chicago Press, 2005). For the opposition simmering at this time, see, for example, Lettre de Riquet à Colbert, ACM 20-59.

50 Père Matthieu de Mourgues, *Relation de la seconde navigation du Canal Royal, 1683*," quoted in Phillippe Delvit, "Un Canal au Midi," in *Conseil d'Architecture, d'Urbanisme et de l'Environment de la Haute-Garonne, Canal royal de Languedoc: le partage des eaux* (Caue: Loubatières, 1992), 204–24. Mourgues's position as inspecteur du canal when he wrote this "relation" is described in François de Dainville, *Cartes Anciennes du Languedoc XVIe–XVIIIe S* (Montpellier: Société Languedocienne de Géographie, 1961), 53.

51 The inscription was a list of the august individuals associated with the presumed great work. These included "Pierre de Riquet, INVENTEUR de ce gd. Ouvrage." Quoted in "Notes sur l'Histoire du Canal de Languedoc par les Descendents de Riquet," 1803, ACM uncataloged, typescript, 66.

52 See Maistre, *Le Canal des Deux-Mers*, 50–51, 57–58. In spite of the original deadline, the contract for the second enterprise gave Riquet more time to finish the canal. This is why in 1679 the États were still negotiating with the king about what and when to pay for the canal. This included indemnification. "Extrait du Cayer de deliberations pris par les gens des trois Etats de la province du Languedoc . . . 12 janvier 1680," ACM 96-19; see also the "Extrait" mentioned above, ACM 96-13, from 1674. But already in November 1672, the États were arguing that the canal had taken so long that they no longer had faith that the canal would be completed, and serve the region and its commerce. See "Extrait du Cayer des Deliberations," 18–28 Novembre 1672, ACM 96-11.

53 *Avis a messieurs les Capitouls de la Ville de Tolose . . . , par Arquier, Doyen de anciens Capitouls. Et Response a cet Avis, Article par Article par Jean de Nivelle, Ancien Capitaine Chasseauant du Canal dans l'atelier de Mr. Sagadenes*, 1667, ACM 01-16.

54 Ibid.

55 Ibid. (emphasis added).

56 Rolt, *From Sea to Sea*, 69.

57 *Avis a messieurs les Capitouls de Tolose* (emphasis added).

58 Ibid.

59 "Memoire de mes remarques du Canal, M. de Seguelay," Octobre 1670, ACM 13–05. For Riquet's critics and Colbert, see Inès Murat, "Les Rapports de Colbert et de Riquet: Méfiance pour un homme ou pour un système?" in *Le Canal du Midi*, ed. Jean-Denis Bergasse (Cessenon: J.-D. Bergasse, 1984), 3:105–22. For the use of templates in construction, see Turnbull, *Masons, Tricksters, and Cartographers*.

60 Arrest of 1666, ACM 3-10, 10.

61 *Registre contenant les ordres, instructions et lettres expediees par Monseigneur Colbert, touchant les fortifications des ponts et chausseees, canal de communication des mers et mines de Languedoc en l'annee 1665*, EPC.

62 To build cofferdams, they pounded long wooden stakes or pilings into the ground close to each other below the waterline to form a barricade to keep out water. The interior space created by the pilings could first be pumped out, and then excavated below ground level down to rock or at least deep heavy clay. After the excavation, the whole interior would be filled with stones and cement, creating a foundation in wet conditions that was both heavy and deep.

63 Vitruvius described how to do this: "[A] coffer dam . . . [is] formed of oaken stakes with ties between them . . . driven down into the water and firmly propped there; the lower surface, under the water, must be leveled off and dredged, . . . and concrete . . . must be heaped up until the empty space . . . is full up by the wall." Vitruvius, *The Ten Books on Architecture*, chapter 12, 162–63.

64 "Sites and Museums in Roman Gaul: Fréjus," *Athena Review* 1, no. 4 (1998): 1.

65 Adgé, "Les Premiers États du Barrage de Saint-Ferréol," 5–22.

66 Books of machines were popular in France in the sixteenth and seventeenth centuries. See Henry Heller, *Labour, Science, and Technology in France, 1500–1620* (Cambridge: Cambridge University Press, 1996), particularly 96–131. For examples of machines with some of the characteristics described by Adgé, see Giorgius Agricola, *De re Metallica; Translated from the First Latin Edition of 1556, with Biographical Introduction, Annotation, and Appendices upon the Development of Mining, Metallurgical Processes, Geology, Mineralogy, and Mining Law from the Earliest Times to the 16th Century*, trans. and ed. Herbert Clark Hoover and Lou Henry Hoover (New York: Dover, 1950): book 5 (103–5) contains simple machines used to raise ore from mines; book 5 (156) also shows a dump truck or cart; book 6 (174–75) illustrates dippers on chains moved by pulleys to raise water from inside a mine; and book 6 (197) also shows a treadmill and machine for both raising ore and pumping water from a mine. For Combe's machine, see Adgé, "Les Premiers États du Barrage de Saint-Ferréol," 12–14.

67 Agricola, *De re Metallica*, book 6, 174, 197.

68 Adgé, "Les Premiers États du Barrage de Saint-Ferréol," 14.

69 Lettre de Riquet à Colbert, 27 Fevrier 1667, ACM 20-28.

70 Lettre de Colbert à Riquet, 1 Avril 1667, ACM 20-32.

71 Lettre de Colbert à Riquet, 27 Avril 1667, ACM 20-36.

72 Lettre de Colbert à Riquet, 12 Avril 1667, ACM 20-33.

73 Lettre de Riquet à Colbert, 19 Avril 1667, ACM 20-34.

74 Lettre de Riquet à Colbert, 26 Juillet 1667, ACM 20-50.

75 Lettre de Colbert à Riquet, 10 Mars 1667, ACM 20-29.

76 Lettre de Colbert à Riquet, 12 Avril 1667, ACM 20-33.

77 Pierre Clément, *Lettres, Instructions et Mémoires de Colbert* (1867; repr., Liechtenstein: Klaus Reprints, 1979), 4:311n1.

78 Sharon Kettering, *Patrons, Brokers, and Clients in Seventeenth-century France* (New York: Oxford University Press, 1986).

79 Lettre de Clerville à Riquet, 6 Juin 1667, ACM 31-18 (emphasis added).

80 Ibid.

81 For the dangers and opportunities of commercial investments for nobles, and the efforts they went to in order to hide their economic ventures, see Samuel Clark, *State and Status: The Rise of the State and Aristocratic Power in Western Europe* (Cardiff: University of Wales Press, 1995), 223–27.

82 Lettre de Clerville à Riquet, 6 Juin 1667, ACM 31-18. Clerville mentioned in a letter on May 28, 1677, a machine that was being used for dredging the harbor. Riquet had argued that Clerville had wanted to build such a thing, and was using Sète as an excuse for trying to develop a machine with uses elsewhere. On Sète, see Josef W. Konvitz, *Cities and the Sea: Port City Planning in Early Modern Europe* (Baltimore: Johns Hopkins University Press, 1978). This was perhaps also the machine developed after the one for Saint Ferréol had failed. In any case, the project at Sète seems also to have failed. See ACM 31-37.

83 ACM 31-15.

84 See, for example, Lettre de Riquet à Colbert 28 Janvier 1668 ACM 21-2, on the problems in Roussillon, their effects on the salt tax, and the need for "assignations."

85 Lettre de Riquet à Colbert, 28 Janvier 1668, ACM 21-1.

86 "Arrêt de conseil portant acceptation des nouvelles offres faites par le Sieur Riquet parachevera les ouvrages de la premiere entreprise dans 4 annees," 20 Aout 1668, ACM 6-12.

87 Compare the role of the New Rome to the "closed world" that was so important to military culture and computer development during World War II; Paul Edwards, *The Closed World: Computers and the Politics of Discourse in Cold War America* (Cambridge, MA: MIT Press, 1996).

CHAPTER 5
Shifting Sands

1 Louis de Froidour, *Lettre a M. Barrillon Damoncourt cotenant la relation & la description des travaux qui se font en Languedoc pour la communication des deux mers* (Toulouse, 1672), 51–52.

2 Paul Sonnino, *Louis XIV and the Origins of the Dutch War* (Cambridge: Cambridge University Press, 1988), pp. 9–28; Anne Blanchard, *Vauban* (Paris: Fayard, 1996), 120–33.

3 Sonnino, *Louis XIV*, 23–28.

4 The French army was increased to 134,000 men in 1668, and decreased to 70,000 quickly after the peace treaty was signed. Ibid., 28; J. A. Lynn, *Giant of the Grand Siècle: The French Army, 1610–1715* (Cambridge: Cambridge University Press, 1997), 46.

5 ACM 13-3. For labor force statistics, see Michel Adgé, "Chronologie des Principaux Évènements de la Construction du Canal, 1662–1694," in *Le Canal Du Midi*, ed. Jean-Denis Bergasse (Cessenon: J.-D. Bergasse, 1985), 4:173–92.

6 Chandra Mukerji, *Territorial Ambitions and the Gardens of Versailles* (Cambridge: Cambridge University Press, 1997), 217.

7 L.T.C. Rolt, *From Sea to Sea: The Canal du Midi* (London: Allen Lance, 1973), 76–77.

8 Lettre de Riquet à Colbert, 5 Mai 1667, ACM 22-14.

9 Lettre de Riquet à Colbert, 14 Janvier 1668, ACM 21-1.

10 Ibid.

11 Ibid.

12 Ibid.

13 Ibid.

14 Ibid.

15 Ibid.

16 Ibid.

17 Ibid.

18 Ibid.

19 ACM 17-10.

20 Cambacérès (apparently Jacques) was the directeur des gabelles du haut Languedoc. See Henri de Cazals, "Armorial du Canal," in *Le Canal Du Midi*, ed. Jean-Denis Bergasse (Cessenon: J.-D. Bergasse, 1984), 3:161.

21 Lettre de Riquet à Colbert, 14 Janvier 1668, ACM 21-1.

22 Ibid.

23 André Maistre, *Le Canal des Deux-Mers: Canal Royal du Languedoc, 1666–1810* (Toulouse: E. Privat, 1968), 69–71.

24 Memoire donné à M. Bezons, Février 1669, ACM 13-3. It includes a discussion of repairs on the locks near Toulouse, work at Saint Ferréol, and completion of the channel from Naurouze to Trèbes. It also described the work being done to expand the rigole de la plaine to make it navigable to Revel and the problem of the earth filled with small stones along the Garonne that made it so hard to secure good footing for walls.

25 "Extraict des registres du conseil d'estat," 10-7-1668, ACM 18-1, in which Riquet is given the direction of mines and forges to furnish materials for the canal at reasonable prices.

26 Maistre, *Le Canal des Deux-Mers*, 69–72. See the "Bail et Adjudication des ouvrages a faire pour la continuation du Canal et du Port de Sète" signed by Claude Bazin, seigneur de Bezons, 20 Août 1668, and ratified in June by Augustin de Solas, a conseiller du Roy in Montpeller, ACM 07-12.

27 Jean-Baptiste Colbert, *Registre contenant les ordres, instructions et lettres expediees par Monseigneur Colbert, touchant les fortifications des ponts et chausseees, canal de communication des mers et mines de Languedoc en l'annee 1669*, 1669, EPC, 20.

28 Lettre de Colbert à Riquet, 9 Juin 1669, ACM 22-18. Also in Colbert, *Registre contenant les orders*, 14–21.

29 Anne Blanchard, Michel Adgé, and Jean-Denis Bergasse, "Les Ingénieurs du Roy," in *Le Canal du Midi*, ed. Jean-Denis Bergasse (Cessenon: J.-D. Bergasse, 1984), 3:183–85; Blanchard, *Vauban*, 138, 187.

30 Pierre Clément, *Lettres, Instructions et Mémoires de Colbert* (1867; repr., Liechtenstein: Kalus Reprints, 1979), 4:325–26.

31 Letter from Riquet to Colbert, quoted in ibid., 4:324. See also Blanchard, Adgé, and Bergasse, "Les Ingénieurs du Roy," 187–88.

32 Clément, *Lettres*, 4:325–26.

33 Ibid., 4:325–26; Blanchard, Adgé, and Bergasse, "Les Ingénieurs du Roy," 186.

34 Instruction pour de la Feuille, Ingénieur allant en Languedoc, 9 Juin 1669, in Clément, *Lettres*, 4:327–29.

35 Lettre de Colbert à de la Feuille, 19 Juillet 1669, in Colbert, *Registre contenant les orders*, 53.

36 Blanchard, Adgé, and Bergasse, "Les Ingénieurs du Roy," 186. Riquet described going into the mountains to study the work there with de la Feuille in Lettre de Riquet à Colbert, 30 Octobre 1669, ACM 22-27. He showed some tenderness toward his new superior, who had been helping him with both the engineering and politics.

37 Lettre de Colbert à Riquet, 27 Juin 1669, in Colbert, *Registre contenant les orders*, 34–35.

38 Riquet wrote to Colbert about the problems with the locks in a letter about a verification he made to Bezons, M. de Saint Papoul, and Clerville just before de la Feuille's arrival. Lettre de Riquet à Colbert, 27 May 1669, ACM 22-16.

39 Lettre de Colbert à Riquet, 27 Juin 1669, in Colbert, *Registre contenant les orders*, 35.

40 Lettre de Colbert à Riquet, 27 Juin 1669, in Colbert, *Registre contenant les orders*, 35.

41 For some of Colbert's doubts about Clerville's financial dealings with Riquet, see Lettre de Colbert à Clerville, 8 Mai 1669, in Colbert, *Registre contenant les orders*, 3–4. For some reservations about Clerville's technical advice, see lettre de Colbert à la Feuille, 30 Août 1669, in Colbert, *Registre contenant les orders*, 78–78.

42 Blanchard, Adgé and Bergasse, "Les Ingénieurs du Roy," 184, 188; Blanchard, *Vauban*, 138, 187.

43 Blanchard, *Vauban*, 138, 187.

44 Ibid., 88, 122–25, 133–39, 162.

45 Jean Mesqui, *Les ponts en France avant le temps des ingénieurs* (Paris: Fayard, 1980), 140, 298; Blanchard, *Vauban*, 138, 187.

46 Jacob de la Feuille was a quite well-known cartographer in the Netherlands at the turn of the eighteenth century. See Jacob de la Feuille (1668–1719) [Denmark], *Dania Regnum, Ducatus Holsatia et Slesvicum Insulæ Danicæ et Provinciæ Iutia Scania, etc.* (Amsterdam: ca. 1710); *Totius Danubii Nova & Accuratiss Tabula Universam Simul Turciam Europæam, Hungariam Magnam Germaniæ Partem Totam Pene Italiam, Atque una Moream cum Sicilia & Candia Complectens* (Amsterdam: de la Feuille, ca. 1710).

47 Blanchard, Adgé, and Bergasse, "Les Ingénieurs du Roy," 184. For some of his few letters from Colbert, see Colbert, *Registre contenant les orders*. And for a few details of his life, see Mesqui, *Les ponts en France avant le temps des ingénieurs*, 140, 298.

48 See Blanchard, Adgé, and Bergasse, "Les Ingénieurs du Roy," 184–85.

49 Archives de la Marine R21 no. 34.

50 Lettre de Colbert à Bezons, 12 Juillet 1669, in Colbert, *Registre contenant les orders*, 47. Colbert expressed the same trust to Bezons expressed in Clément, *Lettres*, 4:333n1. This was in spite of Louis XIV's explicit orders that Clerville make these visits with de la Feuille and continue to verify the work. See ibid., 4:327.

51 Lettre de Colbert à de la Feuille, 26 July 1669, in Colbert, *Registre contenant les orders*, 53.

52 Pons de la Feuille, "Remarques faictes au voiage de Flandres et Hollande en octobre, novembre et décembre 1670, sur les canaux, la construction des écluzes, ponts, jettées en mer et digues, moulins pour l'évacuation des caves et machines pour le nettoyement des canaux et des portz," Paris, 31 Janvier 1671, CCC, no. 448.

53 Lettre de Clerville à Riquet, n.d., ACM 31-37.

54 Lettre de Colbert à Riquet, 10 Octobre 1670, quoted in Clément, *Lettres*, 4:351–52.

55 Bertrand Gabolde, "Les Ouvriers du Chantier," in *Le Canal du Midi*, ed. Jean-Denis Bergasse (Cessenon: J.-D. Bergasse, 1984), 233–38.

56 See, for example, "Employés du Canal pendant sa Construction," ACM 816-01. The report on work by different contrôleurs (27-10-1669, ACM 17-08) suggests this system was already in place by that date.

57 "Memoire des choses nécessaires," 6 Juin 1671, ACM 17-12; lettre de M. Madron criticizing the old work structure, and signed by Gillade and other supervisors, ACM 17-7.

58 Gabolde, "Les Ouvriers du Chantier," 233–38.

59 Maistre, *Le Canal des Deux-Mers*, 72–75; Gabolde, "Les Ouvriers du Chantier," 233–39, particularly the *Affiche* about work conditions (239).

60 ACM 45-4, 46-1, 46-2, and 23-51.

61 Blanchard, Adgé, and Bergasse, "Les Ingénieurs du Roy," 187–88.

62 Riquet, quoted in Rolt, *From Sea to Sea*, 83–84.

63 Olivier de Serres, *Théâtre d'agriculture et mesnages des champs* (Geneva: Mat Hiev Berjon, 1611), 28–29.

64 ACM 31-36.

65 For Clerville's call to war, and a letter he wrote to Riquet expressing anxiety about it, see ACM 31-36. Inès Murat, "Les Rapports de Colbert et de Riquet: Méfiance pour un homme ou pour un système?" in *Le Canal du Midi*, ed. Jean-Denis Bergasse (Cessenon: J.-D. Bergasse, 1984), 108. De la Feuille was first brought to the project as part of the commission that verified Riquet's proposal. He then was asked by Colbert in June 1669 to go to Languedoc to oversee the work since the minister did not trust Riquet's written reports. ACM 22-18. The verification of the project made in May by Bezons, Clerville, and De Saint Papoul had made visible some of the problems on the canal. See ACM 22-16. Riquet was clearly anxious about the verification, although he needed it to get funds. See his letter to Colbert from May 6, 1669, ACM 22-14. Just before the review, Riquet also wrote to Colbert to confess to the lock wall collapses. See his letter from May 27, 1669, ACM 22-16. One month after the review was completed, de la Feuille was sent to oversee the work. De la Feuille was not just a spy for Colbert, although this was part of his job. He was also another engineer, who could bring new ideas to the canal project. This is why he was also sent to the Netherlands and Italy to look at the techniques of canal construction there. ACM 10-70. The canal remained a collaborative enterprise— just new voices were added to the mix. Rolt (*From Sea to Sea*, 83–87) describes the conflict over the routing of the second enterprise and the importance of both Mourgues and de la Feuille in supporting Riquet (76, 86–87, 93) against Clerville's opposition.

66 "Bail et adjucation des ouvrages à faire pour la continuation du canal et du port de Sète," 1668," approved 1669, ACM 07-12.

67 Josef W. Konvitz, *Cities and the Sea: Port City Planning in Early Modern Europe* (Baltimore: Johns Hopkins University Press, 1978), 73–89, particularly 78–81; Alain Degage, "Un Nouveau port en Languedoc, de la fin du XVIe siècle au début du XVIIIe," in *Histoire de Sète*, ed. Jean Sagnes (Toulouse: Privat, 1987), 47–52.

68 In 1669, some men from Montpellier set up a trading company to the Levant in Sète, breaking the monopoly that Marseille had enjoyed prior to that time. See Louis Dermigny, *Sète de 1666–1880* (Montpellier: Actes de l'Institut, 1955), 11. Because Colbert opposed their protectionist policies, he was supportive of this new venture. Because trading with the Levant brought the threat of importing the plague, Sète erected a hospital to quarantine those coming from the area. The terrible plague in Marseille worked to the benefit of Sète. See Alain Degage, "Quarantaine et lazaret à Sète aux XVIIe et XVIIIe siècles," in *Navigation et migrations en Méditerranée de la Préhistoire à nos jours* (Paris: Éditions du Centre National de la Recherche Scientifique, 1990), 1–17.

69 Degage, "Un Nouveau port en Languedoc."

70 Margaret McGowan, *Ideal Forms in the Age of Ronsard* (Berkeley: University of California Press, 1985); David Thompson, *Renaissance Paris* (Berkeley: University of California Press, 1984), chapter 1; Lewis Mumford, *The Culture of Cities* (New York: Harcourt Brace, 1938), chapter 2; Anne Blanchard, *Les Ingénieurs du Roy de Louis XIV à Louis XVI: Étude du corps des fortifications* (Montpellier: Université Paul-Valéry, 1979), 42–54. For an interestingly nuanced reading of Leon Battista Alberti, his architecture, and his city planning that shows his pragmatism as well as idealism, see Anthony Grafton, *Leon Battista Alberti* (New York: Hill and Wang, 2000), 264–65, 269–71, 293–330.

71 On ideal and practical urban planning and the port designs of the period, see Konvitz, *Cities and the Sea*, 3–69. For the plan for Louisville, see Conseil d'Architecture, d'Urbanisme et de l'Environment de la Haute-Garonne, *Canal royal de Languedoc: le partage des eaux* (Caue: Loubatières, 1992).

72 For the port at Frontignan and its demise, see Alain Degage, "Le Port de Sète: Proue Méditerreanéenne du Canal de Riquet," in *Le Canal du Midi*, ed. Jean-Denis Bergasse (Cessenon: J.-D. Bergasse, 1985), 4:265–306, particularly 268; Degage, "Un Nouveau port en Languedoc," 47; Achille Munier, *Frontignan* (Montpellier: C. Coulet, 1874), reprinted by Lacour, 1993, 240–41. There was a long history of illness associated with the closure of the étang from the sea in Frontignan, so building a permanent grau to the sea was a matter of continual interest here.

73 This visit of experts to the area is discussed in Degage, "Le Port de Sète," 268–70. Clerville's plan for Sète is discussed in Konvitz, *Cities and the Sea*, 79–80; Degage, "Un Nouveau port en Languedoc," 50–51; François de Dainville, *Cartes Anciennes du Languedoc XVIe–XVIIIe S* (Montpellier: Société Languedocienne de Géographie, 1961), 56–62. Clerville's successor as military engineer in the region, Niquet, was convinced that Clerville had pecuniary as well as technical interests in the project, hoping to siphon off the funds for his own purposes. Clearly, Clerville wanted to invent a dredging machine or system that could be used on other French harbors, too. That seems to explain his efforts to build an aqueduct into Sète with enough water to flush silt out of the

harbor. He also discussed machines for dredging, but with less enthusiasm. Lettre de Clerville à Riquet, 29 Octobre 1676, ACM 31-31; Lettre de Clerville à Riquet, 2 Janvier 1677, ACM 31-34; Lettre de Clerville à Riquet, 22 Janvier 1677, ACM 31-41; Lettre de Clerville à Riquet, unreadable number Février 1677, ACM 31-36. Still, it is not clear that Clerville's interest in a dredging machine explains why he chose Sète. Dainville argues that between 1634 and 1661, huge sums had been spent to try to fix the port of Agde without success, and this is why the proposal for Sète was tried. Vauban, Dainville contends, continued to prefer Agde, and this was perhaps why his protégé, Niquet, held the same opinion, but this did not mean that more money would have been used to develop this port rather than Sète. Dainville, *Cartes Anciennes du Languedoc*, 56, 63.

74 Riquet was given permission to quarry at Sète as part of the contract for finishing the canal. See Claude Bazin, "Bail et Adjudication des ouvrages à faire pour la continuation du Canal et du port du Cette, 20 Août 1668," 1669, 9, sec. x, ACM 07-12, 27.

75 There was some question of how good a military engineer Clerville actually was. The intimations of his inadequacy are hard to assess since they were raised in part by Vauban with his rise to power. Vauban was a more learned man and a more careful engineer, and Clerville clearly less so. But Colbert still trusted Clerville to work on the devis for the Canal du Midi and Sète, and continued to consult with him even after his humiliation in 1667. Colbert was a conservative man and not likely to do this unless Clerville was a solid engineer, but perhaps he didn't know how to judge. See Konvitz, *Cities and the Sea*, 78.

76 For the various designs for the port on Mediterranean side, see Degage, "Un Nouveau port en Languedoc," 52–53; ACM 7-12, 7–14.

77 Degage, "Un Nouveau port en Languedoc," 47–52; Dainville, *Cartes Anciennes du Languedoc*, 58–59.

78 The opening ceremonies at Sète were described in *Notes sur l'histoire du Canal du Languedoc par les descendents de Riquet*, 1803, ACM uncataloged.

79 For Riquet in Roussillon and his problems setting up his tax farm there, see, for example, ACM 20-1; ACM 21-2.

80 Degage, "Un Nouveau port en Languedoc," 52.

81 Lettre de Riquet à Colbert du 18-01-1670, ACM 23-4.

82 Ibid.

83 Degage, "Le Port de Sète," 272.

84 For a map showing all the plans proposed for solving the problems at Sète, and who developed them, see Degage, "Un Nouveau port en Languedoc," 53. For a description of the two "jetties"—one extending from the rock face of the mountain to protect the harbor, and the other extending from the beach to control the sand—see "Memoire sur le Canal Royal de la jonction des Mers de Languedoc," 1 Mars 1672, ACM 13-15, 5–6. The first had an elbow to enclose the harbor. The other was originally straight, but then designed to have an elbow, too, that filled with sand. There were also a number of other configurations that were tried as the sea tore down what the workers erected. For his consultation with workers, see Froidour, *Lettre a M. Barrillon Damoncourt*, 55–57; Lettre de Riquet à Colbert du 18-01-1670, ACM 23-4. The devis for the second enterprise also listed changes for Sète that show what was discussed and tried. See ACM 07-05.

85 Degage, "Le Port de Sète," 273.

1 Louis de Froidour, *Mémoire du Pays et des États de Bigorre*, intro. and notes Jean Boudette (Paris: H. Champion, Tarbes, Baylac, 1892); Chandra Mukerji, "Women Engineers and the Culture of the Pyrenees: Indigenous Knowledge and Engineering in 17th-century France," in *Knowledge and Its Making in Europe, 1500–1800*, ed. Pamela Smith and Benjamin Schmidt (Chicago: University of Chicago Press, 2008).

2 ACM 13-3; Michel Adgé, "Les Premiers États du Barrage de Saint-Ferréol," *Les cahiers d'histoire de Revel* 7 (2001): 5–22 ; Bernard Gabolde, "Les Ouvriers du Chantier," in *Le Canal du Midi*, ed. Jean-Denis Bergasse (Cessenon: J.-D. Bergasse, 1984).

3 De la Feuille was introduced to these problems early. On the failed locks at Castanet, see Riquet à Colbert, 24 Decembre 1669, ACM 22-38.

4 M. Picard was the exception. He understood that true measures of distance needed to compensate for elevation changes, and he also understood that building water systems like the one for Versailles was dependent on using elevation to advantage. M. Picard, *Traité du Nivellement* (Paris: Estienne Michallet, 1684). For the national survey that began in this period, see Monique Pelletier, *La Carte de Cassini: l'Extraordinaire Aventure de la Carte de France* (Paris: Presses de l'École nationale des ponts et chausses, 1990).

5 Picard was at least able to make reliable enough elevation studies to understand what plans for building the water system for Versailles would or would not work. Ironically, he showed that Riquet's proposal for this was not feasible. Picard, *Traité*.

6 Lettre de Riquet à Colbert, 24 Decembre 1669, ACM 22-38; "Mémoire, 1 Mars 1672," ACM 13-15, 5.

7 "Mémoire, 1 Mars 1672," ACM 13-15, 5.

8 Chandra Mukerji, "Tacit Knowledge and Classical Technique in Seventeenth-century France: Hydraulic Cement as Living Practice among Masons and Military Engineers." *Technology and Culture* 47 (2006): 713–3.

9 Ibid.

10 Bernard Forest de Belidor provides an illustration of the technique for making a jetty out of carpentry. See Bernard Forest de Belidor, "De l'Usage des Eaux a la Guerre," in *Architecture Hydraulique seconde partie qui comprend l'Art de diriger les eaux des la Mer & des Rivieres à l'avantage de la défense des places, du Commerce & de l'Agriculture. par M. Belidor, Colonel d'Infanterie* (Paris: Jombert, 1753), 2:104–5. For the specifications for completing the seawall at Sète, see ACM 7-12, 6–14. For some sense of Riquet's logic for using more reparable methods, see Lettre de Riquet à Monsieur Daguesseau, ACM 29-14.

11 Lettre de Clerville à Riquet, 3 Juin 1677, ACM 31-39; Claude Bazin, "Bail et Adjudication des ouvrages à faire pour la continuation du Canal et du port du Cette, 20 Août 1668," 1669, 9, sec. x, ACM 07-12, 6–14.

12 Allain Manesson Mallet, *Les Travaux de Mars ou l'Art de la Guerre* (Amsterdam: Jan et Gillis Janson à Waesbergue, 1684), 70–81; Vitruvius, *The Ten Books on Architecture*, trans. Morgan (1914; repr., New York: Dover, 1960), 46–49, 51–58; Rabun Taylor, *Roman Builders: At Study in Architectural Process* (Cambridge: Cambridge University Press, 2003), 92–115. For Riquet's use of pozzolana, see

Michel Adgé, "L'Art de l'hydraulique," in *Conseil d'Architecture, d'Urbanisme et de l'Environment de la Haute-Garonne, Canal royal de Languedoc: le partage des eaux* (Caue: Loubatières, 1992), 202–3. See also the 1684 document calling for "la plus grande quantité du lad. pourcelane qu'il pouvroit porter" brought to the province of Languedoc from Italy for repairs at Sète. ACM 18-17.

13 Lettre de Colbert à Riquet, 6 Août 1672, ACM 14-80; Bazin, "Bail et Adjudication."

14 "Devis des ouvrages de Massonerie . . . a la construction d'un pont a la Riviere de Repudre," Octobre 1677, ACM uncataloged paper.

15 Gustavo Giovannoni, "Building and Engineering," in *The Legacy of Rome*, ed. Cyril Bailey (1923; repr., Oxford: Clarendon Press, 1940), 429–74. For pozzolana in the pont-canal, see "Devis des ouvrages de Massonerie . . . a la construction d'un pont a la Riviere de Repudre," Octobre 1677, ACM uncataloged paper.

16 "Devis des ouvrages de Massonerie . . . a la construction d'un pont a la Riviere de Repudre, Octobre 1677, ACM uncataloged paper 2A, 2.

17 "Devis des ouvrages de Massonerie . . . a la construction d'un pont a la Riviere de Repudre," Octobre 1677, ACM uncataloged paper. For the new work structure, see "Mémoire des choses necessaries," ACM 17-12. For templates in construction, see David Turnbull, *Masons, Tricksters, and Cartographers* (Amsterdam: Harwood, 2000).

18 The terms of the contract for the pont-canal and the locks that were bundled with it were typical for civil engineering projects in the period.

19 Lesley Adkins and Roy Adkins, *Handbook to Life in Ancient Rome* (New York: Oxford University Press, 1994), 135–36.

20 Ibid., 135.

21 On the trip from Provence into the Pyrenees, see Strabo, *The Geography of Strabo*, book 4, trans. Horace Jones (Cambridge, MA: Harvard University Press, 1966–1970), 183–87. For archaeology on one of the colonies, see Pierre Aupert, Raymond Monturet, and Christine Dieulafait, *Saint-Bertrand-de-Comminges: Les thermes de forum* (Pessac: Éditions de la Fédération Aqvitania, 2001). Froidour spoke of a Roman-style water fountain as a wonder, not a piece of antiquity. See Louis de Froidour, *Les Pyrenées centrales au XVIIe siècle: lettres par M. de Froidour . . . à M. de Haericourt . . . et à M. de Medon . . . publiées avec des notes par Paul de Casteran* (Auch: G. Foix, 1899), 112–13.

22 Mukerji, "Women Engineers and the Culture of the Pyrenees."

23 On October 20, 1668, Riquet wrote to Colbert that he was staying on in Perpignan to recruit large numbers of workers for the canal. ACM 21-18. He wrote later about finding them in Bigorre. ACM 30-65.

24 Isaure Gratacos, *Femmes pyrénéennes, un statut social exceptionnel en Europe* (Toulouse: Éditions Privat, 2003), 105–15.

25 Froidour, *Les Pyrenées centrales au XVIIe siècle*, 96–97.

26 The recruitment of laborers in 1668 was timed to employ seasonal agricultural laborers after the harvest. See Lettre de Riquet à Colbert, 1668, ACM 07-02.

27 Lettre de Riquet à Colbert, 30 Octobre 1669, ACM 22-27.

28 See also Michel Adgé, "Chronologie des Principaux Évènements de la Construction du Canal, 1662–1694," in *Le Canal du Midi*, ed. Jean-Denis Bergasse (Cessenon: J.-D. Bergasse, 1985), 173–92. He describes the labor force during the first enterprise: January 1667, two thousand *ouvriers*, not by gender; July 1668, eight hundred ouvriers; November 1668, seven to eight thousand ouvriers, not

by gender; March 1669, seven to eight thousand *hommes*; September 1669, three thousand hommes; October 1669, sixty-five hundred hommes, and five to six hundred *femmes*; and December 1669, nine thousand hommes, and fifteen hundred femmes. Just as the number of women in the labor force was growing rapidly, the account books changed, and Adgé stopped this gender inventory. Unfortunately, the important role of women in the second enterprise was not visible to him as a result.

29 Richard Biernacki, *Fabrication of Labor: Germany and Britain, 1640–1914* (Berkeley: University of California Press, 1988).

30 ACM 1072. These account records with the high *liasse* numbers are in the "uncatalogued" boxes at the ACM. They are not consistently numbered.

31 Lettre de Riquet à Colbert, 30 Octobre 1669, ACM 22-27.

32 ACM 13-3.

33 ACM 1072-18; ACM 1072-40.

34 For an example of the cultural complexities of epithets and identity by practices of othering in a regional context, see Marisol de la Cadeña, *The Politics of Race and Culture in Cuzco, Peru, 1919–1991* (Durham, NC: Duke University Press, 2000). This work helps to substantiate the claim that the term held deep significance, but does not provide a means to illuminate what this might have meant in seventeenth-century Languedoc.

35 See Pierre Clément, *Lettres, Instructions et Mémoires de Colbert* (1867; repr., Liechtenstein: Kalus Reprints, 1979), 4:338–39; Lettre de Riquet à Colbert, 30 Octobre 1669, ACM 22-27.

36 For accounts from the Somail region, see ACM 1071; ACM 1072. On Campmas, Roux, and the alimentation system, see ACM 12-02. For more information on Contigny and his work on the Saint Ferréol dam, see Henri de Cazals, "Armorial du Canal," in *Le Canal du Midi*, ed. Jean-Denis Bergasse (Cessenon: J.-D. Bergasse, 1984), 3:151–79, particularly 167. See also Bertrand Gabolde, "Les Ouvriers du Chantier," in *Le Canal du Midi*, ed. Jean-Denis Bergasse (Cessenon: J.-D. Bergasse, 1984), 3:236; ACM 21-18; ACM 30-65. For the working patterns of Pyrenean women, see also Gratacos, *Femmes pyrénéennes*, 105–15. Some women may have stayed by the canal into the eighteenth century, and created their own towns there. In spring 1787, when Thomas Jefferson visited the south of France for his health and took a trip along the Canal du Midi, he reported that (to his astonishment) in some towns women took on roles ordinarily given to men. They were blacksmiths, carpenters, fishermen, butchers, and laborers, and the men of the towns did more domestic labor. Thomas Jefferson, *Le Voyage de Thomas Jefferson sur le Canal du Midi*, ed. Pierre Gérard (Portet-sur-Garonne: Éditions de Laboutière, 1995), 58–60.

37 ACM 12-02.

38 ACM 13-3.

39 There were many problems left to be solved ca. 1672, when the work was mainly left to women. See ACM 13-7.

40 Memoire de l'Etat des travaux du Canal fait par M. Riquet, n.d. (but from the description of the work, apparently the 1670s), ACM 13-14; Gratacos, *Femmes pyrénéennes*, 52.

41 Jean-François Soulet, *La Vie Quotidienne dans les Pyrénées sous l'ancien régime* (Paris: Hachette, 1974), 177–78.

42 For the survey of French mineral waters and Riquet's role in this project, see Lettre de Colbert à Riquet, 12 Janvier 1670, ACM 23-3; letter de Colbert à Riquet, 15 Février 1670, ACM 23-11. For Riquet's views on mountain women, see ACM 21-17, where Riquet discusses keeping the men and women of this area apart by controlling the towns. For Riquet's knowledge of the Midi-Pyrenees as part of his tax farming, see lettre de Colbert à Riquet, 31 Avril 1671, ACM 24-20, where he mentioned many of the areas in the mountains that had Pyrenean canal systems: Bigorre, Foix, and the Neste valley.

43 L. Malavialle, "Une Excursion dans la Montagne Noire," part 4, *Bulletin de la Société Languedocienne de Géographie* (1892): 15:184–86.

44 When Riquet went to recruit laborers in Bigorre in 1673, it was after Froidour had made his observations, and this perhaps encouraged Riquet to find more workers there.

45 Sextus Julius Frontinus, *The Two Books on the Water Supply of the City of Rome of Sextus Julius Frontinus, Water Commissioner of the City of Rome, A.D. 97* (New York: Longmans, Green, 1913).

46 Froidour, *Mémoire du Pays*, 20–22.

47 Frontinus, *Two Books on the Water Supply of the City of Rome.*

48 Giovannoni, "Building and Engineering," 429–74; particularly for Roman aqueducts and their need for continual repairs, see ibid., 465–68.

49 For another indication of the water engineering in the mountains in the period, see Claude Louis Hiacinthe de la Boulaye, "Proces Verbal de visites des Bois du Pays de l'Abour [Adour]," AHMV, SH no. 278, 3, 8. These maps from the Pyrenees (the area of the Adour River) show a dam along the river and what looks like the use of a diversionary canal to irrigate fields. For evidence of the Moorish approach to irrigation channels, see the levadas of Madeira, which had their counterparts southeast of Perpignan. And for the comparative lack of sophistication about the design of dams in the Roman tradition, see Norman Smith, "The Roman Dams of Subiaco," *Technology and Culture* 11, no. 1 (1970): 58–68.

50 Froidour, *Mémoire du Pays*, 115.

51 Ibid., 38–39.

52 Jean-François Soulet, *Les Pyrénées au XIXe siècle* (Bordeaux: Sud Ouest, 2004), 37–56.

53 Froidour, *Mémoire du Pays*, 20.

54 Froidour notes that the forests of the high mountains of Bigorre were held in common. Froidour, *Mémoire du Pays*, 20. The mines were also a central part of Pyrenean culture, as the mountains themselves were known as a site of treasures, and ones needing protection from outsiders who continually came to find these riches. For the patterns of collective management of natural resources in different parts of the Pyrenees, see Soulet, *Les Pyrénees au XIXe siècle*, 37–56, 71–83. See also Olivier de Marliave, *Trésor de la Mythologie Pyrénéene* (Luçon: Éditions Sud Ouest, 1996), 33–41.

55 Froidour, *Mémoire du Pays*, 20–21 (emphasis in original).

56 See, for example, Marliave, *Trésor de la Mythologie Pyrénéene*, 85–91, 150, 161.

57 Froidour, *Mémoire du Pays*, 31–32.

58 The Pyrenees themselves were female in the origin story often told here. The princess Pyrène was seduced by Hercules, and was devoured by beasts afterward. Hercules was upset, and built her a mausoleum that was the Pyrenees

mountains. Marliave, *Trésor de la Mythologie Pyrénéene*, 13. The apparitions of the virgin take many forms. She is associated with some of the mountain sources, but also appears in trees and is associated with the sun. See ibid., 21, 150, 161, 170. See also Gratacos, *Femmes pyrénéennes*, 131–83, particularly 143–71 (which describes the fairies of the mountains) and 177–80 (which describes how the traditional sighting of fairies of the region was turned into apparitions of the Virgin). For the most complete discussion of fairies, gender culture, and the power of women in the mountains, however, see Peter Sahlins, *Forest Rites: The War of the Demoiselles in Nineteenth-Century France* (Cambridge, MA: Harvard University Press, 1994), 40–60. Sahlins recounts stories told in the Midi-Pyrenees of fairies who were taken as wives, showing the fine line between ordinary women and these magical creatures in this area. Ibid., 44–45.

59 We know that the recruitment of laborers in 1668 was timed to employ seasonal agricultural laborers after the harvest. See the document sent to Colbert from Riquet, 1668, ACM 07-02. For the gender division of labor and women's work, see Gratacos, *Femmes pyrénéennes*, 105–8. The women of the Pyrenees were left to do labor in the valleys while the shepherd men traveled up into the meadows. This pattern was as true of Basque areas as the Catalan ones, and the areas of the central Pyrenees. The physical evidence supporting the association of women with these water systems is important, too. The canal systems were also unusually directed toward town water supplies, domestic water uses, and public laundries. The domestic power of these women was well-known. The public laundries might have seemed the most mundane and unimportant feature of these systems, but the fairies of the Pyrenees were always found while they were doing their laundry, and good luck was associated with carrying home their white garments. Female powers, then, were importantly connected to laundries, and the springs that fed them with waters. See ibid., 146–70.

60 Froidour, *Mémoire du Pays*, 31–32. There is also a spiritual significance associated with the development of domestic water supplies. At the summer solstice with the feast of Saint Jean, water was run from sources to houses in order to bless them, and it was also tapped to assure the success of agriculture that season. Marliave, *Trésor de la Mythologie Pyrénéene*, 57. Gratacos shows how the solstice festival was derived from pre-Christian sources that focused on fecundity and the powers of women. See Gratacos, *Femmes pyrénéennes*, 139–40, 143–46.

61 Froidour, *Mémoire du Pays*, 57–58.

62 A woman from this town who saw me photographing the mills in Ustou asked me if I was taking a picture of a particular, whitish stone. She explained that the stone had been her mother's laundry stone. This channel for bringing water to this mill was also the town's collective laundry.

63 Adkins and Adkins, *Handbook to Life in Ancient Rome*, 135.

64 Sahlins, *Forest Rites*, 40–60; Gratacos, *Femmes pyrénéennes*, 143–84.

65 De la Feuille's comments about the women were reported in lettre de Riquet à Colbert, 30 Octobre, 1669, ACM 22-27. The same letter reports that Riquet and de la Feuille were going into the mountains before meeting Clerville in Carcassone. See also the devis from intendant Bezons, also signed by M. Tournier in Carcassone in January 1667, ACM no. 17-2.

66 ACM 1072-1; ACM 1072-3.

67 For the first enterprise, see ACM 17-7. For the second one, see ACM 17-10.

68 For one of the requests for gunpowder, see Lettre de Riquet à Colbert, 4 Juin 1669, ACM 22-19.

69 ACM 1072; ACM 1073.

70 See Michel Adgé, "Le Canal du Midi, ou la jonction des mers en Languedoc," *Mappemonde* 1 (1992): 44–48, particularly 48.

71 L.T.C. Rolt, *From Sea to Sea: The Canal du Midi* (London: Allen Lance, 1973), 89.

72 For uncataloged accounts for the Somail region, see ACM 1071; ACM 1072. For the appointment of Campmas and Roux to the alimenation system, see ACM 12-02. For Contigny's work on the Saint Ferréol dam, see Cazals, "Armorial du Canal," 151–80.

73 These brothers did not have the surname of any of the early workers in the Montagne Noire, so apparently they were not from that area. See Adgé, "Les Premiers États du Barrage de Saint-Ferréol," 7–15, particularly 11.

74 Rolt, *From Sea to Sea*, 24–26. Borel was a respected thinker of his day. He entered the Académie Royale des Sciences as a chemist, and wrote on Cartesian science. See Joseph Frederick Scott, *The Scientific Work of René Descartes, 1596–1650* (Ann Arbor: University of Michigan Press, 2006), 84.

75 See Simon Singh, *Fermat's Enigma* (New York: Anchor Books, 1997), 35–44. Fermat was not at the Académie, but was a judge in Castres. Singh writes as though Marin Mersenne was the only mathematician of interest in the period, and attributes great importance to Fermat's meeting with this man. But the Huguenot Académie at Castres was internationally recognized in the period. See, for example, Thomas Hobbes, *Correspondence of Thomas Hobbes*, ed. Noel Malcolm (Oxford: Oxford University Press, 1994), 2:853.

76 See "Memoire sur le Canal royal de la jonction des Mers de Languedoc, 1er Mars 1672," ACM 13-15, which includes cautions against getting the elevation measures wrong, and the problems it could produce in locks, along with a description of the process of filling the cavities for locks with water to test the volume before doing the masonry for step locks. There was also discussion of the importance of knowing the characteristics of the soil in which the locks were being built. For descriptions of the multiple locks that had problems, and had to be rebuilt in part or whole, see "l'Estat auquel le Chevalier de Clerville a trouvé les ouvrages du canal de la jonction des mers," n.d., ACM 13-12. This document lists all the locks for the second enterprise, including a list of multiple locks in need of repair: St. Roch, Gay, Le Vivier, Foucault, Villaudy, and Puilaurier. The Académie at Castres was well-known in the period. See, for example, Hobbes, *Correspondence*, 2:853; Singh, *Fermat's Enigma*, 35–44.

77 ACM 12-07.

78 See Pierre Saliès, "De l'Isthme Gaulois au Canal des Deux Mers: Histoire des Canaux et Voies Fluviales du Midi," in *Le Canal du Midi*, ed. Jean-Denis Bergasse (Cessenon: J.-D. Bergasse, 1985), 4:55–97, particularly 67. Similar practices for floating logs were used on rivers in other parts of France, too, and similar step-type locks were used on some rivers, especially in connection with mills. Mills were best set in areas where the rivers had rapids, where the water could be diverted into the mill and fall down to the lower level of the river. In these areas, some rivers were also outfitted with locks used to make it easier to navigate down the rapids, or get past rocky areas or weirs when water was low. See Belidor, "De l'Usage des Eaux a la Guerre," 2:324. There was also a multilock staircase on the Canal de Briare, but the engineering was not as difficult because the hillside was

not as large or rocky so it could be made into a perfect incline plane, and the locks could be made the same size and shape. Hubert Pinsseau, "Du Canal de Briare au Canal des Deux Mers: Origines et conséquences d'un système inédit de navigation artificielle," in *Le Canal du Midi*, ed. Jean-Denis Bergasse (Cessenon: J.-D. Bergasse, 1985), 4:27–54, particularly the picture on 53.

CHAPTER 7
Thinking Like a King

1 Lettre de Riquet à Colbert, 27 Mars 1670, ACM 23-16.
2 Lettre de Riquet à Colbert, 1668, ACM 29-8.
3 Frank Lestringant, *Bernard Palissy, 1510–1590, l'écrivain, le réforme, le céramiste* (Coédition Associon Internationale des Amis d'Agrippa d'Aubigné-Saintes, Abbaye-aux-dames: Éditions SPEC, 1990), 167–80; Bernard Palissy, *A Delectable Garden, by Bernard Palissy*, trans. and ed. Helen Morganthau Fox (Peekskill, NY: Watch Hill Press, 1931), 7; Olivier de Serres, *Théâtre d'agriculture et mesnages des champs* (Geneva: Mat Hiev Berjon, 1611), 217.
4 Mack P. Holt, *The French Wars of Religion, 1562–1629* (Cambridge: Cambridge University Press, 2005), 99–101.
5 Ibid., 100.
6 Lettre de Colbert au sieur de la Feuille, 8 Novembre 1669, in Pierre Clément, *Lettres, Instructions et Mémoires de Colbert* (1867; repr., Liechtenstein: Klaus Reprints, 1979), 4:338–39.
7 Edwin Hutchins, *Cognition in the Wild* (Cambridge, MA: MIT Press, 1995).
8 Lettre de Riquet à Colbert, 27 Mars 1670, ACM 23-16.
9 In Goffman's terms, he redefined the situation and this allowed him to redefine himself in ways that were less socially damaging. See Erving Goffman, *The Presentation of Self in Everyday Life* (New York: Doubleday, 1959); Erving Goffman, *Behavior in Public Places* (Glencoe, IL: Free Press, 1963). Goffman provides nice tools for thinking about politics and the self that are too often dismissed.
10 Jean Bodin, *Six Books of the Commonwealth* (New York: Barnes and Noble, 1967).
11 James Scott, *Seeing Like a State* (New Haven, CT: Yale University Press, 1998).
12 "Edit du Roy pour la construction d'un canal du communication des deux mers, Océane & Méditerrannée," 14 Octobre 1666, ACM 03-10, 5–7.
13 Michel de Certeau, *The Writing of History*, trans. Tom Conley (New York: Columbia University Press, 1988), 8–11. Compare to Bodin, *Six Books of the Commonwealth*; Roland Mousnier, *The Institutions of France under the Absolute Monarchy, 1598–1789: Society and the State*, trans. Brian Pearce (Chicago: University of Chicago Press, 1979), 645–720.
14 Arquier, *Avis a messieurs les Capitouls de la Ville de Tolose, par Arquier, Doyen de anciens Capitouls. Et Response a cet Avis, Article par Jean de Nivelle, Ancien Capitaine Chasseauant du Canal dans l'atelier de Mr. Sagadenes*, 1667, ACM 01-16.
15 Michel Adgé, "Chronologie des Principaux Évènements de la Construction du Canal, 1662–1694," in *Le Canal du Midi*, ed. Jean-Denis Bergasse (Cessenon: J.-D. Bergasse, 1985), 182–87.

16 Lettre de Colbert à Riquet, 27 Juin 1669, in Clément, *Lettres*, 4:332; Adgé, "Chronologie des Principaux Évènements," 184.

17 "Instruction pour le sieur de Vos, charpentier," 1 Septembre 1671, in Clément, *Lettres*, 4:359–60.

18 Adgé, "Chronologie des Principaux Évènements," 186.

19 Lettre de Colbert à Clerville, 10 Juin 1673, in Clément, *Lettres*, 5:84.

20 Lettre de Riquet à Colbert, 12 Avril 1667, ACM 20-33.

21 Lettre de Riquet à Bezons, 6 Janvier 1671, ACM 29-11.

22 Carlo Ginzburg, *The Cheese and the Worms: The Cosmos of a Sixteenth Century Miller* (Baltimore: Johns Hopkins University Press, 1980).

23 Lettre de Riquet à Colbert, 13 Mars 1669, ACM 22-8 (emphasis added).

24 Riquet wrote, "The work of the man who made the model [for the fountain] will be known to you. He is a very good sculptor hidden in this village and as a consequence not well-known. But according to my taste, he excels in his art. I believe he is capable of the execution of this grand design. He lived for a long time in Rome, working for twelve years under the Cavalier Berniny [Bernini]. And he is the same person who discovered the beautiful white marbles that we use in our pavements and come down the Garonne. That is why I find in this person the man for the job, and in the Pyrenees the material for realizing the design" (Lettre de Riquet à Colbert, 13 Mars 1669, ACM 22-08).

25 "Questions faites a Mr. Riquet par monsieur de Besons [Bezons] intendant en Languedoc, sur le sujet du Canal de communication des mers," 27 Avril 1670, ACM 13-02. This made the port roughly 83.6 meters square.

26 Lettre de Colbert à M. de Bezons, 26 Juin 1669, in Clément, *Lettres*, 4:330–31.

27 L.T.C. Rolt, *From Sea to Sea: The Canal du Midi* (London: Allen Lance, 1973), 63–65; Lettre de Colbert à M. de Bezons, 26 Juin 1669, in Clément, *Lettres*, 4:330–31; Bertrand Gabolde, "Revel: Des eaux du Sor à la rigole de la plaine," in *Le Canal du Midi*, ed. Jean-Denis Bergasse (Cessenon: J.-D. Bergasse, 1985), 241–44; Adgé, "Chronologie des Principaux Évènements," 181.

28 Lettre de Riquet à Colbert, 4 Août 1671, ACM 24-29.

29 Ibid. Colbert must have had some inkling of this since he asked de la Feuille to go to these rivers, and see what work was done there. See Anne Blanchard, Michel Adgé, and Jean-Denis Bergasse, "Les Ingénieurs du Roy," in *Le Canal du Midi*, ed. Jean-Denis Bergasse (Cessenon: J.-D. Bergasse, 1984), 186.

30 "Memoire de mes remarques du Canal, 9 Octobre 1670," by M. de Seguelay, ACM 13-05; Lettre de Riquet à Colbert, 13 Mars 1669, ACM 22-08. See also Louis de Froidour, *Lettre a M. Barrillon Damoncourt cotenant la relation & la description des travaux qui se font en Languedoc pour la communication des deux mers* (Toulouse: Dominique Camusat, 1672).

31 Adgé, "Chronologie des Principaux Évènements," 182–84; "Extrait des registres du Conseil d'Estat," ACM 11-01.

32 Joseph de Lalande, *Des canaux de navigation, et spécialement du canal de Languedoc* (Paris: Veuve Desaint, 1778), 21.

33 Arnaud Ramière de Fontanier, "Du Grand Canal au Grand Bassin: Fêtes sur le canal du Midi aux XVIIe et XVIIIe siècles," in *Le Canal du Midi et les Voies Navigables dans le Midi de la France*, actes du congrès des fédérations histo- riques languedonciennes (Carcassone: Société d'Études Scientifiques de l'Aude, 1998), 241–48, quote from 243.

34 "Extrait des registres du Conseil d'Etat," 1671, ACM 11-01; Lalande, *Des canaux de navigation*, 21.

35 "Extrait des registres du Conseil d'Etat," 1671, ACM 11-01; Lalande, *Des canaux de navigation*, 21. Interestingly, the route to Castelnaudery required some contour cutting, and along the canal in this region, there were a number of public laundries, perhaps suggesting that Pyrenean women did part of this work.

36 Lettre de Riquet à Colbert, 18 Janvier 1670, ACM 23-4.

37 Ibid.

38 Ibid.

39 Alain Degage, "Le Port de Sète: Proue Méditerranéenne du Canal de Riquet," in *Le Canal du Midi*, ed. Jean-Denis Bergasse (Cessenon: J.-D. Bergasse, 1985), 272.

40 Lettre de Riquet à Colbert, 14 Octobre 1671.

41 Lettre de Riquet à Colbert, 29 Octobre 1676, ACM 31-31.

42 Lettre de Clerville à Riquet, n.d., ACM 31-37.

43 Blanchard, Adgé, and Bergasse, "Les Ingénieurs du Roy," 188.

44 Rolt, *From Sea to Sea*, 113.

45 Ibid., 91, 94–95; "Notes sur l'Histoire du Canal de Languedoc par les Descendents de Riquet," 1803, uncataloged manuscript, ACM.

46 Rolt, *From Sea to Sea*, 93–94.

47 Martine Schwaller, *Ensérune* (Paris: Editions du patrimoine, Imprimerie nationale, 1994), 22–27.

48 Rolt, *From Sea to Sea*, 94–95.

49 Riquet was given permission to join a consortium to acquire and exploit mines in the province, and build forges for refining the ore, led by a Swedish mining expert, the sieur de Besche. Colbert had been told there might be sources of silver there, and was very interested in it. These were ancient mines that had fallen into disrepair, but when they were reopened, they had little left in the way of ore. Mines went broke, and workers went unpaid and soon left their posts. Riquet argued that he should be given some of the mines because he needed metal for hardware on lock doors, and for mills and other buildings. Having his own supplies of materials held the promise of reducing his costs. Colbert had misgivings about Riquet's investment in new ventures when he was so far behind with the canal itself. De la Feuille cautioned him about this as well, but Riquet's proposal did seem a reasonable way to reduce costs, so Colbert approved. "Extrait des registres du Conseil d'Estat," 10 Juillet 1668, ACM 18-01. For the history of efforts in mining in Languedoc, see Clement, *Lettres*, 4:413–46, cxxv–cxxxi.

50 Janice Langins, *Conserving the Enlightenment: French Military Engineering from Vauban to the Revolution* (Cambridge, MA: MIT Press, 2004); Sébastien Le Prestre de Vauban, *A Manuel of Siegecraft and Fortification*, trans. (Ann Arbor: University of Michigan Press, 1968). For the inconclusiveness of the evidence on Clerville's view of Malpas, see Blanchard, Adgé, and Bergasse, "Les Ingénieurs du Roy," 187.

51 Rolt, *From Sea to Sea*, 93.

52 Rolt says Riquet was cut off financially by Colbert for this disobedience, but there were still funds for the canal being paid by the treasury. See Rolt, *From Sea to Sea*, 95. Compare to Jules Guiffrey, ed., *Comptes des Bâtiments du Roi. Sous le Règne de Louis XIX*, 1st ed. (Paris: Imprimerie Nationale, 1881), 943,

1018, 1120. What seems to have happened is that the money for the contract for the second enterprise was spent, and after this point, the only funds for the canal came from the taxes designated for this purpose.

53 Clément, *Lettres*, 4:xciv; Lettre de Colbert à Clerville, 10 Juin 1673, in Clément, *Lettres*, 5:84; Blanchard, Adgé, and Bergasse, "Les Ingénieurs du Roy," 187.

54 Colbert's letter to d'Aguesseau in 1677, in Rolt, *From Sea to Sea*, 91–92.

55 Nivelle, in "Avis a messieurs les Capitouls de la Ville de Tolose, par Arquier, Doyen des anciens Capitouls 1667," ACM 01-16.

56 Supplement of 1,575,452 livres, 13 sols, and 4 deniers in 1672–1673; a base payment on December 7, 1672 of 126,041 livres, 13 sols, and 4 deniers; a regular payment in 1674 of 564,416 livres, 13 sols, and 4 deniers; a regular payment of 54,175 livres in 1675; an extraordinary payment for 1673–1675 of 1,235,242 livres, and from 1677 until the end of the 1670s, 566,666 livres, 13 sols, and 4 deniers based in part on augmentations from the gabelles and the États du Languedoc. See Guiffrey, *Comptes des Bâtiments*, 316–17, 369, 392, 479, 554, 680, 682, 735–36, 804, 877, 943, 1018, 1120.

57 Émile Durkheim, *Suicide: A Study in Sociology*, ed. John Spaulding and George Simpson (London: Routledge, 1952).

58 Inès Murat, *Colbert*, trans. Robert Francis Cook and Jeannie Van Asselt (Charlottesville: University Press of Virginia, 1984), 74–112.

59 Bodin, *Six Books of the Commonwealth*.

60 Chandra Mukerji, *Territorial Ambitions and the Gardens of Versailles* (Cambridge: Cambridge University Press, 1997), 269–72.

61 Lettre de Colbert à d'Aguesseau, 1677, in Rolt, *From Sea to Sea*, 91–92.

62 Michel Foucault, *Order of Things: An Archeology of the Human Sciences* (New York: Pantheon Books, 1970).

63 Lettre de Colbert à Clerville, 10 Juin 1673, in Clément, *Lettres*, 5:84; Chandra Mukerji, "The Unintended State," in *New Cultural Materialisms*, ed. Patrick Joyce and Tony Bennett (New York: Routledge, forthcoming).

64 Lettre de Clerville à Riquet, 7 Mai 167[?], ACM 31-43.

65 Riquet's request to change the locks can be found in his letter of March 19, 1670, ACM 23-15. For Riquet's excitement about his new design of the lock doors, see his letters to Colbert, ACM 23-16, 23-47. And for Colbert's opposition to the plan, see Lettre de Colbert à Riquet, 19 Avril 1670, in Clément, *Lettres*, 4:345.

66 "Bail et Adjuducation des ouvrages a faire, 1668," 1669, ACM 07-12; Lettre de Colbert à Riquet, 27 Juin 1669, and lettre de Colbert à Riquet, 13 Octobre 1669, in Clément, *Lettres*, 4:332, 336–37.

67 Holt, *The French Wars of Religion*, 100–101.

68 Chandra Mukerji, "The Great Forest Survey of 1669–1671," *Social Studies of Science* 37, no. 2 (2007): 227–53.

69 William Beik, *Absolutism and Society in Seventeenth-century France: State Power and Provincial Aristocracy in Languedoc* (Cambridge: Cambridge University Press, 1985); Holt, *The French Wars of Religion*, 100–101.

70 For a different but provocative discussion of governmentality and the built environment, see Patrick Joyce, *Rule of Freedom: Liberalism and the City in Britain* (New York: Verso, 2003).

71 The powers were defined in the "Edit du Roy," 1666, ACM 3-10, 5–7. Requests for state powers are illustrated by a few examples: Lettre de Riquet à Colbert, 7 Decembre 1668, ACM 21-27; lettre de Riquet à Colbert, 12 Janvier 1670, ACM

23-3; lettre de Riquet à Colbert, 12 Fevrier 1670, ACM 23-9; orders from Bazins, ACM 17-2, 17-3, 17-13, 18-9.

72 Lucien Bély, *Dictionnaire de l'Ancien Régime* (Paris: Quadrige PUF, 2005), 510.

73 Lettre de Colbert à Riquet, 14 Decembre 1668, in Clément, *Lettres*, 4:316–17.

74 Riquet was awaiting their decision on October 1, 1670. See lettre de Riquet à Colbert, 1 Octobre 1670, ACM 23-51.

75 For the request for an exemption from duties on wood imported through the Rhône, see lettre de Riquet à Colbert, 1 Mai 1671, ACM 24-19. For the edicts about housing workers (similar to demands for housing soldiers), see ACM 17-2; ACM 17-13.

76 Rolt, *From Sea to Sea*, 76–77.

77 Lettre de Riquet à Colbert, 30 Octobre 1668, ACM 21-20.

78 Ibid.

79 Lettre de Riquet à Colbert, n.d., ACM 21-30.

80 Quoted without citation in Rolt, *From Sea to Sea*, 77.

81 Lettre de Riquet à Colbert, 30 Octobre 1668, ACM 21-20.

82 Rolt, *From Sea to Sea*, 77.

83 Lettre de Colbert à de la Feuille, 19 Julliet 1669, in Jean-Baptiste Colbert, *Registre contenant les ordres, instructions et lettres expediees par Monseigneur Colbert, touchant les fortifications des ponts et chausseees, canal de communication des mers et mines de Languedoc en l'annee 1669*, EPC, 46–47.

84 Boyer and d'Estan's connection with Clerville and work at Rochefort are suggested by some of Clerville's letters. See ACM 31-39; ACM 31-40; ACM 31-43. The soldier clearly had confidence in them before 1678. Papers from Boyer from a dossier on the pont-canal de Répudre are in the ACM uncataloged papers from 1676–1678, starting with "Memoires pour M. le President Riquet," by de la Feuille, n.d., and containing Boyer's testimony that both men worked together at Rochefort. Clerville's verification is in "Estat auquel le Chevalier de Clerville a trouvé les ouvrages du canal de la jonction des mers," n.d., ACM 13-12, 6.

85 Lettre de Clerville à Riquet, 29 Octobre 1676, ACM 31-31.

86 Ibid.; lettre de Clerville à Riquet, 2 Janvier 1677, ACM 31-34; lettre de Clerville à Riquet, 22 Janvier 1677, ACM 31-41; lettre de Clerville à Riquet, [?] Fevrier 1677, ACM 31-36.

87 For payment by volume, see "Devis des ouvrages de Massonerie . . . a la construction d'un pont a la Riviere de Repudre," Oct. 1677, ACM uncataloged paper, 2A, 4. See also "Estat des sommes dont lesr. Emmanuel destan architecti doibt tenir compte sur les ouvrags des Escluses doignon et pont derepudre," 15 Mars 1680, ACM uncataloged paper, 2A, 39-9, 1. It shows that Boyer (jeune) and d'Estan originally took the contract for the locks and the pont-canal together. They separated their work in 1678, Boyer continuing with "hompt et jouarre," and d'Estan continuing with le Repude [Répudre] and "doignon" [d'Oignon]. These sites along the canal were in the same general area, and were described in "Bail et Adjudication des Ouvrages a Faire pour la Continuation du Canal et du Port du Cette [Sète]," 1669, sections XI–XIV, ACM 07-12, 19.

88 "Estat auquel le Chevalier de Clerville a trouvé les ouvrages du canal de la jonction des mers," n.d., ACM 13-12, 6–7.

89 "Memoire pour M. le President Riquet," uncataloged paper, ACM.

90 Jacques Heyman, *The Stone Skeleton: Structural Engineering of Masonry Architecture* (Cambridge: Cambridge University Press, 1995).

91 "Memoires pour M. le President Riquet," uncataloged papers in a dossier about Répudre at the ACM from 1676–1678, starting with "Memoires pour M. le President Riquet," n.d.

92 Letter from Riquet in the uncataloged papers from the dossier on Répudre at the ACM from 1676–1678, starting with "Memoires pour M. le President Riquet," n.d.

93 The receipts for all of the work added up to 42,998 livres, 13 sols, 4 deniers. Since d'Estan did most of the work, he was owed the sum of 32,832 livres 5 sols, 8 deniers, while Boyer was only due 10,166 livres.

94 Documents signed by Pons de la Feuille in 1678 and 1680 in the uncataloged papers from a dossier on Répudre at the ACM from 1676–1678, starting with "Memoires pour M. le President Riquet," n.d.

CHAPTER 8
Monumental Achievement

1 Lettre de Colbert à Riquet, 30 Novembre 1672, in Pierre Clément, *Lettres, Instructions et Mémoires de Colbert* (1867; repr., Liechtenstein: Klaus Reprints, 1979), 4:366–67.

2 Lettre de Clerville à Riquet, 22 Février 1673, ACM 31-28.

3 Anne Blanchard, Michel Adgé, and Jean-Denis Bergasse, "Les Ingénieurs du Roy," in *Le Canal du Midi*, ed. Jean-Denis Bergasse (Cessenon: J.-D. Bergasse, 1984), 3:186–87; lettre de Clerville à Riquet, 1 Juillet 1669, ACM 31-27.

4 Lettre de Clerville à Riquet, 22 Février 1673, ACM 31-28.

5 Lettre de Clerville à Riquet, 25 Juillet 1674, ACM 31-29; lettre de Clerville à Riquet, 11 Mai 1669, ACM 31-25.

6 Pierre Clément, *Lettres, Instructions et Mémoires de Colbert* (1867; repr., Liechtenstein: Klaus Reprints, 1979), 4:373–74; Alain Degage, "Le Port de Sète: Proue Méditerranéenne du Canal de Riquet," in *Le Canal du Midi*, ed. Jean-Denis Bergasse (Cessenon: J.-D. Bergasse, 1985), 4:270–74.

7 L.T.C. Rolt, *From Sea to Sea: The Canal du Midi* (London: Allen Lance, 1973), 93–95.

8 Ibid.

9 Clément, *Lettres*, 4:374–81.

10 Lettre de Colbert à d'Aguesseau, 17 Novembre 1679, in Clément, *Lettres*, 4:388.

11 Lettre de d'Aguesseau à Colbert, 16 Avril 1679, in Clément, *Lettres*, 4:386–87.

12 Lettre de Colbert à d'Aguesseau, 6 Septembre 1679, in Clément, *Lettres*, 4:386.

13 Jules Guiffrey, ed., *Comptes des Bâtiments du Roi. Sous le Règne de Louis XIX*, 1st ed. (Paris: Imprimerie Nationale, 1881), 316–18, 369, 392, 479, 554, 680, 682, 735–36, 804, 877, 943, 1018, 1120.

14 Inès Murat, *Colbert*, trans. Robert Francis Cook and Jeannie Van Asselt (Charlottesville: University Press of Virginia, 1984), 361–76; Robert Villers, "Colbert et les finances publiques," in *Colloque pour le tricentenaire de la mort de Colbert, sous la direction de Roland Mousnier, Un nouveau Colbert*, ed. Jean Favier and Roland Mousnier (Paris: Éditions Sedes, 1985), 177–87.

15 Blanchard, Adgé, and Bergasse, "Les Ingénieurs du Roy," 187–88.

16 Clément, *Lettres*, 4:xcvii.

17 Ibid.

18 ACM 12-10; *Notes sur l'histoire du Canal du Languedoc par les descendents de Riquet*, 1803., ACM uncataloged typescript.

19 Lettre de d'Aguesseau à Colbert, 5 Octobre 1680, and letter de d'Aguesseau à Colbert, 19 Octobre1680, in Clément, *Lettres*, 4:598–99.

20 "Histoire des Descendents," 148–49.

21 Lettre de Colbert à d'Aguesseau, Octobre 1680, in Clément, *Lettres*, 4:389–90.

22 Lettre de Colbert à d'Aguesseau, 17 Novembre 1679, and letter de Colbert à d'Aguesseau, 17 Octobre 1680, in Clément, *Lettres*, 4:388–89.

23 For the sociology of repair work, see Christopher Henke, "The Mechanics of Workplace Order: Toward a Sociology of Repair," *Berkeley Journal of Sociology* 44 (2000): 55–81; Christopher Henke, "Repairing the Nuclear Stockpile: Aging Technology, Tacit Knowledge, and the Future of the Nuclear Weapons Complex" (paper presented at the American Sociological Association meetings, New York, 2007).

24 Anon or François Andreossy(?), "La première navigation sur le canal du Langudedoc, fait par ordre du Roy our la jonction des deux Mers, depuis Toulouse jusques au port de Cette," publiée à Toulouse chez Jean Boudé, ca.1691. In Arnaud Ramière de Fontanier, "Du Grand Canal au Grand Bassin: Fêtes sur le canal du Midi aux XVIIe et XVIIIe siècles," in *Le Canal du Midi et les Voies Navigables dans le Midi de la France*. Actes du congrès des fédérations historiques languedociennes (Carcassone: Société d'Études Scientifiques de l'Aude, 1998), 243–44.

25 Andreossy, "Le première navigation," 243. Selections from the 1681 navigation and an analysis of this event can also be found in Denise Viennet, "Quelques voyages sur le Canal du Midi pendant trois cent ans," in *Le Canal du Midi*, ed. Jean-Denis Bergasse (Cessenon: J.-D. Bergasse, 1982), 2:119–21.

26 Ibid., 119. See also Jean Hugon de Scoeux, *Le Chemin qui Marche* (Toulouse: Loubatières, 1995), 229–30.

27 Andreossy, "Le première navigation," 243.

28 Ibid., 243–44.

29 Ibid., 244.

30 François de Dainville, *Cartes Anciennes du Languedoc XVIe–XVIIIe S* (Montpellier: Société Languedocienne de Géographie, 1961), 50–51.

31 Ibid.

32 Dainville chronicles the difficulties of trying to use maps for "action at a distance." See ibid., 50–55.

33 Ibid., 51–53.

34 Bruno Latour, "Drawing Things Together," in *Representation in Scientific Activity*, ed. Michael Lynch and Steve Woolgar (Cambridge, MA: MIT Press, 1990), 19–68.

35 The problem of language and shifting conceptual understandings of things is described in James Griesemer, "Tracking Organic Processes: Representations and Research Styles in Classical Embryology and Genetics," in *From Embryology to Evo-Devo: A History of Developmental Evolution*, ed. Manfred D. Laubichler and Jane Maienschein (Cambridge, MA: MIT Press, 2007), 375–433. Colbert's search for a language for describing "working" in his own terms demonstrated both the disparity between his patrimonial understandings of service and the path by which he reached at a model of stewardship as service. For the invention of language where there is none, a problem for linguists with some similarities, see, for example, Mark Aronoff, Irit Meir, Carol Padden, and

Wendy Sandler, "The Roots of Linguistic Organization in a New Language," *Interaction Studies* 9, no. 1 (2008): 131–50.

36 Bart Simon, *Undead Science: Science Studies and the Afterlife of Cold Fusion* (New Brunswick, NJ: Rutgers University Press, 2002).

37 The account books after the 1680s documented the shifts in practical authority. The "Etats Sommaire des ouvrages a faire au canal apres la visite de 1681" (uncataloged papers from the ACM, États et projets, 1681–1739), listing all the work still required to perfect the canal, were signed mainly by Andréossy and Gillade. In 1682 and 1683, estimations of the costs for the work were signed by de la Feuille but apparently supervised by Mourgues, and addressed (in the later document) to de Seignelay. And the final accounting of the work needed to complete the canal was written by d'Aguesseau based on measures by Gillade and Mourgues ("Procez Verbal de Verification & Estimation des Ouvrages du Canal par M. Daguesseau," 16 Juillet 1684 ACM 15-3).

38 Blanchard, Adgé, and Bergasse, "Les Ingénieurs du Roy," 185, 188.

39 Clément, *Lettres*, 4:xcix–c (emphasis added).

40 For Vauban's criticisms of Clerville's engineering, see Anne Blanchard, *Vauban* (Paris: Fayard, 1996), 133–39, 155, 162. For Vauban's use of Clerville's plans, see ibid., 265–67; Blanchard, Adgé, and Bergasse, "Les Ingénieurs du Roy," 188–89.

41 Blanchard, Adgé, and Bergasse, "Les Ingénieurs du Roy," 188–91.

42 L. Malavialle, "Une Excursion dans la Montagne Noire," part 1, *Bulletin de la Société Languedocienne de Géographie* (1891), 14:260.

43 Rolt, *From Sea to Sea*, 59; François Gazelle, "Riquet et les eaux de la Montagne Noire: L'idée géniale de l' alimentation du canal," in *Le Canal du Midi*, ed. Jean-Denis Bergasse (Cessenon: J.-D. Bergasse, 1985), 155; Malavialle, "Une Excursion dans la Montagne Noire," part 1, 14:259–61.

44 "Procez Verbal de Verification & Estimation des Ouvrages du Canal par M. Daguesseau," 16 Juillet 1684, ACM 15-3, 2.

45 Ibid.

46 For an engineering drawing of one of the aqueduct bridges designed by Vauban, see *Conseil d'Architecture, d'Urbanisme et de l'Environment de la Haute-Garonne, Canal royal de Languedoc: le partage des eaux* (Caue: Loubatières, 1992), 99.

47 Rolt, *From Sea to Sea*, 59–60.

48 "Memoire de l'estat du Canal de jonction des mers, avec l'estimation des reparations qui doivent y este faites prompement par M. R. de Bonrepos," 28 Novembre 1683, unnumbered and uncataloged manuscript, ACM.

49 Blanchard, Adgé, and Bergasse, "Les Ingénieurs du Roy," 189–90.

50 "Fragment d'un mémoire général et historique dressé par le sieur de Rousset," n.d., ACM 484-11.

51 For some of the complicated relationships among these men, see Blanchard, *Vauban*, 250–62; Dainville, *Cartes Anciennes du Languedoc*, 56, 63; Josef W. Konvitz, *Cities and the Sea: Port City Planning in Early Modern Europe* (Baltimore: Johns Hopkins University Press, 1978), 79–80.

52 Gerhard Mercator, *Historia Mundi*, 2nd ed. (London: Michael Starke, 1637), A3.

53 Robin Wagner-Pacifici, *The Art of Surrender: Decomposing Sovereignty at Conflict's End* (Chicago: University of Chicago Press, 2005), 76–66.

54 Chandra Mukerji, "Printing, Cartography, and Conceptions of Place," *Media, Culture, and Society* 28, no. 5 (2006): 651–69, particularly 655–58; Thomas

Frangenberg, "Chorographies of Florence: The Use of City Views and City Plans in the Sixteenth Century." *Imago Mundi* 46 (1994): 41–64.

55 Map by Andréossy, 1681.

56 "Carte du Canal Royal de la Province de Languedoc sur l'échelle de 5. Lignes pour 100. Toises divisé en 4 parties, subdivisée en 15 feuilles, a été levée par ordre et aux frais des États Généraux . . . sour la Direction de Mr. Garipuy, Ingénieur Directeur des Ponts et Chaussées," 1774.

57 The problematic nature of the classical past, and the debates about purifying or maintaining it, are discussed in Joan Cadden, " 'Nothing Natural Is Shameful': Vestiges of a Debate about Sex and Science in a Group of Late Medieval Manuscripts," *Speculum* 76 (2001): 66–89. For some of the literature used to rewrite the history of the Canal du Midi to give it the appropriate stature for the New Rome, see Bouillet, *Traite des Moyens de render les Rivieres Navigables avec plusieurs desseins de jettées* (Paris: Chez Estienne Michallet, 1693); Bernard Forest de Belidor, "De l'Usage des Eaux a la Guerre," in *Architecture Hydraulique seconde partie qui comprend l'Art de diriger les eaux des la Mer & des Rivieres à l'avantage de la défense des places, du Commerce & de l'Agriculture. par M. Belidor, Colonel d'Infanterie* (Paris: Jombert, 1753); Charles Vallancey, *A Treatise on Inland Navigation or the Art of Making Rivers Navigable: Of Making Canals in All Sorts of Soils, and of Constructing Locks and Sluices* (Dublin: George and Alexander Ewing, 1763); Joseph de Lalande, *Des canaux de navigation, et spécialement du canal de Languedoc* (Paris: Veuve Desaint, 1778).

58 For a description of Schutz's theory and its usefulness for thinking about non-linear time relations in social practices, see Alex Preda, "Where Do Analysts Come from? The Case of Financial Chartism" in *Market Devices*, ed. Michel Callon, Fabian Muniesa, and *Yuval Millo* (Chicago: University of Chicago Press, 2007).

59 Belidor, "De l'Usage des Eaux a la Guerre," 2:233.

60 Ibid., 4:359–60.

61 Ibid., 4:362.

62 Ibid., 4:363.

63 Ibid., 4:364.

64 Ibid., 4:364–65.

65 Lalande, *Des canaux de navigation.*

66 Ibid., iv.

67 Translation by Stephanie Berger of a poem by Pierre Corneille, quoted in ibid., iv–v:

> La Garonne & l'Atax (Aude) dans leurs grottes profondes
> Soupiroient de tout temps pour voir unir leurs ondes;
> Et faire ainsi couler par un heureux penchant
> Les trésors de l'Aurore aux rives Couchant;
> Mais à des voeux si doux, à des flammes si belles
> La nature, attachée à ses loix éternelles,
> Pour obstacles invincible opposoit fièrement
> Des mons & des rochers l'affreux enchaînement.
> France, ton grand Roi parle, & les rochers se sendent;
> La terre ouvre son sein, les plus monts descendent;

Tout cede, & l'eau qui suit les passages ouverts,
Le fait voir tout puissant sur la terre & les mers.

68 M. l'Abbé Maumanet quoted by Lalande in *Des canaux de navigation*, vj.
69 Ibid., 140.
70 Ibid., 30–37.
71 Ibid., 5–6 (emphasis added).
72 Ibid., iij.
73 See, for example, ibid., 25–26, where Lalande describes the inspection in 1681 of the work by de la Feuille, Mourgues, and d'Aguesseau. This inspection and the first official use of the waterway was discussed, he said, in the *Journal des Savans*, June 30, 1681
74 Four of the latter were written before October 1665, when the project seemed full of promise and the patronage ties between the two men were happily being forged. None of the letters mentioned tax problems, or accusations against Riquet by the États or Colbert. The letters from 1667 referred to Riquet's hopes of finishing the first enterprise early. A letter from Colbert dated May 19, 1668, expressed the minister's satisfaction on hearing that a boat had successfully navigated from the "derivation" (perhaps of the Sor) to Naurouze. There was even a copy of the letter from Colbert sent during Riquet's illness, praising the ailing entrepreneur.
75 Lalande, *Des canaux de navigation*, 119–36.
76 Ibid., 136–53.
77 See Rolt's discussion of his sources, two books, one of which was certainly Lalande. Rolt, *From Sea to Sea*, 5.
78 Vallancey, *A Treatise on Inland Navigation*, 109.
79 Ibid., 116–17.
80 Chandra Mukerji, "Cultural Genealogy," *Cultural Sociology* 1, no. 1 (2007).
81 "Histoire du Canal par les descendents de Riquet," 162.

CHAPTER 9
Powers of Impersonal Rule

1 Bruno Latour, *Pandora's Hope: On the Reality of Science Studies* (Cambridge, MA: Harvard University Press, 1999).
2 For the role of technology in political modernism, see Andrew Feenberg, *Questioning Technology* (London: Routledge, 1999).
3 For ongoing political efforts to mask these in effective moments in the struggle to control nature, see Paul Edwards, *The Closed World: Computers and the Politics of Discourse in Cold War America* (Cambridge, MA: MIT Press, 1996); Gabrielle Hecht, *The Radiance of France: Nuclear Power and National Identity after WWII* (Cambridge, MA: MIT Press, 1998); Diane Vaughan, *The Challenger Launch Decision: Risky Technology, Culture, and Deviance at NASA* (Chicago: University of Chicago Press, 1996).
4 Michel Callon, "Some Elements of a Sociology of Translation: Domestication of the Scallops and the Fishermen of St. Brieuc Bay," in Power, Action, and Belief, ed. John Law (London: Routledge and Kegan Paul, 1986), 196–233; see also Michel Callon and Bruno Latour, "Don't Throw the Baby out with the Bath

School!" in *Science as Practice and Culture*, ed. Andrew Pickering (Chicago: University of Chicago Press, 1992), 343–68.

5 Compare to the concept of boundary object in Susan Leigh Star and James R. Griesemer, "Institutional Ecology, 'Translations,' and Coherence: Amateurs and Professionals in Berkeley's Museum of Vertebrate Zoology, 1907–1939," *Social Studies of Science* 19 (1989): 387–420; Joan Fujimura, "Crafting Science: Standardized Packages, Boundary Objects, and 'Translation,'" in *Science as Practice and Culture*, ed. Andrew Pickering (Chicago: University of Chicago Press, 1992), 168–214.

6 See the discussion of Martin Heidegger in Bruno Latour, "Why Has Critique Run out of Steam?" *Critical Inquiry* 30 (2004): 238–39. See also Callon, "Some Elements of a Sociology of Translation"; Michel Serres, *Rome, le livre des foundations* (Paris: Grasset, 1983); Chandra Mukerji, "Intelligent Uses of Engineering and the Legitimacy of State Power," *Technology and Culture* 44, no. 4 (2003): 655–76.

7 Latour, *Pandora's Hope*, 98–108, particularly the diagram on 100. What I try to show in my reappropriation of Latour's model is the salience of working and the logistics of impersonal rule for explaining the knot that Latour leaves ambiguous.

8 Conel Condren, "Liberty of Office and Its Defence in Seventeenth-century Political Argument," *History of Thought* 18, no. (1997): 460–82.

9 This metaphor refers to spreading centers on the seafloor where new crust for the earth is being produced.

10 Compare to Condren, "Liberty of Office."

11 Compare this to the distributed networks cultivated by Steward Brand and his colleagues, described in Fred Turner, *From Counter Culture to Cyber Culture: Stewart Brand, the Whole Earth Network, and the Rise of Digital Utopianism* (Chicago: University of Chicago Press, 2006). The point of this form of organization for members of this counterculture was precisely to avoid organizational capture. In this history of the Canal du Midi, it is clear that this escape is not an escape at all but rather a way of empowering the state through collaborations that politicians cannot control, yet they can certainly exploit. Turner's association of the counterculturalists with the rise of neoliberalism in Washington, DC, makes this point nicely.

12 Norbert Elias, *The Court Society* (New York: Pantheon, 1998).

13 Chandra Mukerji, "The Collective Construction of Scientific Genius," in *Cognition and Communication at Work*, ed. Yrjö Engeström and David Middleton (Cambridge: Cambridge University Press, 1997).

14 James Scott, *Weapons of the Weak: Everyday Forms of Peasant Resistance* (New Haven, CT: Yale University Press, 1985).

15 Chandra Mukerji, "Stewardship Politics and the Control of Wild Weather: Levees, Seawalls, and State Building in 17th-century France," *Social Studies of Science* 37, no. 1 (February 2007): 111–18; Pierre Goubert, *Louis XIV and Twenty Million Frenchmen*, trans Ann Carter (New York: Vintage Books, 1972).

16 Bruno Latour, "Visualisation and Cognition: Thinking with Eyes and Hands," *Knowledge and Society: Studies in the Sociology of Culture Past and Present* 6 (1986): 1–40; Elizabeth L. Eisenstein, *The Printing Press as an Agent of Change: Communications and Cultural Transformations in Early Modern Europe* (New

York: Cambridge University Press, 1979); Adrian Johns, *The Nature of the Book* (Chicago: University of Chicago Press, 1998).

17 Compare this to Joël-Marie Fauquet and Antoine Hennion, *La grandeur de Bach: l'amour de la musique en France au XIXe siècle* (Paris: Fayard, 2000). They argue that the reputation and repertoire associated with Bach were developed by amateurs in the nineteenth century.

18 Stephen D. Krasner, *Sovereignty* (Princeton, NJ: Princeton University Press, 1999). For the power of the Reformation in shifting social relations in Europe in this period, see Robert Wuthnow, *Communities of Discourse: Ideology and Social Structure in the Reformation, the Enlightenment, and Europe Socialism* (Cambridge, MA: Harvard University Press, 1993). The articulation of moral principles during the Reformation was "eventful" in Wuthnow's sense of the term.

19 Claude Rosental, "Making Science and Technology Results Public: A Sociology of Demos," in *Making Things Public: Atmospheres of Democracy*, ed. Bruno Latour and Peter Weibel (Cambridge, MA: MIT Press, 2005), 346–49.

20 Friedrich Nietzsche, *Birth of Tragedy and the Genealogy of Morals*, trans. Francis Golffing (Garden City, NJ: Doubleday, 1956).

21 Michel Foucault, "Governmentality," in *The Foucault Effect: Studies in Governmentality*, ed. Graham Burchell, Colin Gordon, and Peter Miller (Chicago: University of Chicago Press, 1991), 87–104.

22 Chandra Mukerji, "The Unintended State," in *New Cultural Materialisms*, ed. Patrick Joyce and Tony Bennett (New York: Routledge, forthcoming).

23 Louis de Saint-Simon, *The Age of Magnificence*, ed. Ted Morgan (New York: Paragon, 1990).

24 Mukerji, "The Unintended State."

25 Ibid.

26 Ibid.

27 Ibid.

28 Lettre de Colbert à Clerville, 10 Juin 1673, in Pierre Clément, *Lettres, Instructions et Mémoires de Colbert* (1867; repr., Liechtenstein: Klaus Reprints, 1979), 5:84.

29 Mukerji, "The Unintended State."

30 Ibid.

31 Sébastien Le Prestre de Vauban, *La dîme royale*, ed. Le Roy Ladurie (Paris: Imprimerie Nationale, 1992).

32 The character of the threat of this proposal to the patronage system was noted by Louis de Saint-Simon, *Mémoires complets et authentiques de Louis de Saint-Simon duc et pair de France*, ed. Adolphe Chéreual (Paris: Chez Jean de Bonnot, 1966), 5:366–67. See also Vauban, *La dîme royale*; Mukerji, "The Unintended State."

33 Mukerji, "The Unintended State."

34 Callon and Latour, "Don't Throw the Baby out with the Bath School!"

35 Benedict Anderson, *Imagined Communities: Reflections on the Origins and Spread of Nationalism* (London: Verso, 1983).

36 Max Weber, *The Theory of Social and Economic Organization*, ed. A. M. Henderson and Talcott Parsons (New York: Free Press, 1947); Sharon Kettering, *Patrons, Brokers, and Clients in Seventeenth-century France* (New York: Oxford University Press, 1986); Chandra Mukerji, *Territorial Ambitions and the Gardens of Versailles* (Cambridge: Cambridge University Press, 1997). For elements of

power described in table 9.1, see Michel de Certeau, *The Practice of Everyday Life* (Berkeley: University of California Press, 1988); Niccoló Machiavelli, *The Art of War* (New York: Da Capo Press, 1965); Simon Adams, "Tactics or Politics? 'The Military Revolution' and Hapsburg Hegemony, 1525–1648," in *Tools of War*, ed. J. A. Lynn (Urbana: University of Illinois Press, 1990); Perry Anderson, *Lineages of the Absolutist State* (London: New Left Books, 1974); Jean Bodin, *Six Books of the Commonwealth* (New York: Barnes and Noble, 1967); Roland Mousnier, *The Institutions of France under the Absolute Monarchy, 1598–1789: Society and the State*, trans. Brian Pearce (Chicago: University of Chicago Press, 1979); Mukerji, "Intelligent Uses of Engineering"; Patrick Carroll, *Science, Culture, and Modern State Formation* (Berkeley: University of California Press, 2006); Patrick Joyce, *The Rule of Freedom: Liberalism and the City in Britain* (London: Verso, 2003); William Beik, *Absolutism and Society in Seventeenth-century France: State Power and Provincial Aristocracy in Languedoc* (Cambridge: Cambridge University Press, 1985); Sébastien Le Prestre de Vauban, *A Manuel of Siegecraft and Fortification*, trans. George A. Rothrock (Ann Arbor: University of Michigan Press, 1968); Alain Salamagne, "Vauban et les fortifications du Quesnoy," *Revue Historique des Armées*, 162 (1986): 45–51.

37 Scott, *Weapons of the Weak*; Certeau, *The Practice of Everyday Life*.

38 Michael Mann, in a materialist turn, argues the importance of logistics in empires, but does not see it as a separate form of power. See *The Sources of Social Power*. vol. 1. (Cambridge: Cambridge University Press, 1986). See also Ann Swidler, "Culture in Action: Symbols and Strategies," *American Sociological Review* 51 (1986): 273–86.

39 Michel Foucault, *Discipline and Punish*, trans. Alan Sheridan (New York: Pantheon, 1977).

40 Immanuel Wallerstein, *The Modern World-system* (New York: Academic Press, 1974); James Scott, *Seeing Like a State* (New Haven, CT: Yale University Press, 1998); Joyce, *The Rule of Freedom*; Carroll, *Science, Culture, and Modern State Formation*; Mann, *Sources of Social Power*.

41 Callon, "Some Elements of a Sociology of Translation"; Geoffrey Bowker and Susan Leigh Star, *Sorting Things Out* (Cambridge, MA: MIT Press, 1999); Karin Knorr-Centina, *Epistemic Cultures: How the Sciences Make Knowledge* (Cambridge, MA: Harvard University Press, 1999); Mukerji, "Intelligent Uses of Engineering."

42 Bowker and Star, *Sorting Things Out*.

43 Serres, *Rome*.

44 Ibid.

45 Charles Tilly, *The Politics of Collective Violence* (Cambridge: Cambridge University Press, 2003).

46 Max Weber, *From Max Weber: Essays in Sociology*, trans. H. H. Gerth and C. Wright Mills (New York: Oxford University Press, 1946); Tilly, *The Politics of Collective Violence*.

47 Serres, *Rome*, 57.

48 With its emphasis on group material life, what Serres describes is a counterpart—albeit a much gloomier one—to Howard Becker's art worlds. Physical interventions nested in cultural practices are impersonal in both their social origins and material effects—just like the glories of Rome. See Howard Becker, *Artworlds* (Berkeley: University of California Press, 1982).

49 Marcel Hénaff and Anne-Marie Feenberg, "Of Stones, Angels, and Humans: Michel Serres and the Global City," *SubStance* 26, no. 2 (1997): 59–80.

50 The empirical plausibility of Serres's theory of material power is hard to deny. As Luc Boltanski and Laurent Thevenot demonstrate with their research, the grand elements of consciousness such as law, religion, and language are often best understood in terms of usage in modest lifeworlds. Practices of people and things are the infrastructures of group process—ones that may be unspoken but still can sustain an empire. Luc Boltanski and Laurent Thevenot, *De la Justification* (Paris: Gallimard, 1991); Luc Boltanski and Laurent Thevenot, *Justesse et Justice dans le Travail* (Paris: Presses Universitaires de France, 1989).

51 Michel Serres, *Troubadour of Knowledge* (Ann Arbor: University of Michigan Press, 1997); Michel Serres, *Parasite* (Baltimore: Johns Hopkins University Press, 1982).

52 In some ways, Serres's ideas resemble those of George Herbert Mead, who also saw consciousness as deriving from our biological nature. Social process is improvised activity constituting the symbolic and material world in which people live; it is not premeditated but neither is it thoughtless. It is cultural work. George Herbert Mead, *Mind, Self, and Society* (Chicago: University of Chicago Press, 1934).

53 The Fréjus water system is a good example. The aqueduct that served this town was clearly reworked on a number of occasions when leaks developed in the superstructure of the arches or the rock facings to the causeway were impaired. See Jean-Marie Michel, Cherine Gébara, and Jean-Louis Guendon, *L'aqueduc romain de Fréjus. Sa description, son histoire et son environnement* (Lyons: Broché, 2001).

54 Pamela Smith, *The Body of the Artisan* (Chicago: University of Chicago Press, 2004).

55 Edwin Hutchins and Brian Hazlehurst, "Learning in the Cultural Process," in *Artificial Life II,* ed. C. Langdon et al. (Boston: Addison-Wesley, 1991); Edwin Hutchins and Brian Hazlehurst, "How to Invent a Lexicon: The Development of Shared Symbols in Interation," in *Artificial Societies: The Computer Simulation of Social Life,* ed. Nigel Gilbert and Rosaria Conte (London: UCL Press, 1995).

56 Louis de Froidour, *Mémoire du Pays et des États de Bigorre,* intro. and notes Jean Boudette (Paris: H. Champion, Tarbes, Baylac, 1892), 21.

57 Pamela O. Long, *Openness, Secrecy, Authorship: Technical Arts and the Culture of Knowledge from Antiquity to the Renaissance* (Baltimore: Johns Hopkins University Press, 2001).

58 Ibid., 29–45.

59 Froidour, *Mémoire du Pays et des États de Bigorre,* 115, 21.

60 François de Dainville, *Géographie des humanistes* (Geneva: Slatkin Reprints, 1969), 80, 85, 88–93. For a sense of his science and its practice, see also Marguarite Boudon-Duaner, *Bernard Palissy: le potier du roi* (Carrières-sous-Poissy: La Cause, 1989), 44–48.

61 Henry Morely, *Palissy the Potter: The Life of Bernard Palissy, of Saintes, His Labours and Discoveries in Art and Science* (London: Chapman and Hall, 1852), 2:319.

62 Bennett Berger, *The Survival of a Counterculture* (Berkeley: University of California Press, 1981).

63 Bernard Forest de Belidor, "De l'Usage des Eaux a la Guerre," in *Architecture Hydraulique seconde partie qui comprend l'Art de diriger les eaux des la Mer & des Rivieres à l'avantage de la défense des places, du Commerce & de l'Agriculture. par M. Belidor, Colonel d'Infanterie* (Paris: Jombert, 1753); Joseph de Lalande, *Des canaux de navigation, et spécialement du canal de Languedoc* (Paris: Veuve Desaint, 1778).

64 Lyn Spillman, *Nation and Commemoration: Creating National Identities in the United States and Australia* (New York: Cambridge University Press, 1997); Barry Schwartz, *Abraham Lincoln and the Forge of National Memory* (Chicago: University of Chicago Press, 2000); Nietzsche, *Birth of Tragedy and the Genealogy of Morals*; Jeffrey Minson, *Genealogies of Morals* (New York: St. Martin's Press, 1985).

65 Ibid.; Frances Yates, *The Art of Memory* (Chicago: University of Chicago Press, 1966), 1–49.

66 Mary J. Carruthers, *The Book of Memory: A Study of Memory in Medieval Culture* (New York: Cambridge University Press, 1990), 33–37, 71–79.

67 Henri Michel, "Le Canal de Languedoc vu par un astronome 'philosophe,'" in *Le Canal du Midi*, ed. Jean-Denis Bergasse (Cessenon: J.-D. Bergasse, 1982), 1:197–216; Jean-Denis Bergasse, "Le Cult de Riquet en Languedoc au XIXe Siècle," in Le Canal du Midi, ed. Jean-Denis Bergasse (Cessenon: J.-D. Bergasse, 1982), 1:217–51.

BIBLIOGRAPHY

Manuscripts

L'Arrêt d'adjudication des ouvrages à faire pour le canal de communication des Mers en Languedoc est promulgué. Ce même jour, le Roi "fait bail et délivrance à M. de Riquet des ouvrages contenues au Devis" préalablement défini sous l'autorité du Chevalier de Clerville. Reprinted in "Edit du Roy pour la construction d'un canal du communication des deux mers, Océane & Méditerrannée," 14 Octobre 1666, ACM 03-10.

Avis a messieurs les Capitouls de la Ville de Tolose, par Arquier, Doyen de anciens Capitouls. Et Response a cet Avis, Article par Jean de Nivelle, Ancien Capitaine Chasseauant du Canal dans l'atelier de Mr. Sagadenes. 1667, ACM 01-16.

Bazin, Claude. "Bail et Adjudication des ouvrages à faire pour la continuation du Canal et du port du Cette, 20 Août 1668." June 3, 1669 9, sec. x, ACM 07-12.

Colbert, Jean-Baptiste. *Registre contenant les ordres, instructions et lettres expediees par Monseigneur Colbert, touchant les fortifications des ponts et chausseees, canal de communication des mers et mines de Languedoc en l'annee 1669.* 1669. EPC.

de la Feuille, Pons. "Remarques faictes au voiage de Flandres et Hollande en octobre, novembre et décembre 1670, sur les canaux, la construction des écluzes, ponts, jettées en mer et digues, moulins pour l'évacuation des caves et machines pour le nettoyement des canaux et des portz." Paris, 31 Janvier 1671. CCC, no. 448.

Felibien. *Relation de la Feste de Versailles du 18e Juillet 1668.* Paris: Pierre le Petit. BNF Lb37. 360.

Hiacinthe de la Boulaye, Claude Louis. "Proces Verbal de visites des Bois du Pays de l'Abour [Adour]." AHMV, SH no. 278.

Notes sur l'histoire du Canal du Languedoc par les descendents de Riquet. 1803. ACM uncataloged.

Early Sources and Reprints

Agricola, Giorgius. *De re Metallica; Translated from the First Latin Edition of 1556, with Biographical Introduction, Annotation, and Appendices upon the Development of Mining, Metallurgical Processes, Geology, Mineralogy, and Mining Law from the Earliest Times to the 16th Century.* Translated and edited by Herbert Clark Hoover and Lou Henry Hoover. New York: Dover, 1950.

Alberti, Leon Battista. *De re aedificatoria.* Florentiae: Nicolaus Laurentii, 1485.

Andréossy, Antoine. *Histoire du canal du Midi: connu précédement sous le nom de Languedoc.* Paris: F. Buisson, 1800.

Basville, Nicolas de Lamoignon de. *Mémoire sur la province de Languedoc de 1697.* Reprinted and edited in Françoise Moreil, *L'Intendance de Languedoc à la fin du XVIIe Siècle.* Paris: C.T.H.S., 1985.

Belidor, Bernard Forest de. "De l'Usage des Eaux a la Guerre." In *Architecture Hydraulique seconde partie qui comprend l'Art de diriger les eaux des la Mer & des Rivieres à l'avantage de la défense des places, du Commerce & de l'Agriculture. par M. Belidor, Colonel d'Infanterie.* Paris: Jombert, 1753.

Bodin, Jean. *Six Books of the Commonwealth.* New York: Barnes and Noble, 1967.

Bouillet. *Traite des Moyens de render les Rivieres Navigables avec plusieurs desseins de jettées.* Paris: Chez Estienne Michallet, 1693.

Cassini, Jacques. *De La Grandeur et de la Figure de la Terre. Suite des Memoires de l'Academie Royale des Sciences, Annee MDCCXVIII.* Paris: Imprimerie Royale, 1720.

Clément, Pierre. *Lettres, Instructions et Mémoires de Colbert.* Liechtenstein: Kalus Reprints, 1979. First published 1867.

Columella, L.J.M. *Columella on Agriculture 1-Iv.* Edited and translated by Harrison Boyd Ash. Cambridge, MA: Harvard University Press, [1941] 2001. First published 1794.

Estienne, Charles, and Jean Liebault. *L'agriculture et la maison rustique de Charles Estienne . . . ; paracheuee premierement, puis augmentee par M. Iean Liebaut.* Paris: Chez Iacques Du-Puys, 1570.

———. *La Maison Rustique of the Country Farme, Compiled in the French Tongue by Charles Stevens and Jean Liebault, Doctors of Physicke, and Translated into English by Richard Surflet, Practitioner in Physicke.* London: Arnold Hatsfield for John Norton and John Bill, 1606.

Félibien, André. *Receuil de descriptions de peintures et d'autres ouvrages fait pour le roy.* Paris: Veuve de S. Mabre-Cramoisy, 1689.

Froidour, Louis de. *Lettre a M. Barrillon Damoncourt cotenant la relation & la description des travaux qui se font en Languedoc pour la communication des deux mers.* Toulouse: Dominique Camusat, 1672.

———. *Mémoire du Pays et des États de Bigorre.* Introduction and notes by Jean Boudette. Paris: H. Champion, Tarbes, Baylac, 1892.

———. *Les Pyrenées centrales au XVIIe siècle: lettres par M. de Froidour . . . à M. de Haericourt . . . et à M. de Medon . . . publiées avec des notes par Paul de Casteran.* Auch: G. Foix, 1899.

Frontinus, Sextus Julius. *The Two Books on the Water Supply of the City of Rome of Sextus Julius Frontinus, Water Commissioner of the City of Rome, A.D. 97.* New York: Longmans, Green, 1913.

Guiffrey, Jules, ed. *Comptes des Bâtiments du Roi. Sous le Règne de Louis XIX.* 1st ed. Paris: Imprimerie Nationale, 1881.

Hobbes, Thomas. *Correspondence of Thomas Hobbes.* Edited by Noel Malcolm. Vol. 2. Oxford: Oxford University Press, 1994.

Hume, David. *Political Essays.* Edited by Knud Haakonssen. Cambridge: Cambridge University Press, 1994.

Jefferson, Thomas. *Le Voyage de Thomas Jefferson sur le Canal du Midi.* Edited by Pierre Gérard. Portet-sur-Garonne: Éditions de Laboutière, 1995.

La Hire, Philippe de. *L'ecole des arpenteurs: ou l'on enseigne toutes les pratiques de geometrie, qui sont necessaires a un arpenteur: On y a ajoute un abrege du nivellement, & les proprietez des eaux, & les manieres de les juger.* Paris: Thomas Moette, 1689.

Lalande, Joseph de. *Des canaux de navigation, et spécialement du canal de Languedoc.* Paris: Veuve Desaint, 1778.

Lawson, George. *Politica Sacra et Civilis*. Edited by Conal Condren. Cambridge: Cambridge University Press, 1992. First published 1678.

Le Clerc, Sébastian. *Practique de la geometrie sur le papier and sur le terrain*. Paris: Chez Thomas Jolly, 1669.

Machiavelli, Niccoló. *The Art of War*. New York: Da Capo Press, 1965.

———. *The Prince*. Edited and translated by D. Donno. New York: Bantam Dell, 1996.

Malavialle, L. "Une Excursion dans la Montagne Noire." *Bulletin de la Société Languedocienne de Géographie*, vols. 4, 5, 7, 8, 9, 14, 15 (1891–1892).

Mallet, Allain Manesson. *Les Travaux de Mars ou l'Art de la Guerre*. Amsterdam: Jan et Gillis Janson à Waesbergue, 1684.

Mercator, Gerhard. *Historia Mundi*. 2nd ed. London: Michael Starke, 1637.

Moreil, Françoise. *L'Intendance de Languedoc al la fin du XVIIe Siècle*. Paris: C.T.H.S., 1985. Edited edition of Nicolas de Lamoignon de Basville, *Mémoire sur la province de Languedoc de 1697*.

Morely, Henry. *Palissy the Potter: The Life of Bernard Palissy, of Saintes, His Labours and Discoveries in Art and Science*. London: Chapman and Hall, 1852.

Mourgues, Père Matthieu de. "Relation de la seconde navigation du Canal Royal, 1683." Quoted in Philippe Delvit. "Un Canal au Midi." In *Conseil d'Architecture, d'Urbanisme et de l'Environment de la Haute-Garonne, Canal royal de Languedoc: le partage des eaux*, 206. Caue: Loubatières, 1992.

Munier, Achille. *Frontignan*. Montpellier: C. Coulet, 1874. Reprinted by Lacour, 1993.

Palissy, Bernard. *A Delectable Garden, by Bernard Palissy*. Translated and edited by Helen Morganthau Fox. Peekskill, NY: Watch Hill Press, 1931.

———. *Recepte véritable*. Geneva: Froz, 1988.

Perrault, Charles. *Parallèle des anciens et des modernes en ce qui regarde les arts et les sciences*. Munich: Eidos Verlag, 1964.

Picard, M. *Traité du Nivellement*. Paris: Estienne Michallet, 1684.

Saint-Simon, Louis de. *The Age of Magnificence*. Edited by Ted Morgan. New York: Paragon, 1990.

———. *Mémoires complets et authentiques de Louis de Saint-Simon duc et pair de France*. Edited by Adolphe Chéreual. 20 vols. Paris: Chez Jean de Bonnot, 1966.

Serres, Olivier de. *The Perfect Vse of Silk-wormes*. New York: Da Capo, 1971.

———. *Théâtre d'agriculture et mesnages des champs*. Geneva: Mat Hiev Berjon, 1611.

Strabo. *The Geography of Strabo*. Book 4. Translated by Horace Jones. Cambridge, MA: Harvard University Press, 1966–1970.

Vallancey, Charles. *A Treatise on Inland Navigation or the Art of Making Rivers Navigable: Of Making Canals in All Sorts of Soils, and of Constructing Locks and Sluices*. Dublin: George and Alexander Ewing, 1763.

Vauban, Sébastien Le Prestre de. *La dîme royale*. Edited by Le Roy Ladurie. Paris: Imprimerie Nationale, 1992.

———. *A Manuel of Siegecraft and Fortification*. Translated by George A. Rothrock. Ann Arbor: University of Michigan Press, 1968.

Veryard, Ellis. *An Account of Divers Choice Remarks as well as Geographical, Historical, Political, Mathematical, and Moral; Taken in a Journey through the Low Countries, France, Italy, and Part of Spain with the Isles of Sicily and Malta*. London: S. Smith and B. Walford, 1701.

Vitruvius. *The Ten Books on Architecture.* Translated by M. H. Morgan. New York: Dover, 1960. First published 1914.

Contemporary Sources

Adams, Julia. "The Rule of the Father: Patriarchy and Patrimonialism in Early Modern Europe." In *Max Weber's Economy and Society: A Critical Companion*, edited by Charles Camic, Philip Gorski, and David Trubek, 237–66. Stanford, CA: Stanford University Press, 2005.

Adams, Simon. "Tactics or Politics? 'The Military Revolution' and Hapsburg Hegemony, 1525–1648." In *Tools of War*, ed. J. A. Lynn (Urbana: University of Illinois Press, 1990).

Adgé, Michel. "L'Art de l'hydraulique." In *Conseil d'Architecture, d'Urbanisme et de l'Environment de la Haute-Garonne, Canal royal de Languedoc: le partage des eaux*, 202–3. Caue: Loubatières, 1992.

———. "Le Canal du Midi, ou la jonction des mers en Languedoc." *Mappemonde* 1 (1992): 44–48.

———. "Chronologie des Principaux Évènements de la Construction du Canal, 1662–1694." In *Le Canal du Midi*, edited by Jean-Denis Bergasse, 4:173–92. Cessenon: J.-D. Bergasse, 1985.

———. "Les Premiers États du Barrage de Saint-Ferréol." *Les cahiers d'histoire de Revel* 7 (2001): 5–22.

Adkins, Lesley, and Roy Adkins. *Handbook to Life in Ancient Rome.* New York: Oxford University Press, 1994.

Alexander, Jeffrey. *The Meanings of Social Life.* Oxford: Oxford University Press, 2003.

Alliès, Paul. *L'invention du territoire.* Grenoble: Presses Universitaire de Grenoble, 1980.

Anderson, Benedict. *Imagined Communities: Reflections on the Origins and Spread of Nationalism.* London: Verso, 1983.

Anderson, Perry. *Lineages of the Absolutist State.* London: New Left Books, 1974.

Apostolidès, Jean-Marie. 1981. *Le Roi-Machine: spectacle et politique au temps de Louis XIV.* Paris: Éditions de Minuit, 1981.

Aronoff, Mark, Irit Meir, Carol Padden, and Wendy Sandler. "The Roots of Linguistic Organization in a New Language." *Interaction Studies* 9, no. 1 (2008): 131–50.

Aupert, Pierre, Raymond Monturet, and Christine Dieulafait. *Saint-Bertrand-de-Comminges: Les thermes de forum.* Pessac: Éditions de la Fédération Aqvitania, 2001.

Barbet, L. A. *Les Grandes Eaux de Versailles: Installations Méchaniques et Éstangs Artificiels, Descriptions des Fontaines et de leurs Origins.* Paris: H. Dunod et E. Pinat, 1907.

Baumgartner, Frederic. *France in the Sixteenth Century.* New York: St. Martin's Press, 1995.

Bechmann, Roland. *Villard de Honnecourt.* Paris: Picard, 1993.

Becker, Howard. *Artworlds.* Berkeley: University of California Press, 1982.

Becker, Howard, and Robert Faulkner. "Studying Something You Are Part Of : The View from the Bandstand." Paper presented at the international Ethnographies of Artistic Work conference, Sorbonne, Paris, September 2006.

Beik, William. *Absolutism and Society in Seventeenth-century France: State Power and Provincial Aristocracy in Languedoc.* Cambridge: Cambridge University Press, 1985.

Bellah, Robert, et al. *Habits of the Heart.* Berkeley: University of California Press, 1988.

Bély, Lucien. *Dictionnaire de l'Ancien Régime.* Paris: Quadrige PUF, 2005.

Bergasse, Jean-Denis. "Le Cult de Riquet en Languedoc au XIXe Siècle." In *Le Canal du Midi*, edited by Jean-Denis Bergasse, 1:217–51. Cessenon: J.-D. Bergasse, 1982.

———, ed. *Le Canal du Midi*, vols. 1–4. Cessenon: J.-D. Bergasse, 1982–1985.

Berger, Bennett. *The Survival of a Counterculture.* Berkeley: University of California Press, 1981.

Bernet, G. "Riquet intime dans ses residences de Revel, Bonrepos, et Toulouse avant 1668." In *Le Canal du Midi*, edited by Jean-Denis Bergasse, 3:52. Cessenon: J.-D. Bergasse, 1984.

Biernacki, Richard. *Fabrication of Labor: Germany and Britain, 1640–1914.* Berkeley: University of California Press, 1988.

Blanchard, Anne. *Les Ingénieurs du Roy de Louis XIV à Louis XVI: Étude du corps des fortifications.* Montpellier: Université Paul-Valéry, 1979.

———. *Vauban.* Paris: Fayard, 1996.

Blanchard, Anne, Michel Adgé, and Jean-Denis Bergasse. "Les Ingénieurs du Roy." In *Le Canal du Midi*, edited by Jean-Denis Bergasse, 3:181–94. Cessenon: J.-D. Bergasse, 1984.

Boltanski, Luc, and Laurent Thevenot. *Justesse et Justice dans le Travail.* Paris: Presses Universitaires de France., 1989

———. *De la Justification.* Paris: Gallimard, 1991.

Boudon-Duaner, Marguarite. *Bernard Palissy: le potier du roi.* Carrières-sous-Poissy: La Cause, 1989.

Bourdieu, Pierre. *Distinction: A Social Critique of the Judgment of Taste.* Translated by Richard Nice. London: Routledge and Kegan Paul, 1984.

———. *Déchiffrer le France.* Paris: Éditions des Archives Contemporaines, 1988.

Bowker, Geoffrey, and Susan Leigh Star. *Sorting Things Out.* Cambridge, MA: MIT Press, 1999.

Braudel, Fernand. *L'identité de la France.* Paris: Arthaud, 1986.

———. *The Mediterranean and the Mediterranean World in the Age of Philip II.* New York: Harper and Row, 1975.

Brewer, John. *The Sinews of Power: War, Money, and the English State, 1688–1783.* New York: Knopf, 1989.

Buisseret, David, ed. *Monarchs, Ministers, and Maps.* Chicago: University of Chicago Press, 1992.

Burke, Peter. *Fabrication of Louis XIV.* New Haven, CT: Yale University Press, 1992.

Burlats-Brun, P., and Jean-Denis Bergasse. "L'Oligarchie gabelière, soutien financier de Riquet." In *Le Canal du Midi*, edited by Jean-Denis Bergasse, 3:123–41. Cessenon: J.-D. Bergasse, 1984.

Cadden, Joan. "'Nothing Natural Is Shameful': Vestiges of a Debate about Sex and Science in a Group of Late Medieval Manuscripts." *Speculum* 76 (2001): 66–89.

Cadeña, Marisol de la. *The Politics of Race and Culture in Cuzco, Peru, 1919–1991.* Durham, NC: Duke University Press, 2000.

Callon, Michel. "Some Elements of a Sociology of Translation: Domestication of the Scallops and the Fishermen of St. Brieuc Bay." In *Power, Action, and Belief*, edited by John Law, 196–233. London: Routledge and Kegan Paul, 1986.

Callon, Michel, and Bruno Latour. "Don't Throw the Baby out with the Bath School!" In *Science as Practice and Culture*, edited by Andrew Pickering, 343–68. Chicago: University of Chicago Press, 1992.

Carroll, Patrick. *Science, Culture, and Modern State Formation*. Berkeley: University of California Press, 2006.

Carruthers, Mary J. *The Book of Memory: A Study of Memory in Medieval Culture*. New York: Cambridge University Press, 1990.

Cartwright, Nancy. "Evidence-Based Policy: What's Evidence." Lecture, Colloquium in Science Studies, University of California at San Diego, February 11, 2008.

Casti, Emmanuela. "State, Cartography, and Territoriality in Renaissance Veneto and Lombardy." *Imago Mundi* 3 (2001):1–73.

Cazals, Henri de. "Armorial du Canal." In *Le Canal du Midi*, edited by Jean-Denis Bergasse, 3:151–80. Cessenon: J.-D. Bergasse, 1984.

Certeau, Michel de. *The Practice of Everyday Life*. Berkeley: University of California Press, 1988.

———. *The Writing of History*. Translated by Tom Conley. New York: Columbia University Press, 1988.

——— *The Mystic Fable*. Translated by Michael B. Smith. Chicago: University of Chicago Press, 1992.

Cipolla, Carlo. *Guns, Sails, and Empires*. New York: Pantheon, 1965.

Clark, Elizabeth. *History, Theory, Text: Historians and the Linguistic Turn*. Cambridge, MA: Harvard University Press, 2004.

Clark, Samuel. *State and Status: The Rise of the State and Aristocratic Power in Western Europe*. Cardiff: University of Wales Press, 1995.

Cole, Charles W. *Colbert and a Century of French Mercantilism*. Hamden, CT: Archon Books, 1964.

Cole, Michael. *Cultural Psychology: A Once and Future Discipline*. Cambridge, MA: Harvard University Press, 1996.

Cole, Michael, and Sylvia Scribner. *Culture and Thought*. New York: Wiley, 1974.

Cole, Michael, and James Wertsch. *Contemporary Implications of Vygotsky and Luria*. Worcester, MA: Clark University Press, 1996.

Colyar, H. A. de. "Jean-Baptiste Colbert and the Codifying Ordinances of Louis XIV." *Journal of the Society of Comparative Legislation*, new series 13, no. 1 (1912): 56–86.

Condren, Conel. "Liberty of Office and Its Defence in Seventeenth-century Political Argument." *History of Thought* 18, no. (1997): 460–82.

———. "The Politics of the Liberal Archive." *History of the Human Sciences* 12 (1999): 35–49.

Conseil d'Architecture, d'Urbanisme et de l'Environment de la Haute-Garonne. *Canal royal de Languedoc: le partage des eaux*. Caue: Loubatières, 1992.

Corvisier, André. "Colbert et la guerre." In *Colloque pour le tricentenaire de la mort de Colbert, sous la direction de Roland Mousnier, Un nouveau Colbert*, edited by Jean Favier and Roland Mousnier, 287–308. Paris: Éditions Sedes, 1985.

Corvol, Andrée. *L'Homme et l'Arbre sous l'Ancien Régime*. Paris: Economica, 1984.

Cosgrove, Denis. "Mapping New Worlds: Culture and Cartography in Sixteenth-century Venice." *Imago Mundi* 44 (1992): 65–89.

Cronon, William. *Nature's Metropolis*. New York: W. W. Norton, 1991.

Dainville, François de. *Cartes Anciennes du Languedoc XVIe–XVIIIe S.* Montpellier: Société Languedocienne de Géographie, 1961.

———. *Géographie des humanistes.* Geneva: Slatkin Reprints, 1969.

Degage, Alain. "Un Nouveau port en Languedoc, de la fin du XVIe siècle au début du XVIIIe." In *Histoire de Sète*, edited by Jean Sagnes, 47–52. Toulouse: Privat, 1987.

———. "Le Port de Sète: Proue Méditerranéenne du Canal de Riquet." In *Le Canal du Midi*, edited by Jean-Denis Bergasse, 4:265–306. Cessenon: J.-D. Bergasse, 1985.

———. "Quarantaine et lazaret à Sète aux XVIIe et XVIIIe siècles." In *Navigation et migrations en Méditerranée de la Préhistoire à nos jours*, 1–17. Paris: Éditions du Centre National de la Recherche Scientifique, 1990.

Dent, Julian. *Crisis in Finance: Crown Financiers and Society in Seventeenth-century France.* New York: St. Martin's Press, 1973.

Dermigny, Louis. *Sète de 1666–1880.* Montpellier: Actes de l'Institut, 1955.

Devèze, M. "Une Admirable Réforme Administrative: La Grande Réformation des Forêts Royales sous Colbert (1662–1680)." In *Annales de L'École Nationale des Eaux et Forêts et de la Station de Recherches et Expériences.* Nancy: École Nationale des Eaux et Forêts, 1962.

DiMaggio, Paul. "Culture and Cognition." *Annual Review of Sociology* 23 (1997): 263–88.

Drayton, Richard. *Nature's Government: Science, Imperial Britain, and the 'Improvement' of the World.* New Haven, CT: Yale University Press, 2000.

Durkheim, Émile. *Suicide: A Study in Sociology.* Edited by John Spaulding and George Simpson. London: Routledge, 1952.

Duvall, Marguerite. *The King's Garden.* Translated by Annette Tomarken and Claudine Cowen. Charlottesville: University Press of Virginia, 1982.

Edwards, Paul. *The Closed World: Computers and the Politics of Discourse in Cold War America.* Cambridge, MA: MIT Press, 1996.

Eisenstein, Elizabeth L. *The Printing Press as an Agent of Change: Communications and Cultural Transformations in Early Modern Europe.* New York: Cambridge University Press, 1979.

Elias, Norbert. *The Civilizing Process.* New York: Urizen Books, 1978. First published 1939.

———. *The Court Society.* New York: Pantheon, 1998.

Engrand, Charles. "Clients du roi. Colbert et l'État. 1661–1715." In *Colloque pour le tricentenaire de la mort de Colbert, sous la direction de Roland Mousnier, Un nouveau Colbert*, edited by Jean Favier and Roland Mousnier, 85–97. Paris: Éditions Sedes, 1985.

———. "Colbert et les commissaires du roi. Grands jours et intendants. La réduction des officiers a l'obéissance." In *Colloque pour le tricentenaire de la mort de Colbert, sous la direction de Roland Mousnier, Un nouveau Colbert*, edited by Jean Favier and Roland Mousnier, 133–44. Paris: Éditions Sedes, 1985.

Enjalbert, Henri. "Les Hardinesses de Riquet: Données Géomorphologiques de la Région que Traverse le Canal du Midi." In *Le Canal du Midi*, edited by Jean-Denis Bergasse, 4:127–42. Cessenon: J.-D. Bergasse, 1985.

Fabre, François. "La lutte des descendents Andréossy et Riquet pour la paternité du canal." In *Le Canal du Midi*, edited by Jean-Denis Bergasse, 4:341–66. Cessenon: J.-D. Bergasse, 1985.

Fauquet, Joël-Marie, and Antoine Hennion. *Le grandeur de Bach: l'amour de la musique en France au XIXe siècle*. Paris: Fayard, 2000.

Favier, Jean, and Roland Mousnier, eds. *Un nouveau Colbert*. Paris: Éditions Sedes, 1985.

Feenberg, Andrew. *Questioning Technology*. London: Routledge, 1999.

Ferras, Robert. "Le Canal Royal et ses cartes: des images d'une province à la connaissance d'un espace géographique." In *Le Canal du Midi*, edited by Jean-Denis Bergasse, 2:271–308. Cessenon: J.-D. Bergasse, 1983.

Fontanier, Arnaud Ramière de. "Du Grand Canal au Grand Bassin: Fêtes sur le canal du Midi aux XVIIe et XVIIIe siècles." In *Le Canal du Midi et les Voies Navigables dans le Midi de la France*. Actes du congrès des fédérations historiques languedociennes. Carcassone: Société d'Études Scientifiques de l'Aude, 1998.

Foucault, Michel. *Discipline and Punish*. Translated by Alan Sheridan. New York: Pantheon, 1977.

———. "Governmentality." In *The Foucault Effect: Studies in Governmentality*, edited by Graham Burchell, Colin Gordon, and Peter Miller, 87–104. Chicago: University of Chicago Press, 1991.

———. *Madness and Civilization*. Translated by Richard Howard. New York: Vintage Books, 1965.

———. *The Order of Things: An Archeology of the Human Sciences*. New York: Pantheon Books, 1970.

Frangenberg, Thomas. "Chorographies of Florence: The Use of City Views and City Plans in the Sixteenth Century." *Imago Mundi* 46 (1994): 41–64.

Fujimura, Joan. "Crafting Science: Standardized Packages, Boundary Objects, and 'Translation.'" In *Science as Practice and Culture*, edited by Andrew Pickering, 168–214. Chicago: University of Chicago Press, 1992.

Gabolde, Bertrand. "Les Ouvriers du Chantier." In *Le Canal du Midi*, edited by Jean-Denis Bergasse, 3:233–38. Cessenon: J.-D. Bergasse, 1984.

———. "Revel: Des eaux du Sor à la rigole de la plaine." In *Le Canal du Midi*, edited by Jean-Denis Bergasse, 4:241–44. Cessenon: J.-D. Bergasse, 1985.

———. "Riquet à Versailles Vu par le Conteur Charles Perrault." In *Le Canal du Midi*, edited by Jean-Denis Bergasse, 1:185–89. Cessenon: J.-D. Bergasse, 1982.

Gazelle, François. "Riquet et les eaux de la Montagne Noire: L'idée géniale de l' alimentation du canal." In *Le Canal du Midi*, edited by Jean-Denis Bergasse, 4:143–67. Cessenon: J.-D. Bergasse, 1985.

Gébara, Chérine, and Jean-Marie Michel. "L'Aqueduc romain de Fréjus: sa description, son histoire et son environnement." In *Revue Archéologique de Narbonnaise*, supplement 33. Montpellier: Éditions de l'Association de la Revue Achéologique de Narbonnaise, 2002.

Ginzburg, Carlo. *The Cheese and the Worms: The Cosmos of a Sixteenth Century Miller*. Baltimore: Johns Hopkins University Press, 1980.

Giovannoni, Gustavo. "Building and Engineering." In *The Legacy of Rome*, edited by Cyril Bailey, 429–74. Oxford: Clarendon Press, 1940. First published 1923.

Goffman, Erving. *Behavior in Public Places*. Glencoe, IL: Free Press, 1963.

———. *The Presentation of Self in Everyday Life*. New York: Doubleday, 1959.

Gorski, Philip S. *The Disciplinary Revolution: Calvinism and the Rise of the State in Early Modern Europe*. Chicago: University of Chicago Press, 2003.

Goubert, Pierre. *Louis XIV and Twenty Million Frenchmen*. Translated by Ann Carter. New York: Vintage Books, 1972.

Grafton, Anthony. *Leon Battista Alberti*. New York: Hill and Wang, 2000.

Gratacos, Isaure. *Femmes pyrénéennes, un statut social exceptionnel en Europe*. Toulouse: Éditions Privat, 2003.

Grewal, Interpal, and Caren Kaplan. *Scattered Hegemonies*. Minneapolis : University of Minnesota Press, 1994.

Griesemer, James. "Tracking Organic Processes: Representations and Research Styles in Classical Embryology and Genetics." In *From Embryology to Evo-Devo: A History of Developmental Evolution*, edited by Manfred D. Laubichler and Jane Maienschein, 375–433 Cambridge, MA: MIT Press, 2007.

Hall, Stuart. *The Question of Cultural Identity*. London: Polity Press, 1990.

Haraway, Donna. *Primate Visions: Gender, Race, and Nature in the World of Modern Science*. New York: Routledge, 1990.

Harvey, P.D.A. *The History of Topographical Maps*. London: Thames and Hudson, 1980.

Hecht, Gabrielle. *The Radiance of France: Nuclear Power and National Identity after WWII*. Cambridge, MA: MIT Press, 1998.

Heitland, E. "Agriculture." In *The Legacy of Rome*, edited by Cyril Bailey, 475–512. Oxford: Clarendon Press, 1940. First published 1923.

Heller, Henry. *Labour, Science, and Technology in France, 1500–1620*. Cambridge: Cambridge University Press, 1996.

Hénaff, Marcel, and Anne-Marie Feenberg. "Of Stones, Angels, and Humans: Michel Serres and the Global City." *SubStance* 26, no. 2 (1997): 59–80.

Henderson, Kathryn. *On Line and on Paper*. Cambridge, MA: MIT Press, 1999.

Henke, Christopher. "The Mechanics of Workplace Order: Toward a Sociology of Repair." *Berkeley Journal of Sociology* 44 (2000): 55–81.

———. "Repairing the Nuclear Stockpile: Aging Technology, Tacit Knowledge, and the Future of the Nuclear Weapons Complex." Paper presented at the American Sociological Association meetings, New York, 2007.

Heyman, Jacques. *The Stone Skeleton: Structural Engineering of Masonry Architecture*. Cambridge: Cambridge University Press, 1995.

Holt, Mack P. *The French Wars of Religion, 1562–1629*. Cambridge: Cambridge University Press, 2005.

Hutchins, Edwin. *Cognition in the Wild*. Cambridge, MA: MIT Press, 1995.

Hutchins, Edwin, and Brian Hazlehurst. "How to Invent a Lexicon: The Development of Shared Symbols in Interation." In *Artificial Societies: The Computer Simulation of Social Life*, edited by Nigel Gilbert and Rosaria Conte. London: UCL Press, 1995.

———. "Learning in the Cultural Process." In *Artificial Life II*, edited by Christopher Langdon et al. Boston: Addison-Wesley, 1991.

Hutchins, Edwin, and Tove Klausen. "Distributed Cognition in an Airline Cockpit." In *Cognition and Communication at Work*, edited by Yrjö Engeström and David Middleton, 15–34. Cambridge: Cambridge University Press, 1996.

Jasanoff, Sheila. *The Fifth Branch*. Cambridge, MA: Harvard University Press, 1990.

———. *Science at the Bar*. Cambridge, MA: Harvard University Press, 1995.

Johns, Adrian. *The Nature of the Book*. Chicago: University of Chicago Press, 1998.

Joyce, Patrick. *The Rule of Freedom: Liberalism and the City in Britain*. London: Verso, 2003.

Kettering, Sharon. "The Historical Development of Political Clientalism." *Journal of Interdisciplinary History* 18 (1988): 419–47.

————. *Patrons, Brokers, and Clients in Seventeenth-century France*. New York: Oxford University Press, 1986.

Knorr-Cetina, Karin. *Epistemic Cultures: How the Sciences Make Knowledge*. Cambridge, MA: Harvard University Press, 1999.

Konvitz, Josef W. *Cartography in France, 1660–1848*. Chicago: University of Chicago Press, 1987.

————. *Cities and the Sea: Port City Planning in Early Modern Europe*. Baltimore: Johns Hopkins University Press, 1978.

Krasner, Stephen D. *Sovereignty*. Princeton, NJ: Princeton University Press, 1999.

Lamont, Michèle. *How Professors Think: Inside the Curious World of Academic Judgment*. Cambridge, MA: Harvard University Press, 2009.

Langins, Janice. *Conserving the Enlightenment: French Military Engineering from Vauban to the Revolution*. Cambridge, MA: MIT Press, 2004.

Lansing, John. *Priests and Programmers: Technologies of Power in the Engineered Landscape of Bali*. Princeton, NJ: Princeton University Press, 1991.

Latour, Bruno. "Drawing Things Together." In *Representation in Scientific Activity*, edited by Michael Lynch and Steve Woolgar, 19–68. Cambridge, MA: MIT Press, 1990.

————. *Pandora's Hope: On the Reality of Science Studies*. Cambridge, MA: Harvard University Press, 1999.

————. "Visualisation and Cognition: Thinking with Eyes and Hands." *Knowledge and Society: Studies in the Sociology of Culture Past and Present* 6 (1986): 1–40.

————. *We Have Never Been Modern*. Cambridge, MA: Harvard University Press, 1993.

————. "Why Has Critique Run out of Steam?" *Critical Inquiry* 30 (2004): 238–39.

Lee, Carolyn, and Mary Walshok. *A County Level Analysis of California's R&D Activity, 1993–1999*. San Diego: University of California at San Diego Extension, 2002.

Lestringant, Frank. *Bernard Palissy, 1510–1590, l'écrivain, le réforme, le céramiste*. Coédition Associon Internationale des Amis d'Agrippa d'Aubigné- Éditions SPEC, 1990.

Lewis, David. *Counterfactuals*. Malden, MA: Blackwell, 1973.

Lipsitz, George. *Time Passages: Collective Memory and American Popular Culture*. Minneapolis: University of Minnesota Press, 1990.

Long, Elizabeth, ed. *From Sociology to Cultural Studies: New Perspectives*. New York: Blackwell, 1997.

Long, Pamela O. *Openness, Secrecy, Authorship: Technical Arts and the Culture of Knowledge from Antiquity to the Renaissance*. Baltimore: Johns Hopkins University Press, 2001.

Longino, Helen. *The Fate of Knowledge*. Princeton, NJ: Princeton University Press, 2002.

Luria, A. *The Making of Mind*. Edited by Michael Cole and Sheila Cole. Cambridge, MA: Harvard University Press, 1979.

Lynn, J. A. *Giant of the Grand Siècle: The French Army, 1610–1715*. Cambridge: Cambridge University Press, 1997.

Maistre, André. *Le Canal des Deux-Mers: Canal Royal du Languedoc, 1666–1810*. Toulouse: E. Privat, 1968.

Mann, Michael. *The Sources of Social Power*, vol. 1, *A History of Power from the Beginning to A.D. 1760*. Cambridge: Cambridge University Press, 1986.

Marliave, Olivier de. *Trésor de la Mythologie Pyrénéene*. Luçon: Éditions Sud Ouest, 1996.

McGowan, Margaret. *Ideal Forms in the Age of Ronsard*. Berkeley: University of California Press, 1985.

Mead, George Herbert. *Mind, Self, and Society*. Chicago: University of Chicago Press, 1934.

Mesqui, Jean. *Les ponts en France avant le temps des ingénieurs*. Paris: Fayard, 1980.

Mettam, Roger. *Power and Faction in Louis XIV's France*. London: Basil Blackwell, 1988.

Meyer, Jean. "Louis XIV et Colbert: les relations entre un roi et un ministre au XVIIe Siècle." In *Colloque pour le tricentenaire de la mort de Colbert, sous la direction de Roland Mousnier, Un nouveau Colbert*, edited by Jean Favier and Roland Mousnier, 71–84. Paris: Éditions Sedes, 1985.

Michel, Henri. "Le Canal de Languedoc vu par un astronome 'philosophe.'" In *Le Canal du Midi*, edited by Jean-Denis Bergasse, 1:197–216. Cessenon: J.-D. Bergasse, 1982).

Michel, Jean-Marie, Cherine Gébara, and Jean-Louis Guendon. *L'aqueduc romain de Fréjus. Sa description, son histoire et son environnement*. Montpellier: Broché, 2001.

Miller, P. N. *Peiresc's Europe: Learning and Virtue in the Seventeenth Century*. New Haven, CT: Yale University Press, 2000.

Minard, Philippe. *La fortune du Colbertisme: État et industrie dans la France des Lumières*. Paris: Fayard, 1988.

Minson, Jeffrey. *Genealogies of Morals*. New York: St. Martin's Press, 1985.

Morison, Elting E. *From Know-how to Nowhere*. New York: Basic Books, 1974.

Morris, A.E.J. *History of Urban Form*. London: George Godwin Ltd., 1979.

Mosser, Monique, and Georges Teyssot. *The Architecture of Western Gardens*. Cambridge, MA: MIT Press, 1991.

Mousnier, Roland. *The Institutions of France under the Absolute Monarchy, 1598–1789: Society and the State*. Translated by Brian Pearce. Chicago: University of Chicago Press, 1979.

Mukerji, Chandra. "The Collective Construction of Scientific Genius." In *Cognition and Communication at Work*, edited by Yrjö Engeström and David Middleton, 257–78. Cambridge: Cambridge University Press, 1997.

———. "Cultural Genealogy." *Cultural Sociology* 1, no. 1 (2007), 49–71.

———. "Dominion, Demonstration, and Domination: Religious Doctrine, Territorial Politics, and French Plant Collection." In *Colonial Botany: Science, Commerce, and Politics in the Early Modern World*, edited by Londa Schiebinger and Claudia Swan, 19–33. Philadelphia: University of Pennsylvania Press, 2005.

———. "Engineering and French Formal Gardens in the Reign of Louis XIV." In *New Directions in Garden History*, edited by John Dixon Hunt, 22–43. Philadelphia: University of Pennsylvania Press, 2002.

———. "Entrepreneurialism, Land Management, and Cartography during the Age of Louis XIV." In *Merchants and Marvels*, edited by Paula Findlen and Pamela Smith, 248–76. New York: Routledge, 2002).

———. "French *mesnagement* Politics and the Canal du Midi: Labor and Economy in Seventeenth-century France." In *The Mindful Hand: Inquiry and Invention from the Late Renaissance to Industrialization*, edited by Lissa Roberts, Simon

Schaffer, and Peter Dear, 169–88. Amsterdam: Editas and University of Chicago Press, 2007.

———. *A Fragile Power: Scientists and the State*. Princeton, NJ: Princeton University Press, 1990.

———. *From Graven Images: Patterns of Modern Materialism*. New York: Columbia University Press, 1983.

———. "The Great Forest Survey of 1669–1671." *Social Studies of Science* 37, no. 2 (2007): 227–53.

———. "Intelligent Uses of Engineering and the Legitimacy of State Power." *Technology and Culture* 44, no. 4 (2003): 655–76.

———. "Material Practices of Domination and Techniques of Western Power." *Theory and Society* 31 (2002): 1–31.

———. "The Political Mobilization of Nature in French Formal Gardens." *Theory and Society* 23 (1994): 651–77.

———. "Printing, Cartography, and Conceptions of Place." *Media, Culture, and Society* 28, no. 5 (2006): 651–69.

———. "Stewardship Politics and the Control of Wild Weather: Levees, Seawalls, and State Building in 17th-century France." *Social Studies of Science* 37, no. 1 (February 2007): 111–18.

——— "Tacit Knowledge and Classical Technique in Seventeenth-century France: Hydraulic Cement as Living Practice among Masons and Military Engineers." *Technology and Culture* 47 (2006): 713–33.

———. *Territorial Ambitions and the Gardens of Versailles*. Cambridge: Cambridge University Press, 1997.

———. "The Unintended State." In *New Cultural Materialisms*, edited by Patrick Joyce and Tony Bennett. London: Routledge, forthcoming.

———. "Unspoken Assumptions: Voice and Absolutism in the Court of Louis XIV." *Journal of Historical Sociology* 2 (1998): 228–315.

———. "Women Engineers and the Culture of the Pyrenees: Indigenous Knowledge and Engineering in 17th-century France." In *Knowledge and Its Making in Europe, 1500–1800*, edited by Pamela Smith and Benjamin Schmidt, 19–44. Chicago: University of Chicago Press, 2008.

Mumford, Lewis. *The Culture of Cities*. New York: Harcourt Brace, 1938.

Murat, Inès. *Colbert*. Translated by Robert Francis Cook and Jeannie Van Asselt. Charlottesville: University Press of Virginia, 1984.

———. "Les Rapports de Colbert et de Riquet: Méfiance pour un homme ou pour un système?" In *Le Canal du Midi*, edited by Jean-Denis Bergasse, 3:105–21. Cessenon: J.-D. Bergasse, 1984.

Neraudau, J. P. *L'Olympe du Roi Soleil*. Paris: Société des Belles-Lettres, 1986.

Nietzsche, Friedrich. *Birth of Tragedy and the Genealogy of Morals*. Translated by Francis Golffing. Garden City, NJ: Doubleday, 1956.

Orwell, George. *1984*. New York: Signet Classics, 1950.

Parker, David. *The Making of French Absolutism*. London: Edward Arnold, 1983.

Paskvan, Raymond Frank. "The Jardin du Roi: The Growth of Its Plant Collection 1715–1750." PhD diss., University of Minnesota, 1971.

Patterson, Orlando. *Rituals of Blood: Slavery in Two American Centuries*. New York: Basic Books, 1998.

Pelletier, Monique. *La Carte de Cassini: l'Extraordinaire Aventure de la Carte de France*. Paris: Presses de l'École nationale des ponts et chaussées, 1990.

Pinsseau, Hubert. "Du Canal de Briare au Canal des Deux Mers: Origines et con-séquences d'un système inédit de navigation artificielle." In *Le Canal du Midi*, edited by Jean-Denis Bergasse, 4:27–54. Cessenon: J.-D. Bergasse, 1985.

Pinsseau, Pierre. *Le Canal Henri IV, ou Canal de Briare, 1604–1943*. Orléans: R. Houzé, 1944.

Porter, Theodore. *Trust in Numbers: The Pursuit of Objectivity in Science*. Princeton, NJ: Princeton University Press, 1995.

Powell, W. W., and Paul DiMaggio. *The New Institutionalism in Organizational Analysis*. Chicago: University of Chicago Press, 1991.

Preda, Alex. "Where Do Analysts Come from? The Case of Financial Chartism." In *Market Devices*, edited by Michel Callon, Fabian Muniesa, and Yuval Millo. 40–64. Chicago: University of Chicago Press, 2007.

Quenot, Yvette. "Bernard Palissy et Oliver de Serres." In *Bernard Palissy, 1510–1590, l'écrivain, le réforme, le céramiste*, by Frank Lestringant, 93–103. Saint-Pierre-du-Mont: Coédition Associon Internationale des Amis d'Agrippa d'Aubigné- Éditions SPEC, 1990.

Ranum, Orest. *Paris in the Age of Absolutism: An Essay*. Bloomington: Indiana University Press, 1979.

Rinne, Katherine W. "Between Precedent and Experiment: Restoring the Acqua Vergine in Rome, 1560–70." In *The Mindful Hand: Inquiry and Invention from the Late Renaissance to Early Industrialisation*, edited by Lissa L. Roberts, Simon Schaffer, and Peter Dear, 95–116. Amsterdam: Edita—the Publishing House of the Royal Netherlands Academy of Arts and Sciences and the University of Chicago Press, 2007.

Rivet, Bernard. "Aspects Économiques de l'Oeuvre de B. Palissy" et Frank Lestringant, "L'Eden et les Tenèbres Extérieures." In *Bernard Palissy, 1510–1590, l'écrivain, le réforme, le céramiste*, by Frank Lestringant, 167–80. Saint-Pierre-du-Mont: Coédition Associon Internationale des Amis d'Agrippa d'Aubigné- Éditions SPEC, 1990.

Robert, Jean, and Jean-Denis Bergasse. "L'Étrange Destin des Andréossy." In *Le Canal du Midi*, edited by Jean-Denis Bergasse, 3:195–232. Cessenon: J.-D. Bergasse, 1984.

Roberts, Lissa, Simon Schaffer, and Peter Dear, eds. *The Mindful Hand: Inquiry and Invention from the Late Renaissance to Industrialization*. Amsterdam: Edita-the Publishing House of the Royal Netherlands of Academy of Arts and Sciences and University of Chicago Press, 2007.

Rolt, L.T.C. *From Sea to Sea: The Canal du Midi*. London: Allen Lance, 1973.

Rose, Nikolas. *Powers of Freedom: Reframing Political Thought*. Cambridge: Cambridge University Press, 1999.

Roseneau, Helen. *The Ideal City: Its Architectural Evolution*. Boston, MA: Boston Book and Art Shop, 1959.

Rosental, Claude. "Making Science and Technology Results Public: A Sociology of Demos." In *Making Things Public: Atmospheres of Democracy*, edited by Bruno Latour and Peter Weibel, 346–49. Cambridge, MA: MIT Press, 2005.

Roux, Julie. *Les Chemins de Saint-Jacques de Compostelle*. Vic-en-Bigorre: MSM, 1990.

Sagnes, Jean ed. *Histoire de Sète*. Toulouse: Privat, 1987.

Sahlins, Peter. *Boundaries: The Making of France and Spain in the Pyrenees*. Berkeley: University of California Press, 1989.

————. *Forest Rites: The War of the Demoiselles in Nineteenth-Century France*. Cambridge, MA: Harvard University Press, 1994.

Salamagne, Alain. "Vauban et les fortifications du Quesnoy." *Revue Historique des Armées* 162 (1986): 45–51.

Saliès, Pierre. "De l'Isthme Gaulois au Canal des Deux Mers: Histoire des Canaux et Voies Fluviales du Midi." In *Le Canal du Midi*, edited by Jean-Denis Bergasse, 4:55–97. Cessenon: J.-D. Bergasse, 1985.

Sangallo, Antonio. *The Architectural Drawings of Antonio da Sangallo the Younger and His Circle*. Cambridge, MA: MIT Press, 1994.

Santerre, H. Gallet de. "L'empreinte Romaine." In *Histoire du Languedoc*, edited by Philippe Wolff. Paris: Privat, 2000.

Sargent, Arthur J. *The Economic Policy of Colbert*. London: Longmans, Green, and Co., 1899.

Schaffer, Simon. " 'The Charter'd Thames': Naval Architecture and Experimental Spaces in Georgian Britain." In *The Mindful Hand: Inquiry and Invention from the Late Renaissance to Early Industrialisation*, edited by Lissa L. Roberts, Simon Schaffer, and Peter Dear, 279–307. Amsterdam: Edita—the Publishing House of the Royal Netherlands Academy of Arts and Sciences and University of Chicago Press, 2007.

Schwaller, Martine. *Ensérune*. Paris: Éditions du patrimoine, Imprimerie nationale, 1994. Schwartz, Barry. *Abraham Lincoln and the Forge of National Memory*. Chicago: University of Chicago Press, 2000.

Scott, James. *Seeing Like a State*. New Haven, CT: Yale University Press, 1998.

————. 1985. *Weapons of the Weak: Everyday Forms of Peasant Resistance*. New Haven, CT: Yale University Press, 1985.

Scott, Joseph Frederick. *The Scientific Work of René Descartes, 1596–1650*. Ann Arbor: University of Michigan Press, 2006.

Serres, Michel. *Parasite*. Baltimore: Johns Hopkins University Press, 1982.

————. *Rome, le livre des foundations*. Paris: Grasset, 1983.

————. *Troubadour of Knowledge*. Ann Arbor: University of Michigan Press, 1997.

Sewell, William H., Jr. "Language and Practice in Cultural History: Backing Away from the Edge of the Cliff." *French Historical Studies* 21, no. 2 (1988): 241–54.

————. *Logics of History: Social Theory and Social Transformation*. Chicago: University of Chicago Press, 2005.

————. "The Political Unconscious of Social and Cultural History, or, Confessions of a Former Quantitative Historian." In *The Politics of Methods in the Human Sciences: Positivism and Its Epistemological Others*, edited by George Steinmetz, 173–206. Durham, NC: Duke University Press, 2005.

Shapin, Steve. *A Social History of Truth: Civility and Science in Seventeenth-century England*. Chicago: University of Chicago Press, 1994.

Shapin, Steve, and Simon Schaffer. *Leviathan and the Air-Pump: Hobbes, Boyle, and the Experimental Life*. Princeton, NJ: Princeton University Press, 1985.

Simon, Bart. *Undead Science: Science Studies and the Afterlife of Cold Fusion*. New Brunswick, NJ: Rutgers University Press, 2002.

Singh, Simon. *Fermat's Enigma*. New York: Anchor Books, 1997.

Smith, Norman. "The Roman Dams of Subiaco." *Technology and Culture* 11, no. 1 (1970): 58–68.

Smith, Pamela. *The Body of the Artisan*. Chicago: University of Chicago Press, 2004.

Smith, Pamela, and Paula Findlen, eds. *Merchants and Marvels: Commerce, Science and Art in Early Modern Europe*. New York: Routledge, 2002.

Sonnino, Paul. *Louis XIV and the Origins of the Dutch War*. Cambridge: Cambridge University Press, 1988.

Soulet, Jean-François. *Les Pyrénées au XIXe siècle*. Bordeaux: Sud Ouest, 2004.

———. *La Vie Quotidienne dans les Pyrénées sous l'ancien régime*. Paris: Hachette, 1974.

Spillman, Lyn. *Nation and Commemoration: Creating National Identities in the United States and Australia*. New York: Cambridge University Press, 1997.

Star, Susan Leigh, ed. *Cultures of Computing*. Cambridge, MA: Blackwell, 1995.

———, ed. *Ecologies of Knowledge: Work and Politics in Science and Technology*. Albany: State University of New York Press, 1995.

Star, Susan Leigh, and James R. Griesemer. "Institutional Ecology, 'Translations,' and Coherence: Amateurs and Professionals in Berkeley's Museum of Vertebrate Zoology, 1907–1939." *Social Studies of Science* 19 (1989): 387–420.

Stroup, Alice. *A Company of Scientists: Botany, Patronage, and Community at the Seventeenth-century Parisian Royal Academy of Sciences*. Berkeley: University of California Press, 1990.

Swidler, Ann. "Culture in Action: Symbols and Strategies." *American Sociological Review* 51 (1986): 273–86.

Taylor, Rabun. *Roman Builders: A Study in Architectural Process*. Cambridge: Cambridge University Press, 2003.

Thompson, David. *Renaissance Paris*. Berkeley: University of California Press, 1984.

Tilly, Charles. *Coercion, Capital, and European States, 1000-1999*. Oxford: Blackwell, 1990.

———. *The Contentious French*. Cambridge, MA: Harvard University Press, 1986.

———. *The Politics of Collective Violence*. Cambridge: Cambridge University Press, 2003.

Turnbull, David. *Masons, Tricksters, and Cartographers*. Amsterdam: Harwood, 2000.

Turner, Fred. *From Counter Culture to Cyber Culture: Stewart Brand, the Whole Earth Network, and the Rise of Digital Utopianism*. Chicago: University of Chicago Press, 2006.

Vaughan, Diane. *The Challenger Launch Decision: Risky Technology, Culture, and Deviance at NASA*. Chicago: University of Chicago Press, 1996.

Vérin, Hélen. *La Gloire des Ingéneurs: L'intelligence techniques du XVIe au XVIIIe siècle*. Paris: Albin Michel, 1993.

Vidal, Mary. *Watteau's Painted Conversations: Art, Literature, and Talk in 17th- and 18th -century France*. New Haven, CT: Yale University Press, 1992.

Villers, Robert. "Colbert et les finances publiques." In *Colloque pour le tricentenaire de la mort de Colbert, sous la direction de Roland Mousnier, Un nouveau Colbert*, edited by Jean Favier and Roland Mousnier, 177–87. Paris: Éditions Sedes, 1985.

Virville, Davy de. *Histoire de la Botanique en France*. Paris: Société d'Édition d'Enseignement Supérieure, 1954.

Wagner-Pacifici, Robin. *The Art of Surrender: Decomposing Sovereignty at Conflict's End*. Chicago: University of Chicago Press, 2005.

Wallerstein, Immanuel. *The Modern World-system*. New York: Academic Press, 1974.

Weber, Max. *From Max Weber: Essays in Sociology*. Translated by H. H. Gerth and C. Wright Mills. New York: Oxford University Press, 1946.

———. *The Theory of Social and Economic Organization*. Edited by A. M. Henderson and Talcott Parsons. New York: Free Press, 1947.

Wilford, John Noble. *The Mapmakers*. New York: Vintage, 1981.

Woodward, David. *Maps as Prints in the Italian Renaissance: Makers, Distributors, and Consumers*. London: British Library, 1996.

Wuthnow, Robert. *Communities of Discourse: Ideology and Social Structure in the Reformation, the Enlightenment, and Europe Socialism*. Cambridge, MA: Harvard University Press, 1993.

———. *All in Sync: How Music and Art Are Revitalizing American Religion*. Berkeley: University of California Press, 2003.

Wuthnow, Robert, and John Evans. *The Quiet Hand of God: Faith-Based Activism and the Public Role of Mainline Protestantism*. Berkeley: University of California Press, 2002.

Yates, Frances. *The Art of Memory*. Chicago: University of Chicago Press, 1966.

Zerubavel, E. *The Seven Day Circle*. Chicago: University of Chicago Press, 1985.

INDEX

absolutism (*see* state)

Académie Royale des Sciences 21, 28–29, 32, 43, 102, 119

Adam and Eve 223

Adgé, Michel 85, 102

Agde 94, 166, 195, 199, 220–21

Agoût River 43–44, 159, 200

Agricola, Giorgius 85

Aguesseau, Henri d' 158, 167–70, 178–81, 189, 201

Aigues-Mortes 50, 94 163

ainesse absolut, see Pyrenees

Alzau River 47, 53, 200

ancients and moderns 6, 8, 60–61, 199

Andréossy, François 47–48, 53, 56, 107, 186, 193–95, 200, 211

Anglure de Bourlemont, Charles-François d', bishop of Castres, archbishop of Toulouse 17, 27, 35, 44, 47–48, 52, 78

anomie 157, 168, 174

Antibes 111

Apollo 194, 211

aqueducts, 60, 65, 118, 121, 123–25, 139, 153, 162–63, 189, 195; Le Répudre 121, 123–25; classical 123, 125–26; proposed by Vauban 189

archival records of the canal 128–30

Arcons, César d' 113

Argelès-Gazost 135

Argeliers 144

Aristotle 225

arpenteurs-géometres (see surveyors)

Arquier, M. 79–80

artisans, 9, 12, 13, 62–65, 85, 95, 117–18, 144, 158, 163, 192, 202–3, 207–8; and tacit knowl-edge 38; and classical engineering 12, 118, 120–25; and standards of practice 173–74; erasure of 192, 201–2, 223–24; in the building trades 12, 38, 50, 74–75, 77, 118

Atlantic economy 28, 62, 111, 114, 200

Atlantic Ocean 8, 16–17, 60

Aude River 37, 42–43, 45, 53–54, 56, 61, 67 95, 109, 110, 130, 143, 163, 200

Aulus-les-Bains 137

authority, formal structure of, 70, 108; moral 22–25, political 2, 12, 15–22, 25–27, 36, 60–62, 68, 88, 90, 113; social 12, 61, 70, 108; technical 23–24, 30–31, 36, 61, 68–70, 79–80, 87–88, 105, 108. *See also* leader-ship, finances

authorship 69, 154–70, 180, 186–93, 196, 202, 204, 207–8, 221, 223–26. *See also* Louis XIV, Riquet

Avessens, Jospeph, sieur de Tarabel 50, 131

Bachelier, Nicolas 42

Bagnère de Bigorre, *see* Bigorre

Bazin de Bezons, Claude 49, 54, 54, 59, 77 94, 97, 99–102, 106, 155, 159, 172

Beaucaire 102

Belidor, Bernard Forest de 198–99

Bernassonne River 53

Béziers 38, 49, 107, 110, 117, 129–30, 142–43, 149–52, 163, 166, 182, 199; Jesuit college in 151

Bigorre 127–28, 131, 134, 136–37, 146, 221

Blanchard, Anne 102

Bodin, Jean 168

Bonrepos 27, 40, 45

Bonzi, Pierre de 182–83, 196

bookkeeping 70, 76–77, 99–100, 103, 150, 158, 224; and contractual obligations 172–74, 181, 212; and gender 127–29; and job classifications 70, 76–77, 128–30, 222–23; and records of work 100, 107, 128–30, 150–51; and system of oversight 70, 76–77, 79–80, 100, 212; and hidden expenditures 85, 97; and Riquet's finances 97, 99–101, 178–82; changing methods of 100–1, 107, 128–30. *See also* finances

Borel, Pierre 43, 45, 48, 150, 159, 22

Boutheroue, Hector, sieur de Bourgneuf
50–53, 55, 81, 192, 200
Boutheroue, Guilaume 43
Bowker, Geoffrey 216
Boyer, André 125, 172–74
Bradley, Hugh 42
Brest 111
building techniques, *see* construction
methods

Caesar, Julius 60–61
calculations, *see* measures
Callon, Michel 204, 216
Cambacérès, Jacques (?) 96
Cammazes 55–56, 165, 187
Campmas, Pierre 45–48, 53, 55, 130, 191, 200
Campmases 72
Canal de Bouridou 94
Canal de Briare 4–5, 50–51, 169, 177
Canal de Silveréal 94, 95
canal(s): and classical symbolism 70–72,
75–78, 191, 193–95; commerce on 6, 18–19,
42, 199–200; traffic on 4–5, 51; political
significance of 16, 70, 176–80, 191; scale
of 220
canal features 195–97, 220; classical
techniques 63, 65–66, 70–72, 75–77, 126;
banks 51; bridges 6, 95, 197, 223; channel
65–66, 89, 95 157; drains 51, 65, 118, 126,
146, 189, 220; tow path 71, 123; instabilities of 92; laundries 65, 147, 220; levees
57, 71, 126, 204; mills 65; overflows 118,
126; skimmers/silt barriers 126, 145–46;
underpasses 95, 118, 220; wooden reinforcements 58, 190. *See also* aqueducts,
locks, materials, Malpas tunnel, ports,
and water features, engineered.
Capestang 110, 204
Carcassone 42, 45, 48–49, 53, 160, 201
Carroll, Patrick 216
Cassini, Jacques 32–33
Castelnaudery 157, 160–62, 182–83, 201
Castres 43–44, 49, 159; Académie in 150
Catalans/Catalonia 69, 74, 135, 170–71, 200
Catholic Church 17–22
Cavalier, Jean 50
Certeau, Michel de 215–16
Cesse River 144
Chambonas Abbey 49

Cheust 146–48, 221
China 200
Cicero 225
Clerville 99, 102, 103, 105, 106, 108, 110, 191,
199–200; and classical engineering 70;
and Colbert 31, 87–89, 98, 172, 212; and
dam design 54–56; and excitement about
canal 12; and fall from royal favor 177,
211–3; and his clients 56, 70, 85, 88–89,
121, 172–74, 177; and logistics 212–3; and
oversight of the canal 49–56, 74, 186–89,
211; and pozzolana 121, 122, 125; and
private enterprises 88–89, 162–63; and
Riquet 87–89, 154, 157, 162–63, 165–66,
169, 172–74, 177–78; and Sète 94, 110, 113,
115, 121, 174; and specifications 56, 66, 69,
74, 81, 125, 160, 219; and standing with the
king 158; and the commission 35, 49–56,
101; and verifications 87, 97, 100, 172–73;
death of 163, 180; role as military engineer
50–51, 191; and Vauban 186–189, 199
coasts 13, 51, 91, 94, 194; instability of 91,
111–16; engineering problems on 117,
162–63
cofferdams, *see* construction methods
cognitive tools 35, *see also* distributed
cognition
Colbert, and administrative policies 25–27,
30, 60, 98, 111, 159, 203, 206, 209, 211;
and canal contracts 66–69, 184–85;
and Clerville 31, 51, 87–89, 102; and
d'Aguesseau 166–70, 178–81, 184 and
Etats 85–87; and fear of Riquet's death
176–80; and forest reform 28, 33, 41; and
intendants 49, 59, 166–70; and La Feuille
98–103, 105–6, 161, 166; and Louis XIV 88,
92, 93, 157, 206; and loss of control 154–58,
160–66, 168–70, 174–75, 177–80, 208, 221;
and natural knowledge 25–32, 35; and
New Rome 59–62; and nobility 19, 21, 26,
61–62; and patrimonial politics 18–19, 22,
25–27, 221; and petit academie 60; and
pozzolana 122–23; and propaganda 13, 17,
59–61, 181–82, 223–25; and stewardship
politics 22, 25–27; and territorial politics
22, 25–27, 30, 111, 206; and women
laborers 130; as minister of the king's
household 16, 25, 31; as minister of the
navy 16, 50, 111; as minister of the

domination 2, 5, 204, 214–16. *See also* land,
     Louis XIV, power
dominion, *see* stewardship
Durfort 52
Durkheim 156
duties/tolls 18–19, 68, 162, 170, 200

École des Ponts et Chaussées 197
Edict of Nantes 210
Egypt 200
Encyclopédie 207
engineering books 3, 198–202
engineering; academic/formal 11, 47, 70, 117,
     205; as a living tradition 117, 118–28,
     131–34, 137–40, 153, 201, 208, 220, 224, 227;
     civil 18, 50; classical 6, 11–12, 64, 70–71,
     82–83, 117–18, 125–28, 131–34, 153, 164–65,
     208, 221, 224–25; Dutch 3, 103–6, 113, 116,
     200; German 200; Italian 3, 42, 47–48,
     71, 75, 200, 224; indigenous 38, 118–21,
     123–27, 131–34, 136–42, 153, 192, 205, 219,
     225; limits of 36–38, 69, 80–81, 204, 221;
     military 11, 29–32, 50, 69, 71–72, 75, 121,
     165, 185–90, 197–99; political value of 16,
     193–200; schools of 150–51, 197;
     structural 117–18, 192; threat of 211–14
Ensérune (oppidum) 164, 198
epistemic authority 5–6, 24, 27, 36–37, 41,
     48–50, 59, 79–81, 109; and credibility 27,
     36–38, 48, 70, 79–81, 87–89, 109, 154–55,
     167, 208, 224; and rank 6, 36–37, 52–54,
     59, 79–81, 154–55, 182, 215, 222; and "work-
     ing" 7, 58–59, 64, 154–57, 184–86, 198–207,
     221–24; and the state 15–17, 36, 70, 206
epistemic communities 5, 7, 11, 20–29, 35,
     48, 176, 203
Estan, Immanuel d' 125, 172–174
Estienne, Charles and Jean Liebault 23
étangs 13, 94, 112–13; étang de Montady
     164–65; étang de Thau 54, 114; étang de
     Vendres; *See also* water features
États du Languedoc 38, 49, 61–62, 67–69,
     78, 85–87, 89–90, 93, 96–97, 154, 171,
     181–82, 202, 206
experts 27–35, 38, 50–56, 79–81, 156, 159, 163,
     173, 178, 182, 212, 223

fairies 135, 141
Farrand, Jean 98

femelles, *see* women
Fermat, Pierre de 150
finances 55, 62, 67–69, 77, 85–87, 94–98,
     106–9, 156–58, 161–63, 167–68, 172–74,
     179–81, 199–200
financiers 9, 12, 39, 206
first enterprise 66–70, 72–77, 79–85, 89–90,
     100, 118–19, 128–29, 156–58, 201; specifica-
     tions for 66–67, 201; route for 66–67,
     79–80, 160–61; time limit for 68–69, 79.
     See also contracts, finances
floods 2, 37, 51, 71, 78, 109, 122–23, 143–44,
     146, 185–86, 188–89, 198–99, 204, 208. *See
     also* storms
folklore 48–49, 135, 141
Fonseranes lock staircase 117, 149–53, 195,
     199, 219
Fontesorbes 141
forests, depictions of 33–35, 131–32, 197;
     enterprises in 13; reform of 25, 131–32;
     timbering methods in 152–53
fortresses 50–51, 97, 198–99, 212–14; at
     Brisach 212; at Philippsbourg 212;
     Château de Trompette 212
fossé; counter-fossé 31, 136, 144
Foucault, Michel 216, 218
Fouquet, Nicolas 168
François I 42
Fréjus 82, 84
French Revolution 199, 226
Fresquel River 40, 42–46, 53
Froidour, Louis de 41–42, 72, 76, 116, 126,
     131–34, 136–37
Fronde 20
Frontignan 113
Frontinus 131
funding, *see* finances

gabelle 5, 17, 39, 67–69, 85, 158, 167, 171,
     178; in Roussillon 67–69, 85–87, 171–72;
     resistance to 69, 85, 92, 132–33. *See also*
     Riquet, tax
Galileo 32
Ganges 49
Garipuy, M. 197–198
Garonne River 1, 37, 42–43, 45, 52, 56, 61,
     67, 76–82, 114, 116, 159, 197, 200, 208,
     220
Gaul (old)11–12, 60–63, 79–80, 174, 226

gender 76–77, 127–30, 132, 135–36, 139–41, 151, 192, 194. *See also* women

genius 6, 8–9, 48–49, 153, 182, 184, 191–92, 198–202, 208, 223–24

gentlemen 5, 9–10, 12, 28–29, 32, 36, 43, 49–56, 62, 78, 90, 102, 128, 132, 154–55, 192, 206–7, 222, 227

Gerde 137

Gibraltar 8, 18, 114, 200

Gillade, Dominique 107, 163, 181, 186

Girou River 45

God 2, 5, 7, 19, 23–25, 41, 59, 154–59, 174, 176, 185, 210, 222, 226

goddess worship 135–36, 141

Graissans 40, 49, 57

grau 113

Greece 200

Griesemer, James 209–10

gunpowder 117, 129, 142–43, 158, 163, 198

Guyon, Jacques 43

Hapsburg Empire 13, 16, 18–19, 224

Henri IV 7, 18, 20, 24–25, 43, 177

Hérault River 94, 95, 220

history of the Canal du Midi; canonical 88, 180, 192, 198–202, 209–11, 223–26; Colbert's role in 182–86, 224, 226; and impersonal rule 203–4, 223–25

Hondius 2, 32

Huguenots 17, 18, 22–23, 111, 146, 150, 155, 170

humanism 22, 24, 36–37, 120–21, 198, 201, 211, 218, 224

Hurez, M. 87

Hutchins, Edward 10; and Brian Hazelhurst 219

Huygens, Christian 32–33

hydraulic cement 118, 120–125

hydraulics 2, 3, 5, 10, 15, 17, 39–40, 117, 199, 204, 207, 219–20; and mining 3; and salt production 39–40; as a living tradition 117–28, 131–34, 137–40, 201, 192, 208, 220, 224, 227; classical 12, 54, 126–27, 144, 164–65, 198, 201, 227; Dutch 3, 103–6, 198–99; formal knowledge of 2, 3, 5, 36, 47, 126, 192, 201, 207; in military engineering 50–51, 198–99; indigenous 12, 117, 131–34, 136–41, 144, 150, 152–53, 192, 201, 220; Italian 3; limits on knowledge of 55, 144, 208, 221; scale of 220;

local knowledge of 10, 50, 127, 133. *See also* water systems, and knowledge

ideal city 75, 111

impersonal rule 4–5, 7, 24, 30, 169–70, 174–75, 177–78, 183, 192, 197–99, 202–9, 211–23, 225–27

inclines 48, 53, 119–20, 126–27, 130, 134, 137, 142, 144

indemnifications 25, 72, 78–79, 89, 98, 108, 168, 199

infrastructure 5, 6, 18–19, 24, 31, 70, 92, 110–11, 114–15, 154, 184–86, 214–16. *See also* engineering

intendants 77, 96–99, 100, 102, 158, 178–83, 214

irrigation 3, 12–13, 137, 139, 223

Italy 200, 224

Jacquinot, Etienne 50

Jardin du Roy 28

Jennse, Regnier 113

Jesuits 77–78, 109, 128, 150–51, 181, 186

jetties, *see* La Feuille, Sète

jeu du Canal Royal 196–97

Joyce, Patrick 216

Joyeuse, Cardinal 18

Knorr-Cetina, Karin 216

knot for technoscientific culture 203–6

knowledge: and learning 5, 9–11, 125, 154, 131–32, 222–23, 225; experiential 38–42, 51, 63–65, 70, 80, 108, 117–19, 131–34, 150, 152; formal 6, 11, 36, 42–44, 70, 117–18, 125, 150–51, 192, 201; formalization of 33–35, 192, 197–98, 207–8, 220–23, 225–26; limits of 33–35, 37–38, 55, 112–13, 115–16, 144, 152, 178; local 6, 12–13, 28, 38, 108, 125, 131–34, 152, 201, 207, 221, 225; natural 2–4, 6, 13, 16, 22–35, 38, 78, 80, 116, 118, 136, 141, 155, 191, 197–98, 201, 204–8, 223; peasant 36, 39–42, 45–49, 74–75, 117, 141; provenance of 12, 64, 70–71, 80, 117, 126, 153; tacit 11–12, 38, 63–64, 74, 117–20, 125, 150–53, 225; technical 5, 80, 206

La Feuille, Jacob

La Feuille, Pons de 50, 98–106, 110, 118, 155, 173, 186; and Colbert 98–106, 161–62, 168,

La Feuille, Pons de (cont.)
178; book by 102–6; family 102; engineering knowledge 102–6, 115–16; and women laborers 128, 130
La Hire 2
La Nouvelle 54
La Radelle 94
La Robine 109
La Rochelle 111
labor practices; supervision 70, 77, 106–7; pay scales 76–77, 179, 128; piece work 77, 130, 142; wage labor 62, 107, 179; subcontracts 95, 106–7; changes for second enterprise 106–7
laborers 107, 115, 181; and classical engineering 63, 118, 125–27, 137, 142, 192, 227; and tacit knowledge 63, 127, 137, 142, 201, 203, 206, 208, 227; seasonal 62, 92, 96, 127–28, 181; shortages of 91–92, 103, 181. *See also* artisans, peasants, women
Lalande, Jospeh Jérôme 199–201
land 109, 119; agricultural uses of 22–27, 30–31, 38; cultures of 13, 22–23, 117, 219; destruction of 61, 90, 92, 176, 182, 202, 214, 224; domination of 2–5, 13, 16, 22, 25, 60, 90, 92, 118–20, 126, 193–95, 197–200, 204–5, 214, 218; engineering of 30–31, 90, 109, 117–18, 184–86, 203, 219–20; improvements 6, 9, 12, 22–25, 30–31, 78–79, 133–34, 154–55, 185, 191, 194–95, 199, 201–3, 205, 211, 218–19, 222; rocky 13, 30–31, 117, 126, 142–43; shifting 91–95, 110–16, 163, 204, 208; dry 1–2, 4, 13, 117, 183
Languedoc; as a geographical region 2, 4–7, 9–11, 13 17–27, 40–44, 47–48, 50–51, 62, 70, 88, 91, 93, 103, 191, 197; as an economic region 199–201; as a political region 4, 16, 19, 38, 49–50, 61–62, 68, 86–88, 90, 95–98, 101–2, 170, 179, 181–82, 185, 193, 199, 201; cultural traditions in 9–10, 13–14, 39, 117–21; 207; people/communities of 9–10, 13–14, 62–66, 117–18, 128, 197, 207; popular resistance in 22–23, 61–62, 132–33, 170; physical features of 2, 13–14, 40–43, 47–48, 51–56, 61, 91–92, 109–13, 116–20, 126–28, 152, 179, 185, 191, 193, 197, 201, 203, 208; religion in 17, 155, 170; Roman heritage/image of 61, 63–66,

70–76, 117–20; and independence from monarchy 4–7, 61, 170, 177. *See also* Etats du Languedoc
Lanta 49
Latour, Bruno 203–6, 216
Laudot River 40–41, 54, 57, 72, 74, 189
Lauragais ridge 27, 40–41, 44, 49, 52
Le Puy 49
Le Répudre 123–25, 153, 172–74, 189, 199
Le Somail 107, 129, 142–13
Le Tellier, Michel 16, 60
Le Tellier, François-Michel, *see* Louvois
leadership 9, 12, 24–25, 29–30, 36–39, 49–50, 87, 103, 106–9, 117–118, 154–66, 171–75, 177–78, 180–82, 186–92.
legacy 70; of Canal du Midi 10, 182–85, 193–202, 208; of Rome 6, 16, 19, 70, 185, 199, 208, 221; of Louis XIV 16, 19, 70, 118, 193–94, 198, 200, 208, 224, 226; of Riquet 109, 198, 201, 208, 225–26; purification of 70, 72, 192, 200–2, 209, 223–26
Lers River 42
levees, *see* canal features
locks 2, 3, 6, 27, 51–52 , 65, 66, 78, 80, 94, 100, 157, 172–74, 183, 185, 195, 197, 204; and opening of the canal 183; as innovations 13,42, 70, 81, 149–53, 220; contract for Jouares, Ons and Oignon 173–74; dimensions of 51, 81; doors 81, 83, 92, 158, 103–4, 156, 158, 204, 220; Dutch 81, 103–4, 199–200; failures 69, 81–83, 101, 185; for Sète 113–14; Italian 200; near Toulouse 81–83, 101, 185; on rigoles 159; on the Laudot 189; pressures on 81, 83, 103; oval 82–83, 157, 199; retrofitting of 120, 151–52; round 195, 199, 220–21; staircases of 51, 81–82, 95, 118, 120, 143, 149–53, 169, 191, 199, 219–20; tambours 151, 156, 169, 204, 220; wooden walls for 82
logistics 5–7, 65, 133–34, 211–26
Loire River 4, 43
Lombardy 3, 42, 47
Lorient 111
Louis XIII 206
Louis XIV 2, 19, 25–26, 86, 161, 182, 198; and danger of stewardship politics 61–62, 207, 211–14; and local nobles 19, 25–26, 62, 65, 81, 86, 194–96, 209, 224; and New

Rome 61, 90, 118, 193–96, 199, 224; and
patronage politics 20–21, 26, 61–62,
155–57, 168, 209, 212–13; and Riquet's tax
debts 181; and royal will 15–16, 18, 89, 92,
116, 168, 176, 184, 186, 193–95; and
singularity 15, 156–57, 176, 203, 206,
210–11, 214, 224; and sovereignty 15, 193,
199, 209–10; and war 21, 91–92, 199; as
authorial source of the canal 2, 182, 184,
193, 196, 202, 204, 208, 221, 224–25; as
self-centered 212; geopolitical ambitions
15–16, 210, 224; proposed statue of 158–59
Lourdes 136
Louvois (see Le Tellier) 16, 60, 92, 102
Lully, Jean-Baptiste 60
Lyons 54, 199

Machiavelli 20–21, 26, 61, 157, 172, 207
machines: dirt-lifting 80, 83, 85; dredging
105, 162–63
Malpas 163–66, 169, 178, 180; 184, 189–90,
197–99, 222
Mann, Michael 216
manouvriers 128–29
maps/charts/plans 4–5, 10, 28–35, 50–53, 56,
100, 111, 136, 184, 193–98, 201, 208
Marie-Thérèse d'Autriche 91
Mariotte, sieur 183
Marseillan 95
Marseilles 111
Marx 216
massif central 42
materials 2, 38, 108, 115, 162, 173, 179, 185,
220; and costs 87; clay 2, 39–40, 71, 73–74;
bricks 64; iron 220; stone 30, 63–65,
70–71, 73, 76–77, 82, 189–90, 220; wood 2,
30, 58, 70–71, 73, 75, 82, 85, 190, 220. *See
also* mortar and walls.
Maunenet, Abbé 200
Mazarin, Jules, Cardinal 25
Mazère-de-Neste 137–40
measures/calculations 2, 10–11, 15–16,
28–29, 32, 36–37, 52, 96, 119–20, 162, 201,
206; arguments over 79–81; as impersonal 205–6; finances 55, 172–74; distance
30, 38, 48, 52–53, 58; elevation 2, 3, 17,
30–32, 40–41, 48, 52, 152; inclines 48, 53,
119–20, 126–27, 130; limits of 119–20, 144;

of completion 184–86; of latitude and
longitude 32; of legitimacy 206; of soil
properties 58, 81, 119–20; of volume 120,
126, 151–53; of water loss 2; soundings 30,
182–83. *See also* specifications and
verifications
Medailhes, Michel and Pierre 150–53
Mediterranean: Sea 8, 11, 16–17, 35, 60,
111–16; trade 28
memory 77–78; material memory 10–12,
50–52, 63–64, 78, 118–21, 125–27, 131–32,
134, 136–41, 145–47, 152–53, 193–202, 219;
memory palace 225–27
Mendel, Gregor 210
Merville/Marville 102
mesnagement politics, *see* stewardship
Milan 42
mills 13, 39–40, 45, 52, 137–38, 147, 153, 162;
millstreams 45–46; windmills 49
mineral waters 131
miners 142, 165
mines 3, 39, 45, 85, 165–66, 190
Minocchio 158
Miquelets 69, 85, 170–71
models/mock-ups 17, 27, 30, 36, 76, 106, 201
modernism 5–8, 12, 60–61, 66, 70, 78, 93,
155, 169–70, 182–84, 198–203, 209, 222,
227
Molière 60
Monbel, sieur de 183
Mont Saint Clair 111, 113–15, 121
Montady 164–65, 198
Montagne Noire 13, 40, 43, 45–48, 52–56, 67,
77, 85, 89, 91, 116, 118–19, 129–30, 141–42,
162, 165, 208
Montaigu, D. 184
Montauban 89
Montpellier 68, 89, 98, 178, 181
monument(s) 6, 13, 17, 63, 69–70, 76, 78,
114, 118, 158, 185, 187, 191–92,
195–96, 200
Morison, Elting 64
mortar; ways to make 64–65; in wall
construction 64–65, 81; furnaces for 65,
115; and quarrying 65, 114. *See also*
hydraulic cement
Mourgues, Père Matthieu de 77–78, 109,
128, 150–51, 181, 186, 189

water features (engineered) (cont.)
132, 134, 136–37, 139, 141; drains 51, 65, 118,
126, 146, 189, 220; intakes 126, 145–47, 189,
199, 220; irrigation/drainage systems 3,
12–13, 134, 136–37, 139, 147, 223; laundries
127, 132, 137–41, 147–48, 223, 226; reservoirs 53–56, 69, 72, 74, 100, 126, 134, 141,
152, 157, 199; salt pans 39–40, 71; settling
ponds 126, 141, 145–47, 153, 187, 189, 197,
199, 220–21; weirs 137–38, 141, 146, 189
water features (natural) 38, 197–98; étangs
13, 54, 95, 98, 112–16, 164–65; rivers 13, 19,
28, 30–32, 33, 41–48, 94–95, 113, 153;
sources 45–46, 76, 126–27, 134–37, 141, 160
water supply, and reservoirs 2, 53, 54, 69, 72,
146–48, 157; and silt problems 119,
144–46; for Sète 162–63; for a canal in
Languedoc 2, 4, 27, 42–43, 45–48, 51–56,
58, 69, 89, 141–42; Pyrenean 125–28,
131–32, 134–41, 146–48; success of 158. *See
also*, Naurouze, rigoles, Saint Ferréol
water systems; Moorish 72, 127, 132, 153;
Roman 72, 126–27, 131–32, 153, 221;
Pyrenean 131–34, 136–41, 146–48
watersheds 2, 35, 118, Mediterranean 13, 27,
35, 37, 41, 53, 57, 117, 119, 126, 220; Atlantic
27, 35, 37, 41, 53, 57, 220

weapons of the weak, *see* tactics
weather 118; rainfall patterns; storms 91
Weber, Max 26, 214, 217
weirs 137–38, 141, 46
wind 112, 121
windmills (see mills)
wine 111, 199
women 5, 9, 76–77, 92, 115, 117, 119, 125, 142,
174, 192, 194, 207–8, 222; and classical
hydraulics 117, 125–27, 142, 192, 227;
and silt control 119, 144–46; lower
costs of 76–77, 119; as seasonal laborers
127–28; and the Fonseranes lock staircase
151–53; as femelles 128–30, 222–23; as
indigenous engineers 117, 127, 142, 201,
203, 206, 208, 227; erasure of 192, 201–2,
223–24; in the Le Somail region 143
workforce; 76–77, 91–92, 116, 121 ateliers/
brigades 76–77, 128; supervisory
structure 76–77, 129; and subcontracts
107, 129; and gender 76–77, 128–30;
and skilled workers 76–77, 129; diversity
of 10, 76;
working: as epistemic culture/measure 5,
36, 58, 64, 154–57, 184–86, 192,
198–207, 221–23; as religious
discipline 22

PRINCETON STUDIES IN CULTURAL SOCIOLOGY